Retinal Diseases: From Molecular Mechanisms to Therapeutics

Retinal Diseases: From Molecular Mechanisms to Therapeutics

Guest Editor

Soo-Young Kim

MDPI

Basel • Beijing • Wuhan • Barcelona • Belgrade • Novi Sad • Cluj • Manchester

Guest Editor
Soo-Young Kim
Independent Researcher
Carlsbad, CA
USA

Editorial Office
MDPI AG
Grosspeteranlage 5
4052 Basel, Switzerland

This is a reprint of the Special Issue, published open access by the journal *Life* (ISSN 2075-1729), freely accessible at: https://www.mdpi.com/journal/life/special_issues/4642F92096.

For citation purposes, cite each article independently as indicated on the article page online and as indicated below:

Lastname, A.A.; Lastname, B.B. Article Title. *Journal Name* **Year**, *Volume Number*, Page Range.

ISBN 978-3-7258-5785-2 (Hbk)
ISBN 978-3-7258-5786-9 (PDF)
https://doi.org/10.3390/books978-3-7258-5786-9

Cover image courtesy of Soo-Young Kim

Contents

About the Editor

Soo-Young Kim

Soo-Young Kim studied psychology (BA), life science (BS), immunology and medical engineering (MS), and neurobiology and human genetics (PhD) at Korea University in Seoul, South Korea. She then studied retina development and degeneration at the National Eye Institute, National Institutes of Health, USA, for 5 years, and after she started her retina translational studies within the discipline of nanomedicine and pharmaceutics at the Ophthalmology Department of the Wilmer Eye Institute, Johns Hopkins University and School of Pharmacy, Virginia Commonwealth University, focusing on early target development and validation in the context of inflammatory retina diseases such as age-related macular degeneration and diabetic retinopathy. Her current most prominent interest is gene delivery and protein engineering. Recently, she worked at an exosome biotech company, ExosomePlus Inc., as a project manager to develop exosome therapeutics to protect against ophthalmic diseases, and at AAV biotech company, Neurophth Therapeutics Inc., to be part of AAV therapeutic development for AMD and glaucoma treatment. While researching target molecules and translating the development, she started to develop an interest in protein engineering and structuring and is currently a member of the membrane protein database mpstruc, University of California, Irvine, and an independent researcher and consultant for biopharma startup research and development.

Preface

The retina, one of the sensory organs, is unique in that it is the only sensory organ that directly belongs to the central nervous system. Retinal ganglion cells project their axons into the thalamic region of the brain, providing a direct structural and functional link between the eye and the brain. This unique feature makes retinal research essential not only for understanding vision but also for gaining valuable insights into broader neurological mechanisms and diseases.

This Special Issue brings together contributions from an outstanding editorial team and distinguished researchers across diverse fields of medical and health sciences. The collection aims to advance scientific knowledge, deepen understanding of disease mechanisms, support the development of therapies for retinal diseases, and strengthen the collaborative spirit within the scientific and medical community.

Topics covered in this Reprint include therapeutic development, such as gene- and cell-based therapies; current treatment approaches, including anti-VEGF therapy; strategies in vision rehabilitation and eye exercises for disease prevention and management; diagnostic methodologies such as VESsel GENeration (VESGEN) analysis and application of deep neural networks; and fundamental research on photoreceptor synapse development and transcriptome analyses.

This collection provides readers with a comprehensive, up-to-date overview of retinal research, encompassing therapeutic, diagnostic, and basic studies with implications not only for ophthalmology but also for neuroscience and regenerative medicine.

Soo-Young Kim
Guest Editor

life

MDPI

Article

Murine Retina Outer Plexiform Layer Development and Transcriptome Analysis of Pre-Synapses in Photoreceptors

Soo-Young Kim [1,*], Christine Haewon Park [1], Bo-Hyun Moon [2] and Gail K. Seabold [1]

[1] Neurobiology-Neurodegeneration and Repair Laboratory, National Eye Institute, National Institutes of Health, Bethesda, MD 20892, USA; parkchr@snu.ac.kr (C.H.P.); gail.seabold@nih.gov (G.K.S.)
[2] Department of Oncology and Lombardi Comprehensive Cancer Center, Georgetown University Medical Center, Washington, DC 20057, USA; bm366@georgetown.edu
* Correspondence: gungil@korea.ac.kr; Tel.: +1-(240)-421-2657

Abstract: Photoreceptors in the mammalian retina convert light signals into electrical and molecular signals through phototransduction and transfer the visual inputs to second-order neurons via specialized ribbon synapses. Two kinds of photoreceptors, rods and cones, possess distinct morphology and function. Currently, we have limited knowledge about rod versus (vs.) cone synapse development and the associated genes. The transcription factor neural retina leucine zipper (NRL) determines the rod vs. cone photoreceptor cell fate and is critical for rod differentiation. *Nrl* knockout mice fail to form rods, generating all cone or S-cone-like (SCL) photoreceptors in the retina, whereas ectopic expression of *Nrl* using a cone-rod homeobox (*Crx*) promoter (*CrxpNrl*) forms all rods. Here, we examined rod and cone pre-synapse development, including axonal elongation, terminal shaping, and synaptic lamination in the outer plexiform layer (OPL) in the presence or absence of *Nrl*. We show that NRL loss and knockdown result in delayed OPL maturation and plasticity with aberrant dendrites of bipolar neurons. The integrated analyses of the transcriptome in developing rods and SCLs with NRL CUT&RUN and synaptic gene ontology analyses identified G protein subunit beta (*Gnb*) 1 and p21 (RAC1) activated kinase 5 (*Pak5 or Pak7*) transcripts were upregulated in developing rods and down-regulated in developing SCLs. Notably, *Gnb1* and *Gnb5* are rod dominant, and *Gnb3* is enriched in cones. NRL binds to the genes of *Gnb1*, *Gnb3*, and *Gnb5*. NRL also regulates pre-synapse ribbon genes, and their expression is altered in rods and SCLs. Our study of histological and gene analyses provides new insights into the morphogenesis of photoreceptor pre-synapse development and regulation of associated genes in the developing retina.

Keywords: retina outer plexiform layer; photoreceptor synapse; spherule; pedicle; neural retina leucine zipper; transcriptional regulation; gene expression

1. Introduction

The central nervous system is composed of distinct groups of neurons, which have specific morphologies. Neuronal morphologies represent typical functions and connections, which are controlled by genetic cues and synaptic activity [1,2]. However, there is still more to be learned about the genetic codes that determine the functionality and morphology of each neuron type. The retina belongs to the central nervous system but has fewer types of neurons compared to the brain, providing an easily accessible model for studying mechanisms associated with neuronal differentiation, morphogenesis, synaptogenesis, and plasticity.

The retina possesses highly laminated structures, with three layers of nuclei and two synaptic layers, consisting of five major neuron types: photoreceptor, bipolar, horizontal, amacrine, and ganglion cells. The photoreceptor is the first neuron type, which initiates visual cues and transmits them to the brain via second-order and then ganglion cells, which form the optic nerve. The transmission of visual signals occurs in the retina outer plexiform layer (OPL) by synaptic connections between pre-synapses of photoreceptors and post-synapses in the dendritic fields of over 12 sub-types of bipolar neurons, along with the dendrites and axonal terminals of horizontal cell neurons [3,4].

The mammalian photoreceptors are separated into rods and cones. Rods function under low light conditions, and cones respond to daylight, mediating color vision [5]. Depending on the opsins inside S-, S/M-, or M-cones in the murine retina, different spectral wavelengths of light are detected. Rods and cones differ morphologically in outer and inner segments, nuclear regions, and synapses [6,7]. They have a common pre-synapse structure, called the ribbon, consisting of similar molecules such as ribeye and bassoon. However, the rod pre-synapse terminal (spherule) contains only one ribbon, while the cone pre-synapse terminal (pedicle) contains multiple ribbons inside [8–10], indicating more complex connections with different types of bipolar neurons and distinct regions of horizontal neurons to code color vision [11,12]. However, so far, the genetic codes to govern or manage specific development between spherule versus (vs.) pedicle are mostly elusive.

The neural retina leucine zipper (NRL), a basic motif-leucine zipper transcription factor essential for rod cell differentiation, actively commits photoreceptor cells to rod cell fate [13–15]. In *Nrl*$^{-/-}$ retina, postmitotic photoreceptor precursors do not develop into rods, producing an excess of S-cones or S-cone-like (SCL) cells [15] with multiple ribbons inside [16], and display electroretinographically cone phototransduction and cone bipolar function [13,15,17,18].

Here, we elucidated the details of the retina OPL development in wild-type (WT) and *Nrl*$^{-/-}$ retinas, using individual photoreceptor labeling to understand photoreceptor synapse development, connections, and plasticity. We examined morphological and histological characterization of developing rods and cones and analyzed pre-synapse development between spherule vs. pedicle. Multi-transcriptome datasets [19–21] were used to further categorize NRL-target genes delineating photoreceptor pre-synapse development.

2. Materials and Methods

2.1. Animals

All studies adhered to the Association for Research in Vision and Ophthalmology (ARVO) Statement for the Use of Animals in Ophthalmic and Vision Research and were approved by the Animal Care and Use Committee of the National Eye Institute. WT C57BL/6J, *Nrl*$^{-/-}$ [15], Nrl promoter-driven green fluorescent protein (*Nrl*p-*GFP*) [14], *Nrl*$^{-/-}$/*Nrl*p-*GFP*, *Cone-rod homeobox* (*Crx*) promoter-driven *Nrl* (*Crxp-Nrl*) [13], *clomeleon* (*Clm*)-*GFP* [22], *Clm-GFP/Nrl*$^{-/-}$, and *Clm-GFP/CrxpNrl* mice were used. *Nrl*$^{-/-}$/*Nrl*p-*GFP*, *Clm-GFP/Nrl*$^{-/-}$, and *Clm-GFP/CrxpNrl* mice were established by mating for this study. Timed pregnant C57BL/6J and CD1 mice used for in vivo retina electroporation were obtained from Jackson Laboratories (Bar Harbor, ME, USA) and Charles River (Rockville, MD, USA). The total numbers of animals used are listed in Supplementary Table S9.

2.2. DNA Construction

Nrl promoter-driven enhanced GFP (*Nrl*p-*EGFP*) [23] and 0.5 kb *S-opsin* promoter-driven *tdTomato* (*S-opsin*p-*tdT*) were used to label rod and cone photoreceptors by in vivo electroporation. Previously, the 2.8 kb mouse *Nrl* promoter (−2734 to +119) was cloned into the *pEGFP-N1* vector [23]. For the construction of *S-opsin*p *tdT*, human Ubiquitin C promoter (*pUB*)-*tdT* was made from *pUB-GFP* vector [24] by replacing *GFP* with *tdT*, and 0.5 kb S-opsin promoter (−529 to −1) was placed into *pUB-tdT*, removing *pUB* by *Sal*I digestion. The following PCR primers were used to amplify the S-opsin promoter: forward 5′-ggtcgacggcaggatgcagttgtttctgc-3′ and reverse 5′-ccgtcgactcccgcttgggatgccct-3′. The *Nrl* shRNA construct (target sequence: ggtcctgtctctatggaagg) was previously described [25].

2.3. In Vivo Electroporation

Electroporation was carried out to label photoreceptors, as previously described [25,26], with minor modifications. Newborn WT CD1, C57BL/6J, and *Nrl*$^{-/-}$ pups were anesthetized by chilling on ice. DNA (1–2 μg/μL) in sterile water containing 0.025% fast green with a total volume of 0.2 μL was injected into the subretinal space by a Hamilton syringe. Electrodes (Harvard Apparatus, Holliston, MA, USA) were placed on either side of the head. Five 80 V pulses (50 ms duration and 950 ms interval) were applied to each mouse.

2.4. Immunohistochemistry

Mouse eyes were enucleated and the entire eyeballs were fixed in 4% paraformaldehyde (PFA), or the posterior eyecups were fixed for 30 min. The entire eyes were embedded in agarose and sectioned at 100 μm thickness with vibratome (Leica VT1000S; Leica, Wetzlar, Germany), and the posterior eyecups were cryoprotected in sucrose and embedded in optimal cutting temperature (OCT) compound (Sakura Finetek USA Inc., Torrance, CA, USA) and cut with a cryostat (Thermo Microm HM550; Thermo Fisher Scientific, Kalamazoo, MI, USA). Fluorescent staining of retinal sections and whole mounts was performed as described [27]. The following primary antibodies were used: anti-GFP (Rockland Immunochemicals, Pottstown, PA, USA), anti-DsRed (Rockland Immunochemicals), anti-Ribeye (BD Biosciences Transduction Laboratories, San Jose, CA, USA), anti-Protein Kinase C alpha (PKCα) (Sigma-Aldrich, St. Louis, MO, USA), anti-Calbindin (Calbiochem, La Jolla, CA, USA), anti-Receptor accessory protein 6 (Reep6) [28], anti-Cone arrestin (CAR, Millipore, Billerica, MA, USA), anti-Guanine nucleotide-binding protein alpha subunit (Goα) (Chemicon, Billerica, MA, USA), and anti-M-opsin (Millipore). Relevant secondary antibodies were conjugated with Alexa Fluor 488, 568, 633, or 647 (Life Technologies, Grand Island, NY, USA). Alexa Fluor 594-conjugated Peanut agglutin lectin (PNA) (Invitrogen, Carlsbad, CA, USA) was incubated with secondary antibodies. Images were taken on confocal microscopes (Leica, Zeiss 700, and 780; Leica).

2.5. Image Analysis

For area measurement of synapse terminals, OPL confocal images were serially taken by less than 0.5 μm thickness from the middle location of whole mount retinas (See Supplementary Figure S1). The most distinct and largest synapse terminal among the series of terminals of each photoreceptor was chosen, and the area was measured using Imaging J software (versions 1.46 and 1.53). For measurements of OPL thickness, PKCα tips with ribbons, and relative ribbon distribution, the middle location of vibratome-sectioned retinas with optic nerve heads was used. Statistical analyses were performed using Student's T-test and/or one-way ANOVA (Tukey or Kriskal–Wallis test) using Prism.

2.6. Gene Profile Analyses

The published data of flow-sorted photoreceptor transcriptome datasets (Gene Expression Omnibus (GEO) accession # GSE74660) [20] and NRL CUT&RUN-seq (GEO accession # GSE197420) [21] were used for gene profile analyses. Data samples were obtained via the fasterq dump of the SRA toolkit (accessed from 10 February to 2 March 2024; https://github.com/ncbi/sra-tools/). The sequencing reads (from postnatal day P2 to P28 rods and SCLs) from GSE74660 RNA seq was trimmed using Trimmomatic v0.39 [29] with parameter HEADCROP:10, and the P10 bio-replicate reads from GSE197420 NRL CUT&RUN-seq were trimmed with parameter ILLUMINACLIP:(adapter):2:15:4:4: true SLIDINGWINDOW:4:20, similar to the previous study [21], with a small modification. Each sample was aligned to the mouse genome (mm10) using HISAT2 [30], and BAM files were converted using SAMtools [31]. For the RNA seq data, three Count matrixes of reads with 55,487 genes were generated using a reference gtf file of GRCm38.102 by featureCounts: Total, Rod_P4_P14, and SCL_P10_P28. The Count matrixes of Rod_P4_P14 and SCL_P10_P28 were applied to identify differentially expressed genes (DEGs) of interest (adjusted (adj.) $p < 0.05$, fold changes 1.5 or 2) by DESeq2 [32]. The count matrix of Total (P2 to P28 rods and SCLs datasets) was further normalized using the transcript per million (TPM) method, and TPMs were used to generate heatmaps. MACS2 was used to identify peaks of the P10 NRL CUT&RUN-seq data [33], and bamCoverage of deepTools was used to generate bigwig files with parameter normalization using bins per million mapped reads (BPM) [34]. For the gene annotation, a bed file of the combined peaks of 4 P10 bio-replicates of NRL CUT&RUN-seq was generated using MSPC [35] and annotated using the GREAT annotation tool with its basal parameters [36]. The gene lists of RNAseq and NRL CUT&RUN-seq were integrated using the Venn_Diagram webtool (accessed on 25 February 2024; https://bioinformatics.psb.ugent.be/webtools/Venn/), and heatmaps were generated using ggplot2 from R Studio. To identify candidate synapse genes in photoreceptors, Synaptic Gene Ontologies and Annotations (SynGO) [37] and DAVID gene functional classification analyses [38] were used.

3. Results

3.1. Wild-Type Outer Plexiform Layer Development in Murine Retina

Synaptic morphogenesis and connections in murine retinal OPL occur postnatally, and the number of synaptic connections continues to increase up to 3 weeks after birth [39]. The immunofluorescence staining of vertical retina sections using markers for horizontal neurons (Calbindin), rod bipolar neurons (PKCα), and presynaptic ribbon protein (Ribeye) displays the developing lamination pattern of the retina OPL (Figure 1A,B). As previously described [40], we observed the appearance of the OPL at P6 (Figure 1A, arrowheads). The segregation of axon terminals and dendritic fields of horizontal neurons was visible after P10 (Figure 1A, arrows), and ribbons aligned as clusters in the inner portion of OPL (Figure 1B, arrows), showing rod- and cone-laminated synaptic connections with horizontal and bipolar neurons.

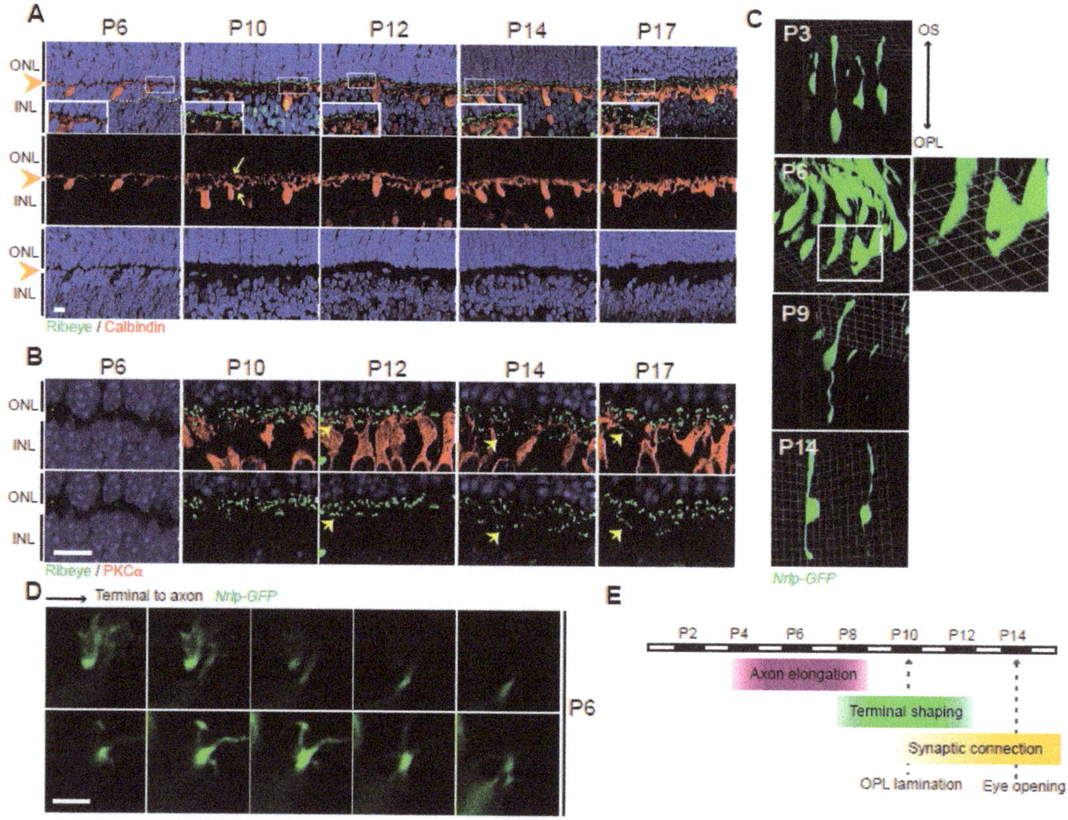

Figure 1. Wild-type outer plexiform layer (OPL) development. (**A**) Developing OPL stained by anti-Ribeye (synaptic ribbons, green) and anti-Calbindin (horizontal neurons, red). Nuclei stained with

DAPI. The boxed areas are shown in insets with higher magnification. OPL (arrowheads) and separate fields of dendrites and axon branches from horizontal neurons (arrows) are shown. (**B**) Developing OPL stained by anti-Ribeye (green) and anti-PKCα (rod bipolar neurons, red). Clusters of pedicle ribbons (arrows) are shown. (**C**) Developing rod photoreceptors. Three-dimensional Volocity converted confocal images labeled by in vivo electroporation of *Nrl*p-*EGFP* plasmids. Insert at P6 is shown with higher magnification. (**D**) Growth cone-like structure at P6 terminals of rod photoreceptors. Confocal images displayed at z-thickness of 0.5 μm from synaptic terminal to axon stalk, labeled by in vivo electroporation of *Nrlp-EGFP*. (**E**) Schematic summary of OPL and photoreceptor synapse development. Abbreviations: ONL, outer nuclear layer; INL, inner nuclear layer; OPL, outer plexiform layer; DAPI, 4′,6-diamidino-2-phenylindole; P, postnatal day; PKCα, protein kinase C alpha. Scale bars, 10 μm in (**A**,**B**) and 5 μm in (**D**).

We further examined the sequential events of GFP-labeled rod photoreceptors by in vivo electroporation to elucidate pre-synapse development of photoreceptor axon terminals since the gross observation of immunofluorescence staining in the sectioned retinas fails to give information of individual photoreceptor development. P0 or P1 retinas were in vivo electroporated with *Nrl*p-*EGFP* plasmids (Supplementary Figure S1), and the labeled photoreceptors were considered to represent the developing steps of rod photoreceptors because the majority of rods are generated around the time of birth [41,42]. Observations of morphology were made at time points of P3, P6, P9, and P14 (Figure 1C) using confocal images and three-dimensional (3D) visualization using Volocity software (version 6.0; Perkin-Elmer, Wattham, MA, USA). At P3, rods displayed an elongated segment from the cell body towards the outer segment but no distinct structure in the opposite direction towards the OPL to form an axon (Figure 1C). At P6, rod axon terminals were forming, and lamellipodia-like structures were seen at the leading edges in Volocity restoration (Figure 1C, boxed insert) and in the z-stack serial optical image data of confocal microscopy (Figure 1D). The lamellipodia-like structures at the terminal tips had the appearance of growth cones, exploring their path and leading axonal growth at the axonal terminals [42]. The observation of these structures was temporal and disappeared quickly. The lamellipodia structures were no longer observed in the whole retinas of P9, indicating that the rod axons were actively elongating from P3 until before P9. At P9, we observed the typical shape of rod photoreceptors from the top of the outer segment to the tip of axon terminals (Figure 1C). We concluded that rod axons elongated between P3 and P9, and the OPL lamination started occurring around P9 and P10. Maturation is accelerated thereafter via pre-synapse morphogenesis and synaptogenesis (Figure 1E) between photoreceptors and second-order neurons, the horizontal and bipolar cells [43,44].

3.2. *Characterization of Individual Spherules and Pedicles*

To further identify spherules and pedicles in mature OPL, the retina vertical sections were double-labeled with antibodies against spherules (Reep6) [28] and pedicles (PNA) [11] (Figure 2A). Staining displayed segregated labeling of spherules (green) and pedicles (red). Double labeling of PNA and CAR in the sectioned retina (Figure 2A) and the OPL of the whole mount retina (Figure 2B) demonstrated the distribution of S-, M/S-, and M-cone pedicles since the PNA labeling is dominant in S-cone pedicles [45,46] and CAR labels M-cone pedicles [47]. The retina stained with CAR and PNA reveals the dominant pedicle type of M/S-cones, along with a smaller number of pure M- (white arrows) and pure S-cone pedicles (arrowheads) (Figure 2A,B). We determined the size of the pedicles by measuring the CAR-stained areas in the OPL of retina whole mounts and found that the synaptic terminal size reached full size before eye opening around P14 in mouse retinas (Figure 2B).

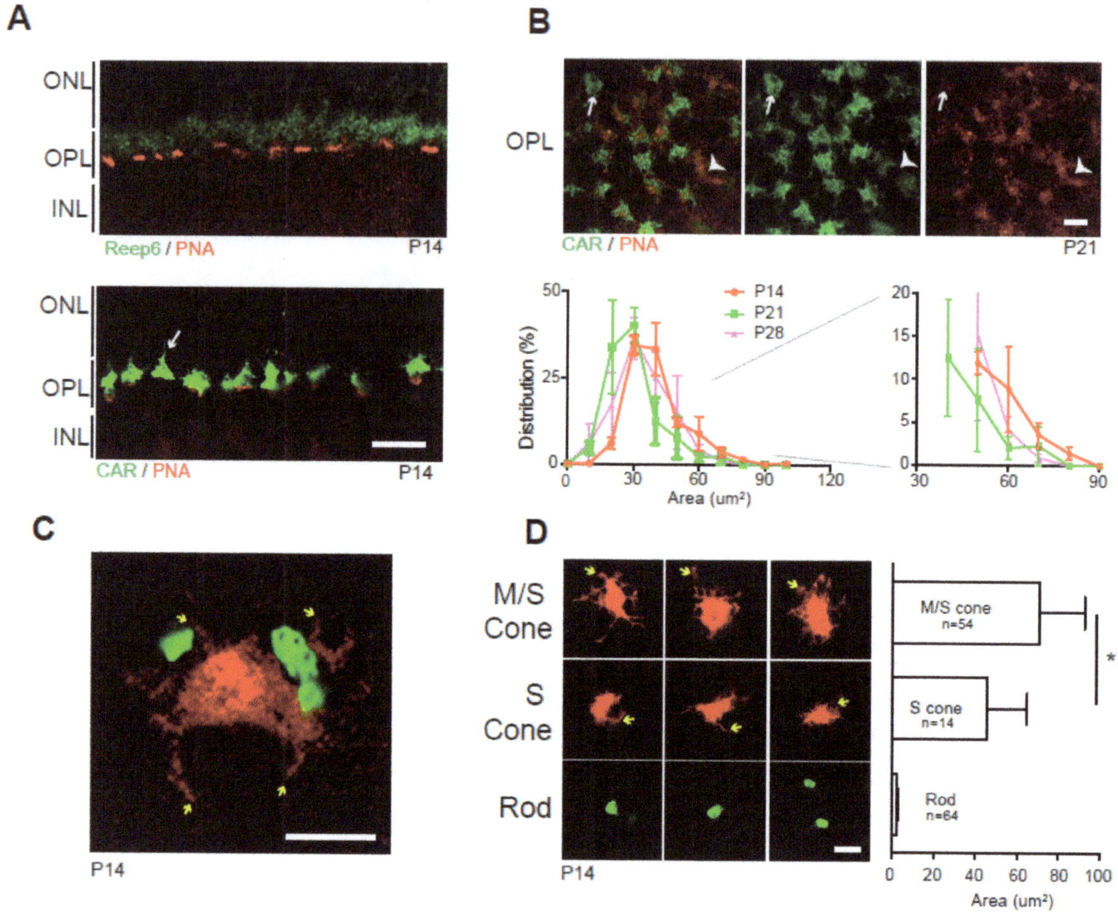

Figure 2. Wild-type spherule and pedicle. (**A**) P14 vertical retina sections stained by anti-Reep6 (spherules; green) and PNA (pedicles; red) (upper panel) or anti-CAR (M-cone, green) and PNA (S-cone, red) (lower panel). (**B**) Horizontal OPL images of retina whole mounts stained by anti-CAR and PNA (upper panels). Pure M-cone (arrows) and S-cone (arrowheads) pedicles are observed. The graph shows the distribution (%, average ± SEM) of CAR pedicle areas in OPL of P14, 21, and 28 whole mount retinas. Over 180 CAR positive pedicles were measured from 3 wild-type C57BL/6J retinas. The area distribution after 30 μm^2 is magnified on the left side. (**C**) Spherules (green) and a pedicle (red) in CD1 retina whole mount labeled by in vivo electroporation of *Nrl*p-*EGFP* and *S-opsinp-tdT*. Telodendria (yellow arrows) are observed. (**D**) Representative images of M/S-, S-cone pedicles and spherules and their area size comparison. Telodendria (yellow arrows) are observed in cones. M-cone and pure S-cone pedicles are segregated by anti-M-opsin staining in the retina whole mounts labeled by *S-opsin*p-*tdT* electroporation. M/S-pedicles (n = 54), S-pedicles (n = 14), and Rod spherules (n = 64) from 3 to 5 wild-type CD1 retinas were measured. The graph displays the average ± SD of each: 70.79 ± 21.48 μm^2 for M pedicles, 45.91 ± 18.97 μm^2 for S pedicles, and 2.64 ± 0.81 μm^2 for rod spherules. * $p \leq 0.05$, two-tailed T-test. Abbreviations: CAR, cone arrestin; PNA, peanut agglutin lectin; *S-opsin* promoter-driven *tdTomato (S-opsinp-tdT)*; P, postnatal day; ONL, outer nuclear layer; OPL, outer plexiform layer; INL, inner nuclear layer. Scale bars, 10 µm in (**A**), and 5 µm in (**B–D**).

To clearly distinguish the morphology and size of individual spherules and pedicles, we labeled rods and cones separately by gene delivery of *Nrl*p-*EGFP* (rods) and *S-opsin*p-*tdT* (cones) plasmids. The confocal image in Figure 2C presents representative spherules and a pedicle in their relative size and morphology, and telodendria on cone pedicles are clearly observed (Figure 2C, yellow arrows). We also stained the retinas with M-opsin antibody to discriminate M-cone (M/S) and pure S-cone pedicles (Supplementary Figure S2). The area measurement of synaptic terminals (Figure 2D) indicated that M-cone pedicles were 1.5-fold bigger than S-cone pedicles, consistent with a previous report [45]. Compared to pedicles, the size of rod spherules was tiny, with over a 20-fold size difference.

3.3. Characterization of Pre-Synapses in Nrl$^{-/-}$ Photoreceptors

NRL is a major transcription factor governing rod differentiation [5], and a study showed that the retina OPL of 18-week-old *Nrl*$^{-/-}$ mice has pedicle-like pre-synapses using electron microscopy [16]. We expanded the previous observation by examining the pre-synaptic terminals of GFP-labeled photoreceptors in WT *Nrl*p-*GFP* and *Nrl*$^{-/-}$/*Nrl*p-*GFP* mice (Figure 3) and postnatally delivered *Nrl*p-*EGFP* and/or *Nrl* short hairpin ribonucleic acid (shRNA) in photoreceptors (Figure 4) at different developing stages. We consistently observed enlarged presynaptic terminals of the original rods in P18 retina vertical sections and whole mounts of *Nrl*$^{-/-}$/*Nrl*p-*GFP* mice, compared to those of WT *Nrl*p-*GFP* (Figure 3A,B). The staining of the ribbon structure protein, Ribeye (red), further revealed the presence of several ribbons inside *Nrl*$^{-/-}$ SCL pre-synapses. However, there were also still some tiny spherule-like pre-synapses in the retina OPL of *Nrl*$^{-/-}$/*Nrl*p-*EGFP* mice at P14 (yellow arrows in Figure 3C), suggesting a delayed commitment of the pedicle-like morphologies in SCLs or delayed size maturation in these pre-synapses compared to those in WT retina. To verify this observation in individual pre-synapses as well, we labeled rods and cones with *Nrl*p-*EGFP* and *S-opsin*p-*tdT* plasmids by in vivo electroporation (Figure 4). The SCLs in *Nrl*$^{-/-}$ retinas labeled by both *Nrl*p-*EGFP* and *S-opsin*p-*tdT* plasmids displayed a larger size of pre-synapse terminals than those of WT spherules but were still smaller than WT S-cone pedicles (Figure 4A–C). The SCL pre-synapse terminals also continued to significantly increase in size from P14 (14.17 ± 8.42 μm^2) to P21 (20.61 ± 10.39 μm^2) (Figure 4C). There was a similar observation in postnatal *Nrl* knockdown photoreceptors. Postnatal *Nrl* knockdown was applied to the CD1 retinas using *Nrl* shRNA by in vivo electroporation. A 2:1 ratio of *Nrl* shRNA and *Nrl*p-*EGFP* plasmids was applied to knockdown NRL expression because this ratio displayed reduced variability in protein expression. (Supplementary Figure S3). The postnatal *Nrl* knockdown demonstrated an increasing size of SCL pre-synaptic terminals by P35 (Figure 4D–F), but the size of SCL pre-synapse terminals did not reach the size of WT S-cone pedicles (Figure 4C). Meanwhile, rod spherule size was not different from P14 to P28 in the retinas of CD1 mice (Figure 4F). We additionally observed that the pedicle size of the M-cone in *Nrl*$^{-/-}$ retinas was smaller than that of WT M-cone pedicles. The smaller size in the M-cone pedicle was maintained from P14 to P21 (Supplementary Figure S4). Our observation indicates that the pedicle-like morphogenesis in pre-synapses of *Nrl*$^{-/-}$ SCLs is gradually determined and that cells originally committed to rods might be switchable, transforming to SCLs.

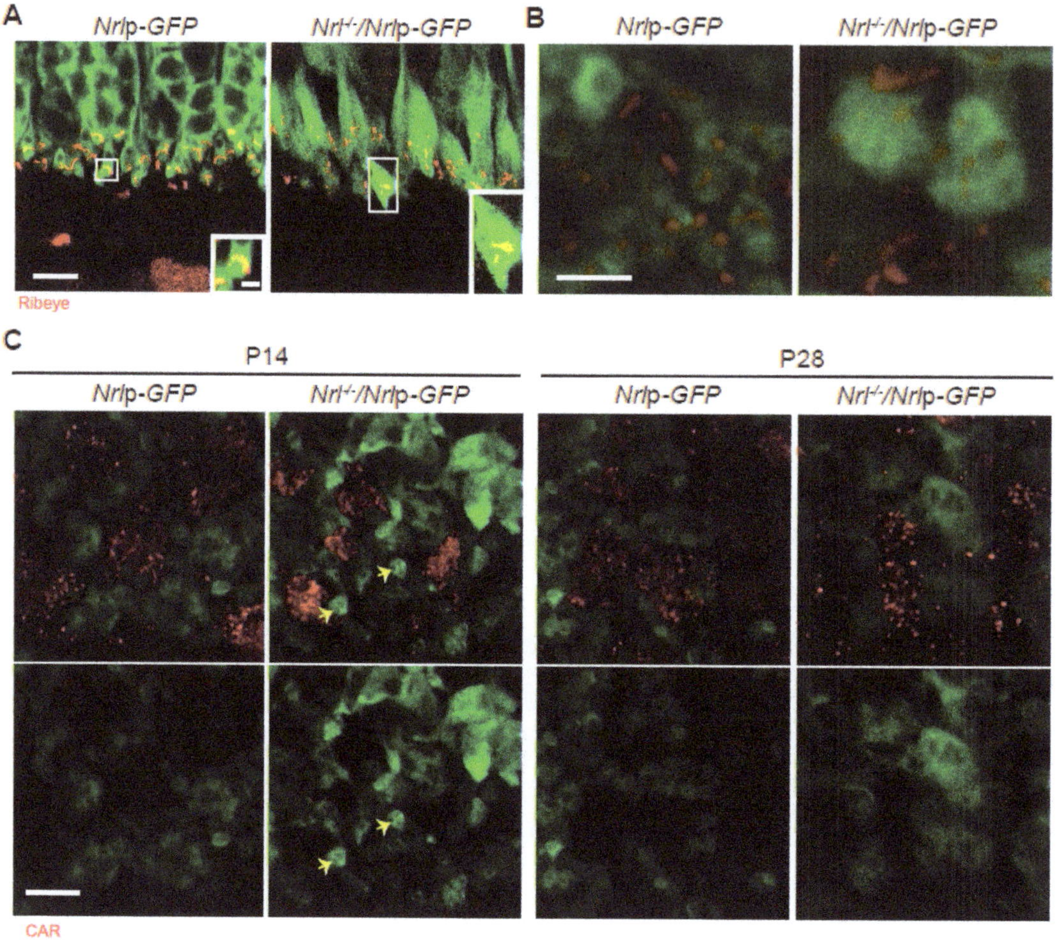

Figure 3. Photoreceptor pre-synapse terminals in *Nrl*$^{-/-}$ retina. (**A**) Vertical retina sections of P18 *Nrl*p-*GFP* and *Nrl*$^{-/-}$/*Nrl*p-*GFP* mice, stained by anti-Ribeye (red). (**B**) Horizontal OPL of retina whole mounts in P18 *Nrl*p-*GFP* and *Nrl*$^{-/-}$/*Nrl*p-*GFP* mice, stained by anti-Ribeye (red). (**C**) Horizontal OPL of retina whole mounts in P14 and P28 *Nrl*p-*GFP* and *Nrl*$^{-/-}$/*Nrl*p-*GFP* mice, stained by anti-CAR (red). Arrows indicate pre-synaptic terminals in small size. Abbreviations: *Nrl*, neural retina leucine zipper; *GFP*, green fluorescent protein; *Nrl*p-*GFP*, *Nrl* promoter-driven *GFP*; P, postnatal day; CAR, cone arrestin. Scale bars, 1 μm in magnified box of (**A**), 5 μm in (**A**,**C**), and 2.5 μm in (**B**).

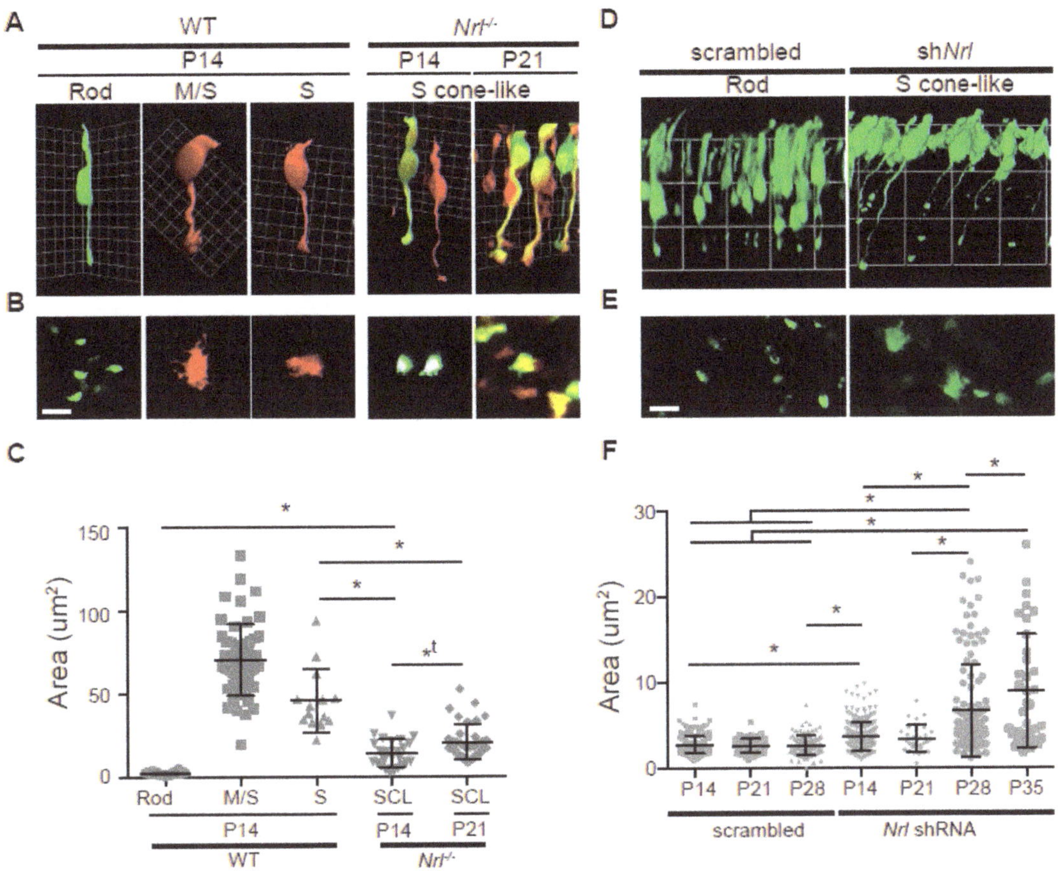

Figure 4. Pre-synapse comparison in wild-type rod, M/S-cone, pure S-cone, and *Nrl*$^{-/-}$ or *Nrl* knockdown S-cone-like (SCL) photoreceptors. (**A**) Representative Volocity 3D images of wild-type rod, M/S-cone, pure S-cone, and *Nrl*$^{-/-}$ SCL photoreceptors, taken from wild-type or *Nrl*$^{-/-}$ retina whole mounts labeled by *Nrl*p-*EGFP*, *S-opsinp-tdT*. Rods (green only), cones (red only) and SCLs (mixed green and red) were imaged. M/S- and pure S-cones were differentiated by staining with an anti-M-opsin antibody. (**B**) Representative confocal images of pre-synapse terminals of wild-type rod, M/S-cone, pure S-cone, and *Nrl*$^{-/-}$ SCL photoreceptors. (**C**) Size distribution of pre-synapses in wild-type rod (n = 64), M/S-cone (n = 54), pure S-cone (n = 14) and *Nrl*$^{-/-}$ SCL (P14, n = 25; P21, n = 38) photoreceptors. (**D**) Representative Volocity 3D images of P28 retina whole mounts labeled by electroporation of scrambled or *Nrl* shRNA plasmid (sh*Nrl*) with *Nrl*p-*EGFP* (2:1 ratio). (**E**) Representative confocal images of pre-synaptic terminals expressing scrambled or *Nrl* shRNA. (**F**) Size distribution of pre-synapses in control (P14, n = 248; P21, n = 64; P28, n = 98) and developing *Nrl* shRNA SCL photoreceptors (P14, n = 246; P21, n = 31; P28, n = 124; P35, n = 36) labeled with *Nrl*p-*EGFP*. Data of area in measurement were analyzed by one-way ANOVA (Tukey or Kriskal–Wallis test) and T-test (two-tailed) in Prism. *, statistically meaningful in one-way ANOVA and T-test; *t, statistically meaningful in T-test. Abbreviations: WT, wild-type; *Nrl*, neural retina leucine zipper; 3D, three-dimensional; *Nrl*p-*EGFP*, *Nrl* promoter-driven enhanced *GFP*; *S-opsin* promoter-driven *tdTomato* (*S-opsinp-tdT*); P, postnatal day; SCL, S-cone-like; shRNA, short hairpin ribonucleic acid. Scale bars, 5 μm in (**B**,**E**).

3.4. $Nrl^{-/-}$ Retina Outer Plexiform Layer Development

To obtain a deeper understanding of developing OPL in $Nrl^{-/-}$ retinas, we stained pre-synapse ribbons (Ribeye), rod bipolar neurons (PKCα), and ON-bipolar neurons (Goα) in WT, $Nrl^{-/-}$ (cone-only), and *CrxpNrl*, (rod-only) mice at P10 to P17 (Figure 5). There were several differences between WT and $Nrl^{-/-}$ retinas during the OPL development (Figure 5A). First, there was the increased thickness of OPL observed in $Nrl^{-/-}$ retinas (Figure 5B), and there were longer, extended dendritic stalks of the rod bipolar neurons in $Nrl^{-/-}$ retinas (Supplementary Figure S5), consistent with a previous report [48]. Third, there was no clear OPL segregation in $Nrl^{-/-}$ retinas, whereas, in the inner portion of WT OPL retinas, clusters of cone ribbons were observed. (Figure 5A, white dashed circles). The ribbon distribution was measured as the relative location within the OPL by numerical assignment from 0 to 1, where 0 corresponds to the location at the border between ONL and OPL and 1 to the location at the border between OPL and INL, as previously reported [49]. The ribbon location of the WT pedicles was not measured at P10 because the clusters of ribbons were not observed in all sections at this stage. However, the spherule and pedicle ribbons started segregating after P10 (Figure 5C). The ribbon locations were measured by the shortest length from the bottom of the ONL (Supplementary Figure S6), and the relative locations were calculated within the OPL. Spherule ribbons were ultimately located within one of the five upper parts of the OPL (0.183 $\pm$ 0.046), while pedicle ribbons were at the middle of the OPL (0.518 $\pm$ 0.077), similar to our previous report [49]. At P17, the average photoreceptor ribbon location in $Nrl^{-/-}$ OPL was between WT spherules and pedicles (0.377 $\pm$ 0.044) (Figure 5C). Next, we measured possible synaptic connections of rod bipolar neurons with photoreceptor pre-synapses by counting the number of stained PKCα tips with ribbons at their top since rod bipolar neurons are stained with PKCα antibody [8] (Figure 5A,D). Over 80% of dendritic tips of rod bipolar neurons had ribbons in WT OPL (Figure 5D), while $Nrl^{-/-}$ OPL had fewer synaptic connections between photoreceptors and rod bipolar neurons at early stages of development (Figure 5A, yellow arrows). However, by P17, the percentage of ribbon tips aligned with rod bipolar dendritic tops was not significantly different (Figure 5D), although there was still a portion of ribbons located under the borderline of the rod bipolar dendritic fields in $Nrl^{-/-}$ OPL, compared to WT (Supplementary Figure S7). Over 20% of the ribbons in $Nrl^{-/-}$ retina were located under the top dendritic borderline of the rod bipolar neurons, while under 5% of the ribbons were in WT retina, indicating that less than 5% of the ribbons in the WT OPL of the vertical retina sections belong to the pedicles, whereas at least over 20% of the ribbons in $Nrl^{-/-}$ retina do not connect with rod bipolar neurons.

To further examine the morphological changes in the dendritic fields and connections of cone bipolar neurons caused by disrupted photoreceptor inputs, we stained the retinas of 1.5-month WT, $Nrl^{-/-}$ (cone-only), and *CrxpNrl* (rod-only) [13] mice with PKCα (green) and Goα (red) antibodies (Figure 6A). In addition, we stained *Clm*-GFP [22], *Clm-GFP/*$Nrl^{-/-}$, and *Clm-GFP/CrxpNrl* mice with Ribeye (red) (Figure 6B). Goα antibody labels all ON-bipolar neurons, including rod bipolar neurons [50,51], and *Clm-GFP* specifically labels type 9 cone bipolar neurons [45], which connect with S-cone photoreceptors. In the WT retina, dendrites of ON-cone bipolar neurons (PKCα negative and Goα positive) were located clearly in the lower portion of OPL, whereas in $Nrl^{-/-}$ and *CrxpNrl* retinas, dendrites of ON-cone bipolar neurons extended to the level of dendritic tips of rod bipolar neurons, indicating that normal OPL lamination was not achieved (Figure 6A). Furthermore, there were differences in synaptic connections between dendritic tips of type 9 cone bipolar neurons and photoreceptor ribbons in WT, $Nrl^{-/-}$, and *CrxpNrl* retinas (Figure 6B). The dendritic tips of type 9 cone bipolar neurons in WT retinas were located in the pedicle layer, the lower portion of OPL, while the dendrites of type 9 cone bipolar neurons in $Nrl^{-/-}$ retinas were extended into the upper portion of OPL. In *CrxpNrl* mice, horizontally prolonged dendritic branches (Figure 6B, asterisks) of type 9 cone bipolar neurons were observed, suggesting that disturbance of sensory input of photoreceptors leads to dendritic rewiring and synaptic plasticity of second-order neurons in the retina.

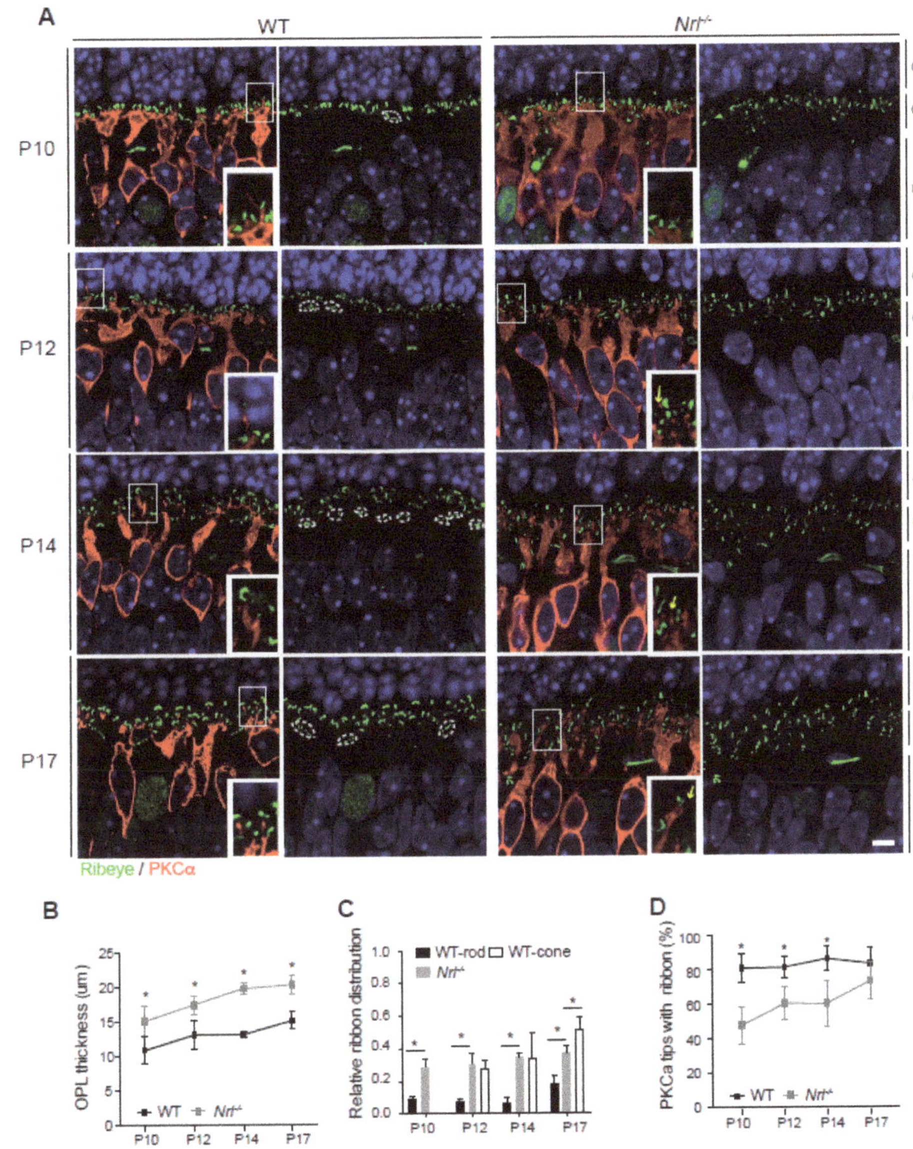

Figure 5. Outer plexiform layer development and synaptic connection in *Nrl*$^{-/-}$ retina. (**A**) Developing (P10 to P17) retinas of wild-type and *Nrl*$^{-/-}$ mice stained by anti-Ribeye (green) and anti-PKCα (red). Clustered pedicle ribbons (white dotted lines) and dendritic tips of rod bipolar neurons without synaptic ribbons (yellow arrows) are observed. (**B**) Comparison of OPL thickness in developing wild-type and *Nrl*$^{-/-}$ retinas. Measurement was quantified on five images of the middle retina (with optic nerve head) from each of three to four animals in different developing stages. Values represent mean ± SD. * $p \leq 0.05$, two-tailed T-test. (**C**) Comparison of the ribbon distribution in OPL. Distance of ribbon location from the ONL bottom when the OPL thickness is considered 1.0. The location of individual ribbons was measured with each OPL thickness in over two images from each of three to four animals. Values represent mean ± SD. * $p \leq 0.05$, two-tailed T-test. (**D**) Number comparison (%)

of rod bipolar neuron dendritic tips with or without ribbons aligned at their tops. Dendritic tips of rod bipolar neurons were measured at P10 (*WT*, n = 363; *Nrl*$^{-/-}$, n = 627), P12 (*WT*, n = 445; *Nrl*$^{-/-}$, n = 953), P14 (*WT*, n = 433; *Nrl*$^{-/-}$, n = 691), and P17 (*WT*, n = 197; *Nrl*$^{-/-}$, n = 269). Values represent mean ± SD. * $p \leq 0.05$, two-tailed T-test. Abbreviations: WT, wild-type; *Nrl*, neural retina leucine zipper; P, postnatal day; PKCα, Protein Kinase C alpha; ONL, outer nuclear layer; OPL, outer plexiform layer; INL, inner nuclear layer. Scale bars, 5 μm in (**A**).

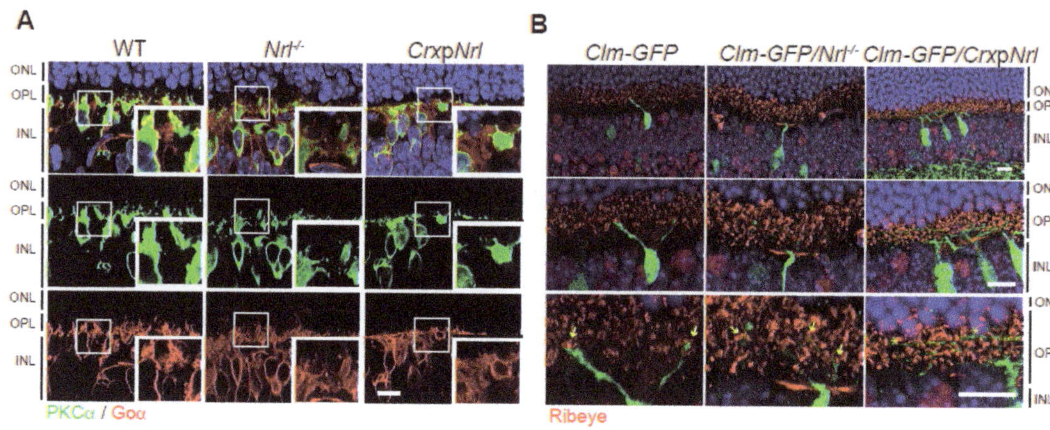

Figure 6. Synaptic plasticity in *Nrl* transgenic retinas. (**A**) 1.5-month retinas of wild-type, *Nrl*$^{-/-}$ (cone-only) and *CrxpNrl* (rod-only) mice stained by anti-PKCα (rod bipolar neurons, green) and anti-Goα (all ON bipolar neurons, red). (**B**) Retinas from 1.5-month-old *Clm-GFP* (type 9 cone bipolar neurons, green), *Clm-GFP/Nrl*$^{-/-}$, and *Clm-GFP/CrxpNrl* mice stained with anti-Ribeye (red). Abbreviations: WT, wild-type; *Nrl*, neural retina leucine zipper; *CrxpNrl*, *Cone-rod homeobox* promoter-driven *Nrl*; Clm, clomeleon; GFP, green fluorescent protein; PKCα, Protein Kinase C alpha; Goα, guanine nucleotide-binding protein alpha subunit; ONL, outer nuclear layer; OPL, outer plexiform layer; INL, inner nuclear layer. Scale bars, 10 μm in (**A**,**B**).

Taken together, our data confirm that disturbed input signals modify dendritic fields of connected neurons and might further lead to the rewiring of whole neural circuitries. Furthermore, synapse maturation would be achieved by functional synaptic connections with partner neurons, which may be one reason for the delayed SCL maturation and OPL alteration in these disturbed photoreceptor models.

3.5. Comparison of Gene Expression in Spherules vs. Pedicles

Next, to identify synaptic molecules that are specific to spherule vs. pedicle development, we took advantage of the published transcriptome profiles of sorted GFP-labeled photoreceptors in developing *Nrl-GFP* and *Nrl*$^{-/-}$/*Nrl-GFP* retinas [20] and NRL CUT&RUN-seq [21]. NRL regulates genes of both rods and cones; it activates rod genes and suppresses cone genes, directly and/or indirectly, in balance with other transcription factors and epigenetic regulators [5,19,21,52,53] (Figure 7A). NRL directly targets nuclear receptor subfamily 2 group E member 3 (*Nr2E3*), a cone gene repressor, and also binds the promoter region of opsin 1 medium-wave-sensitive (*Opn1mw*) [19]. NRL interactions with both rod and cone genes have been reported [21,54]. Here, in our individual rod cell labeling study (Figure 1C–E), we observed initial segment elongation at P3 and axonal elongation around P6. Synaptic morphology appeared around or before P14 in WT rod photoreceptors, but there was delayed synapse development in SCLs (Figures 3 and 4) and OPL alterations afterward (Figures 5 and 6). Thus, to further understand gene networks in spherules vs. pedicles, we extracted gene lists of upregulated genes in WT rods between P14 vs. P4 (upRod, 2-fold changes, adj. $p < 0.05$) and upregulated genes in *Nrl*$^{-/-}$

SCLs between P28 vs. P10 (upSCL, 1.5-fold changes, adj. $p < 0.05$), and CUT&RUN NRL binding genes in P10 retinas. We then combined the DEGs of upRod and upSCL and the NRL-binding genes using Venn diagrams (Figure 7B). We obtained 319 upregulated NRL binding genes in developing WT rods (upRod-NRL) and 83 upregulated NRL binding genes in *Nrl*$^{-/-}$ SCLs (upSCL-NRL) (Supplementary Table S1). We further integrated the upRod by combining down-regulated SCL genes (downSCL) and obtained 34 genes, of which transcripts are upregulated in developing rods but down-regulated in developing SCLs, regardless of NRL binding (Figure 6C, Supplementary Table S1). We then applied SynGO analyses to these identified genes [37]. SynGO is based on a generic conventional synapse model and only uses synapse genes determined by actual experiments, not based on big data prediction. SynGO analyses identified 45 synapse genes from the 319 upRod-NRL list (Supplementary Table S2), highlighting the Synaptic Vesicle Membrane (SVM) in SynGO hierarchy visualization (Figure 7D), and 12 synapse genes from the 34 upRod and downSCLs with SynGO visualization of pre- and post-synapses (Figure 7E). The well-known post-synapse components in conventional neuronal synapses, for example, postsynaptic density protein 95 (PSD-95; discs large MAGUK scaffold protein 4, Dlg4), dystrophin muscular dystrophy (Dmd), and dystroglycan 1 (DG, Dag1), are pre-synapse ribbon components in the retina [55,56]. There were 8 SynGO genes in the 83 upSCLs-NRL list (Figure 7F) and only 11 SynGO genes from the gene list of 600 upSCLs without NRL binding (Supplementary Table S2), suggesting that SynGO ontology might not entirely represent cone pedicles, a highly specialized sensory neuron synapse. On the other hand, there were 66 SynGO synapse genes in the 678 upRod and no-NRL binding group (Supplementary Table S2). Most of the genes in the 45 SynGO upRod-NRL group also displayed upregulation during SCL development, even if the upregulation did not reach the threshold fold-change (Figure 7G), indicating that there could be common synapse components between spherules and pedicles, even though the expression levels and patterns are different. However, G protein subunit beta (*Gnb*) 1 and p21 (rac family small GTPase 1, RAC1) activated kinase 5 (*Pak5*; also called *Pak7*) transcripts were upregulated in rods from P2 to the P28 but down-regulated in developing SCLs (Figure 7G, indicated by a green asterisk). *Gnb1*, one of the NRL-binding genes, is enriched in rod photoreceptors and associated with rod-cone dystrophy [57]. In our analyses, *Gnb3* and *Gnb5* are also NRL-binding genes (Supplementary Figure S7A,B). *Gnb3* is enriched in developing SCLs, and *Gnb5* is upregulated in developing rods (Figure 7H). *Gnb3* is already known to be expressed in the segments and pedicles of cone photoreceptors and is associated with phototransduction [58,59].

To survey more candidate pedicle and spherule genes, DAVID gene functional classification analyses were applied, and more synapse candidate genes were extracted (Tables 1–3 and Supplementary Tables S3–S8). In DAVID analyses of the upRod-NRL genes, there were GOterms such as myelin sheath (GO ID:0043209), neuron projection (GO ID: 0043005), and growth cone (GO ID: 0030426) (Table 1), indicating that our gene analyses properly represent developing rod spherules, including axon elongation. Among them, Neurofascine (Nfasc) is already known as a rod-specific synapse component, and the knockout of *Nfasc* displayed spherule abnormalities and altered rod bipolar dendrites [60]. Previously, we also reported that the knockdowns of *Nfasc*, rod outer segment membrane protein 1 (*Rom1*), Phospholipase C eta 2 (*Plch2*), and Syntrophin alpha 1 (*Snta1*) cause spherule abnormalities [49]. DAVID analyses of upSCL-NRL genes displayed GOterms related to actin binding (GO ID:0003779), actin cytoskeleton (GO ID: 0015629), and microtubule (GO ID:0005874) (Table 2), indicating that pedicles might need additional support or cytoskeleton structuring inside. G protein subunit alpha transducin 2 (*Gnat2*) (GO ID: cell morphogenesis), one of the well-known cone-specific genes, was also identified in the upSCL-NRL gene list. Defects in *Gnat2* cause achromatopsia, or a lack of color vision [61,62]. DAVID analyses of upRod and downSCL added 7 more synapse candidates, as well as 12 SynGo genes (Table 3). Next, we focused on known presynaptic components at photoreceptor ribbon synapses. Synaptic ribbons are specialized presynaptic tethers holding synaptic vesicles and have been described as a conveyor belt to deliver vesicles to the active zone [63]. The differential expressions of several genes, such as Complexin (*Cplx*) 4 and Synaptic vesicle glycoprotein (*Sv*) 2b in spherules and *Cplx3* and *Sv2a* in pedicles, are known [64,65]. In P28 WT and *Nrl*$^{-/-}$ retinas, most ribbon pre-synapse genes (pink indicated in Figure 7I) were expressed higher in SCLs than rod photoreceptors, and just a few genes (complexin4: Cplx4, Receptor accessory protein 6: Reep6, solute carrier family 17: Slc17a7, and unc-119 lipid binding chaperone: Unc119) were highly expressed in rods rather than SCLs (Figure 7I, indicated in green). Interestingly, NRL binds broadly to several pre-synapse ribbon genes (Figure 7I, indicated with bold font).

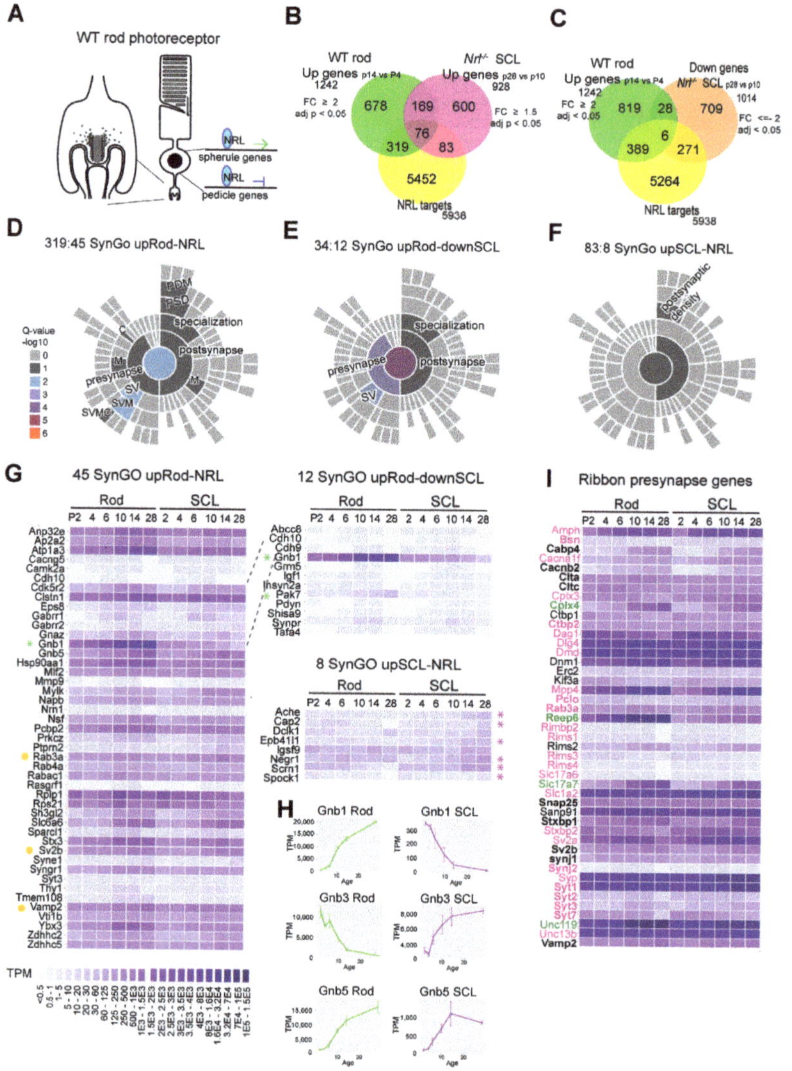

Figure 7. RNA-seq and CUT&RUN-seq analyses of photoreceptor pre-synapse genes. (**A**) Schematic drawing of NRL gene regulation in rod photoreceptors. NRL activates rod genes and suppresses cone genes. (**B**,**C**) Venn diagrams displaying differently expressed genes in rods, SCLs, and NRL-binding genes. (**D**–**F**) SynGO visualizations of upRod-NRL, upRod-downSCL, and upSCL-NRL. (**G**) SynGO

synapse genes, in order from P2 to P28 in transcript per million (TPM) heatmaps: SynGO upRod-NRL (left), SynGO upRod-downSCL (upper right), SynGO upSCL-NRL (down right). Genes in which transcripts are upregulated in developing rods and down-regulated in developing SCLs up to P28 are indicated with a green asterisk, and genes in which transcripts are down-regulated in rods and upregulated in developing SCLs are indicated with a purple asterisk. Known ribbon-associated genes are indicated with a yellow dot. (**H**) *Gnb1*, *Gnb3*, and *Gnb5* expression in developing rods and SCLs measured by RNAseq dataset (TPM). The line plots (average ± SD) from all relevant transcripts in 2 to 4 bio-replicates were generated using ggplot2 from R studio. (**I**) Ribbon synapse genes, in order from P2 to P29 in the TPM heatmap. Gene names in magenta correspond to genes showing higher expression in SCLs than rods at P28. Green gene names correspond to genes showing higher expression in rods than SCLs at P28. Gene names in bold font indicate that NRL binds to the genes. Abbreviations: WT, wild-type; *Nrl*, neural retina leucine zipper; upRod-NRL, upregulated NRL binding genes in developing WT rods; upSCL-NRL, upregulated NRL binding genes in developing S-cone-like photoreceptors; downSCL, down-regulated genes in developing S-cone-like photoreceptors; adj. *p*, adjusted *p*-value; TPM, transcript per million; SynGO, synaptic Gene Ontologies and annotations.

Table 1. DAVID analysis of Rod upregulated NRL targets. This table displays relevant Gene Ontology (GO) terms and genes that could be associated with synapse development.

GOterm	Go ID	Genes
myelin sheath	0043209	NSF, NAPB, HSP90AA1, ENO1B, ATP1A3, THY1, ENO1, COX6A1, NFASC, PKM, GNB1, GNB5, SLC25A12, ALDOA, SLC25A5
synapse	0045202	RAB3A, NRN1, RPLP1, CABP4, CLSTN1, CAMK2A, ATP1A3, TULP1, YBX3, ZDHHC2, LY6C1, EPS8, GABRR2, GABRR1, SYNGR1, SV2B, TMEM108, RABAC1, SH3GL2, SNTA1, CACNG5, PTPRN2, WHRN, CARTPT, MYRIP, USH2A, CDH10, SPARCL1, RPS21, VAMP2, SNTB2
neuron projection	0043005	HSP90AA1, PACRG, WHRN, CAMK2A, RTN4RL2, LSM4, EPS8, GABRR2, GABRR1, PCP4, NFASC, RIT2, SV2B, BBS7, STX3, VAMP2, CDK5R2, RGS6
dendrite	0030425	GNAZ, RCVRN, KCNJ12, KCNIP2, CAMK2A, ZDHHC5, RTN4RL2, THY1, TRAK1, TXN2, SLC6A6, PEX5L, NFASC, GNB1, TMEM108, STX3
Calcium ion binding	0005509	CRB1, SYT3, RCVRN, KCNIP2, CABP4, CLSTN1, GUCA1A, RHOT2, PCP4, EHD4, VSNL1, PITPNM1, CDH10, CDHR1, SPARCL1, PLCH2, SLC25A12, GAS6, NUCB2, SLC25A25, TGM3
vesicle-mediated transport	0016192	NSF, NAPB, RAB4A, SYT3, TULP1, AP2A2, ARL4D, SPIRE2, STX3, VAMP2, HSPA1B, VTI1B, HSPA1A
ion transport	0006811	KCNG1, KCNJ5, KCNE2, KCNK9, KCNJ12, KCNIP2, SLC6A15, ATP1A3, SLC5A2, SLC4A5, GABRR2, GABRR1, NIPAL1, CNGA1, SCN4A
growth cone	0030426	EPS8, WHRN, ENO1B, RASGRF1, STX3, THY1, ENO1, CDK5R2
calmodulin binding	0005516	PCP4, RIT2, CAMK2A, VAMP2, MYLK, SNTA1, SNTB2
GABA-ergic synapse	0030425	SLC6A6, GABRR2, GABRR1, CDH10, CLSTN1, ZDHHC5
postsynaptic membrane	0045211	SLC6A6, GABRR2, GABRR1, CLSTN1, CACNG5, SNTA1, SYNE1
chemical synaptic transmission	0007268	GABRR2, GABRR1, SV2B, CLSTN1, CARTPT
extracellular matrix	0031012	ADAMTSL1, COL26A1, COL3A1, ADAMTS3, COL4A3, RTN4RL2, MMP9
transmembrane transport	005508	KCNG1, SLC25A29, SV2B, CNGA1, SCN4A, SLC5A2, SLC25A5, SLC25A12, SLC25A25, SLC4A5
cell adhesion	0007155	NFASC, PRPH2, CDHR1, CDH10, CLSTN1, ROM1, COL4A3, ADAM9, SPG7, THY1, CD34

Table 1. *Cont.*

GOterm	Go ID	Genes
extracellular region	0005576	CRB1, CCDC126, CDNF, PRCD, CLSTN1, LOXL4, TULP1, PLA2G7, GLB1L2, CST3, ADAMTSL1, FTH1, NRTN, FAM3C, CD34, ST3GAL1, CTSB, CHGA, COL26A1, CARTPT, MMEL1, USH2A, MMP9, POMC, COL3A1, RBP3, QPCT, COL4A3, SPARCL1, GAS6, NUCB2, HSPA1A

129 genes: ADAM9, ADAMTS3, ADAMTSL1, ALDOA, AP2A2, ARL4D, ATP1A3, BBS7, CABP4, CACNG5, CAMK2A, CARTPT, CCDC126, CD34, CDH10, CDHR1, CDK5R2, CDNF, CHGA, CLSTN1, CNGA1, COL26A1, COL3A1, COL4A3, COX6A1, CRB1, CST3, CTSB, EHD4, ENO1, ENO1B, EPS8, FAM3C, FTH1, GABRR1, GABRR2, GAS6, GLB1L2, GNAZ, GNB1, GNB5, GUCA1A, HSP90AA1, HSPA1A, HSPA1B, KCNE2, KCNG1, KCNIP2, KCNJ12, KCNJ5, KCNK9, LOXL4, LSM4, LY6C1, MMEL1, MMP9, MYLK, MYRIP, NAPB, NFASC, NIPAL1, NRN1, NRTN, NSF, NUCB2, PACRG, PCP4, PEX5L, PITPNM1, PKM, PLA2G7, PLCH2, POMC, PRCD, PRPH2, PTPRN2, QPCT, RAB3A, RAB4A, RABAC1, RASGRF1, RBP3, RCVRN, RGS6, RHOT2, RIT2, ROM1, RPLP1, RPS21, RTN4RL2, SCN4A, SH3GL2, SLC25A12, SLC25A25, SLC25A29, SLC25A5, SLC4A5, SLC5A2, SLC6A15, SLC6A6, SNTA1, SNTB2, SPARCL1, SPG7, SPIRE2, ST3GAL1, STX3, SV2B, SYNE1, SYNGR1, SYT3, TGM3, THY1, TMEM108, TULP1, TXN2, USH2A, VAMP2, VSNL1, VTI1B, WHRN, YBX3, ZDHHC2, ZDHHC5. SynGO analyses also provided genes underlined.

Table 2. DAVID analysis of SCL-upregulated NRL targets. This table displays relevant Gene Ontology (GO) terms and genes that could be associated with synapse development.

GOterm	Go ID	Genes
actin binding	0003779	DIAPH3, PARVG, WIPF1, EPB41L1, MYO7A, CAP2
dendrite	0030425	ACHE, OPN4, NEGR1, NR1D1, IGSF9, CNNM4
actin cytoskeleton	0015629	RINL, PARVG, WIPF1, MYO7A
cell morphogenesis	0000902	GREM1, GNAT2, CAP2
ATP-dependent microtubule motor activity, minus-end-directed	0008569	KIFC1, DNAH7C
axon	0030424	ACHE, OPN4, MAP4, IGSF9, DCLK1
amino acid transmembrane transport	0003333	SLC7A8, SLC38A3
microtubule	0005874	KIFC1, MAP4, DNAH7C
calcium ion binding	0005509	HEG1, CALU, SPOCK1, AMY1, DNAH7C
extracellular region	0005576	GREM1, ACHE, ORM1, CALU, SPOCK1, AMY1

25 genes: ACHE, AMY1, CALU, CAP2, CNNM4, DCLK1, DIAPH3, DNAH7C, EPB41L1, GNAT2, GREM1, HEG1, IGSF9, KIFC1, MAP4, MYO7A, NEGR1, NEGR1, OPN4, ORM1, PARVG, RINL, SLC38A3, SLC7A8, SPOCK1, WIPF1. SynGO analyses also provided genes underlined.

Table 3. DAVID analysis of Rod upregulated and SCL down-regulated genes. This table displays relevant Gene Ontology (GO) terms and genes that could be associated with synapse development.

GOterm	Go ID	Genes
synapse	0045202	SYNPR, RNF112, GRM5, CDH10, PDYN, SHISA9, CDH9, INSYN2A
cell-cell junction assembly	0007043	CDH10, NR1H4, CDH9
neuronal dense core vesicle lumen	0099013	IGF1, PDYN
postsynaptic membrane	0045211	GRM5, GRM6, SHISA9, CDH9
glutamatergic synapse	0098978	GRM5, CDH10, IGF1, SHISA9, CDH9
postsynaptic density	0014069	RNF112, GRM5, SHISA9, INSYN2A
calcium-mediated signaling using intracellular calcium source	0035584	GRM5, VCAM1
myotube differentiation	0014902	IGF1, ANKRD2
response to zinc ion	0010043	VCAM1, ABCC8

Table 3. *Cont.*

GOterm	Go ID	Genes
regulation of long-term neuronal synaptic plasticity	0048169	GRM5, AGT
dendrite	0030425	GRM6, GNB1, PDYN, CDH9
extracellular region	0005576	TAFA4, VCAM1, FRZB, IGF1, PDYN, BMP7

19 genes: *ABCC8*, AGT, ANKRD2, BMP7, *CDH9*, *CDH10*, FRZB, *GNB1*, *GRM5*, GRM6, IGF1, *INSYN2A*, NR1H4, *PDYN*, *RNF112*, *SHISA9*, *SYNPR*, *TAFA4*, VCAM1. SynGO analyses also provided genes underlined. A total of *11 genes* (Italic) showed the highest expression in either central nervous system (CNS) embryonic day18 (E18), cerebellum adult, or frontal lobe cortex compared to other tissues in mouse ENCODE transcriptome data.

4. Discussion

Our study displays that NRL loss switches and/or alters synapse morphology, size, ribbon number, ribbon location, and connections with bipolar neurons from rods to cones. *Nrl*$^{-/-}$ retinas exhibited pedicle properties in all mature photoreceptors, although the pedicle size in SCL photoreceptors was smaller than that of WT pedicles, and the maturation was delayed compared to that in WT retinas. Furthermore, the switch in photoreceptor presynaptic types caused dendritic alternation of the second-order bipolar neurons in *Nrl*$^{-/-}$ and *CrxpNrl* retinas. However, in the switched photoreceptors, synaptic input or connections did not cause degeneration of the second-order neurons. Therefore, this study provides a good model for studying photoreceptor adaptations and altered synapse development and how altered gene regulation can govern these events.

To identify photoreceptor synapse candidate genes and understand ribbon genes in the context of spherules and pedicles, we applied our synapse development time course in WT and *Nrl*$^{-/-}$ retinas for big data analyses using datasets of developing rods and SCLs [20], NRL CUT&RUN-seq [21], and GO analysis tools, such as SynGO [37] and DAVID [38]. Spherules and pedicles are quite different in their shapes and functions. When over 12 different types of bipolar neurons are considered [66], there might be a variety of synapse alterations and connections among the different types of photoreceptors, from rods to cones and from M-, M/S-, and S-cones. In spite of their differences in morphology and function, there are a limited number of unique genes expressed in spherules vs. pedicles. Reep6 is one gene that is highly expressed only in spherules [67]. Our analyses also revealed *Gnb1* and *Pak7*, whose transcripts were specific in developing and mature rods. Acetylcholinesterase (*Ache*), Cyclase-associated actin cytoskeleton regulatory protein 2 (*Cap2*), Doublecortin-like kinase 1 (*Dclk1*), Erythrocyte membrane protein band 4.1-like 1 (*Epb41l1*), Neuronal growth regulator 1 (*Negr1*), Secerin 1 (*Scrn1*), Sparc/osteonectin, cwcv, and kazal-like domains proteoglycan 1 (*Spock1*) were specific to developing and mature SCLs, suggesting that they could be specific to cone pedicles. Interestingly, Gnb1 was specific in rods, whereas Gnb3 was specific in SCLs. It is already known that Gnb3 is expressed in WT pedicles [58]. However, the synaptic function of Gnb3 is still unknown. Our analyses also suggested that Cplx3 and Sv2a are expressed higher in pedicles than in spherules, and Cplx4 and Sv2b are more specific in spherules. Interestingly, there was differential expression of known and common ribbon pre-synapse genes between rods and SCLs, and NRL regulation of ribbon synapse genes was confirmed (Figure 7I) [54]. However, caution needs to be taken when interpreting the differential expression of pre-synapse ribbon genes between rods and SCLs because SCLs are not WT cones, although current and previous data analyses suggest that SCLs are close to WT cones [68,69]. It is noteworthy that the switched or hybrid photoreceptors in this current, or a previous [16], study did not cause degeneration of the second-order neurons. In addition, *Nrl* knockout delayed pedicle development compared to the WT, and postnatal *Nrl* knockdown using *Nrl* shRNA delayed the switch between spherule vs. pedicle, suggesting that committed photoreceptors might re-enter differentiation and form another type of photoreceptor cell. This finding may be important when considering clinical trials (NCT05203939; NCT04945772; NCT03326336; NCT04278131) using optogenetic or transcription factor gene delivery to treat retinitis pigmentosa or other degenerative retina diseases. Furthermore, the rod or cone photoreceptor synapse genes revealed in this study could be candidate genes for genetic diseases of the brain and sensory system because synapse genes are common components in different neuron sets. For example, our DAVID analyses provide synapse gene lists of Kyoto Encyclopedia of Genes and Genomes (KEGG) DISEASE pathways such as Parkinson's disease, Huntington's disease, Alzheimer's disease, and long-term depression (Supplementary Tables S3–S8). Finally, this study once more indicates that NRL is broadly involved in the regulation of both spherule and pedicle genes during photoreceptor development and paves the road toward a deeper understanding of synapse genes and their function in rod and cone photoreceptors.

Supplementary Materials: The following supporting information can be downloaded at: https://www.mdpi.com/article/10.3390/life14091103/s1, Figure S1: Retina in vivo electroporation; Figure S2: Pedicle discrimination between M/S and pure S cone photoreceptors; Figure S3: Efficiency of *Nrl* shRNA constructs; Figure S4: Size comparison of cone arrestin; Figure S5:Vertical retina images of WT and *Nrl*$^{-/-}$; Figure S6: OPL thickness and relative ribbon location in WT and *Nrl*$^{-/-}$ retinas; Figure S7: Aberrant synaptic connectivity of rod and cone pre-synapses to distinct biopolar cells; Figure S8: NRL binding to the genes *G protein subunit beta (Gnb)1*, *Gnb3* and *Gnb5*; Table S1: Gene lists of Venn diagram; Table S2: SynGo genes from the gene lists of Venn diagram; Table S3: Upregulated Rod NRL targets—Chart, DAVID; Table S4: Upregulated Rod NRL targets—Clustering, DAVID; Table S5: Upregulated SCL targets—Chart, DAVID; Table S6: Upregulated SCL NRL targets—Clustering, DAVID; Table S7: Upregulated Rod downregulated SCL—Chart, DAVID; Table S8: Upregulated Rod downregulated SCL—Clustering, DAVID; Table S9: Animals used.

Author Contributions: Methodology, S.-Y.K. and G.K.S.; Software, S.-Y.K. and B.-H.M.; Validation, S.-Y.K., C.H.P. and G.K.S.; Formal analysis, S.-Y.K., C.H.P., B.-H.M. and G.K.S.; Investigation, S.-Y.K., C.H.P. and G.K.S.; Data curation, S.-Y.K. and B.-H.M.; Writing—original draft, S.-Y.K.; Writing—review & editing, S.-Y.K., C.H.P., B.-H.M. and G.K.S.; Visualization, S.-Y.K.; Supervision, S.-Y.K. All authors have read and agreed to the published version of the manuscript.

Funding: This research was supported by Intramural research program of the National Eye Institute (EY000450, EY000546).

Institutional Review Board Statement: The animal research is supported by National Institutes of Health Animal study NEI-650.

Informed Consent Statement: Not applicable.

Data Availability Statement: The original contributions presented in the study are included in the article/Supplementary Materials, further inquiries can be directed to the corresponding author/s, and the original data used in this study are already available in NCBI Gene expression omnibus datatbase at accession numbers GES74660 and GSE197420.

Acknowledgments: First of all, we thank Anand Swaroop for the mentoring, support, and laboratory resources; Robert Fariss for technical support, teaching, and advice; and Matthew Brook, Marie-Audrey I. Kautzmann, Juthaporn Assawachananont, Jacob Nellissery, Yide Mi, and Wei Li at the National Eye Institute for technical support and advice. This research was supported by the Intramural Research Program of the National Eye Institute, the National Institutes of Health.

Conflicts of Interest: The authors declare no competing interests.

Abbreviations

3D	three-dimensional
adj. *p*	adjusted *p*-value
ARVO	Association for Research in Vision and Ophthalmology
BPM	bins per million mapped reads
CAR	cone arrestin
Clm	clomeleon
Cplx	complexin

CrxpNrl	Crx promoter
DAPI	4′,6-diamidino-2-phenylindole
DEGs	differentially expressed genes
Dlg4	discs large MAGUK scaffold protein 4
Dmd	dystrophin muscular dystrophy
downSCL	down-regulated genes in developing S-cone-like photoreceptors
KEGG	Kyoto Encyclopedia of Genes and Genomes
GEO	Gene Expression Omnibus
GFP	green fluorescent protein
Gnb	G protein subunit beta
GO	Gene Ontology
Goα	guanine nucleotide-binding protein alpha subunit
Nr2E3	nuclear receptor subfamily 2 group E member 3
NRL	neural retina leucine zipper
Nrlp-EGFP	Nrl promoter-driven enhanced GFP
Nrlp-GFP	Nrl promoter-driven GFP
OCT	optimal cutting temperature
OPL	outer plexiform layer
P	postnatal day
PFA	paraformaldehyde
PKCα	Protein Kinase C alpha
PNA	peanut agglutin lectin
PSD-95	postsynaptic density protein 95
pUB	Ubiquitin C promoter
S-opsinp-tdT	S opsin promoter-driven tdTomato
SCL	S-cone-like
shRNA	short hairpin ribonucleic acid
Sv	Synaptic vesicle glycoprotein
SynGO	Synaptic Gene Ontologies and annotations
TPM	transcript per million
upRod-NRL	upregulated NRL binding genes in developing WT rods
upSCL-NRL	upregulated NRL binding genes in developing S-cone-like photoreceptors
vs.	versus
WT	Wild-type

References

1. Sperry, R.W. Chemoaffinity in the Orderly Growth of Nerve Fiber Patterns and Connections. *Proc. Natl. Acad. Sci. USA* **1963**, *50*, 703–710. [CrossRef]
2. Katz, L.C.; Shatz, C.J. Synaptic activity and the construction of cortical circuits. *Science* **1996**, *274*, 1133–1138. [CrossRef] [PubMed]
3. Peichl, L.; Gonzalez-Soriano, J. Morphological types of horizontal cell in rodent retinae: A comparison of rat, mouse, gerbil, and guinea pig. *Vis. Neurosci.* **1994**, *11*, 501–517. [CrossRef]
4. Ghosh, K.K.; Bujan, S.; Haverkamp, S.; Feigenspan, A.; Wassle, H. Types of bipolar cells in the mouse retina. *J. Comp. Neurol.* **2004**, *469*, 70–82. [CrossRef] [PubMed]
5. Swaroop, A.; Kim, D.; Forrest, D. Transcriptional regulation of photoreceptor development and homeostasis in the mammalian retina. *Nat. Rev. Neurosci.* **2010**, *11*, 563–576. [CrossRef]
6. Kennedy, B.; Malicki, J. What drives cell morphogenesis: A look inside the vertebrate photoreceptor. *Dev. Dyn.* **2009**, *238*, 2115–2138. [CrossRef]
7. Mustafi, D.; Engel, A.H.; Palczewski, K. Structure of cone photoreceptors. *Prog. Retin. Eye Res.* **2009**, *28*, 289–302. [CrossRef]
8. Wassle, H. Parallel processing in the mammalian retina. *Nat. Rev. Neurosci.* **2004**, *5*, 747–757. [CrossRef] [PubMed]
9. Sterling, P.; Matthews, G. Structure and function of ribbon synapses. *Trends Neurosci.* **2005**, *28*, 20–29. [CrossRef]
10. tom Dieck, S.; Altrock, W.D.; Kessels, M.M.; Qualmann, B.; Regus, H.; Brauner, D.; Fejtova, A.; Bracko, O.; Gundelfinger, E.D.; Brandstatter, J.H. Molecular dissection of the photoreceptor ribbon synapse: Physical interaction of Bassoon and RIBEYE is essential for the assembly of the ribbon complex. *J. Cell Biol.* **2005**, *168*, 825–836. [CrossRef]
11. Wassle, H.; Puller, C.; Muller, F.; Haverkamp, S. Cone contacts, mosaics, and territories of bipolar cells in the mouse retina. *J. Neurosci.* **2009**, *29*, 106–117. [CrossRef] [PubMed]
12. Ahnelt, P.; Kolb, H. Horizontal cells and cone photoreceptors in human retina: A Golgi-electron microscopic study of spectral connectivity. *J. Comp. Neurol.* **1994**, *343*, 406–427. [CrossRef] [PubMed]

13. Oh, E.C.; Khan, N.; Novelli, E.; Khanna, H.; Strettoi, E.; Swaroop, A. Transformation of cone precursors to functional rod photoreceptors by bZIP transcription factor NRL. *Proc. Natl. Acad. Sci. USA* **2007**, *104*, 1679–1684. [CrossRef]
14. Akimoto, M.; Cheng, H.; Zhu, D.; Brzezinski, J.A.; Khanna, R.; Filippova, E.; Oh, E.C.; Jing, Y.; Linares, J.L.; Brooks, M.; et al. Targeting of GFP to newborn rods by Nrl promoter and temporal expression profiling of flow-sorted photoreceptors. *Proc. Natl. Acad. Sci. USA* **2006**, *103*, 3890–3895. [CrossRef] [PubMed]
15. Mears, A.J.; Kondo, M.; Swain, P.K.; Takada, Y.; Bush, R.A.; Saunders, T.L.; Sieving, P.A.; Swaroop, A. Nrl is required for rod photoreceptor development. *Nat. Genet.* **2001**, *29*, 447–452. [CrossRef]
16. Strettoi, E.; Mears, A.J.; Swaroop, A. Recruitment of the rod pathway by cones in the absence of rods. *J. Neurosci.* **2004**, *24*, 7576–7582. [CrossRef]
17. Daniele, L.L.; Lillo, C.; Lyubarsky, A.L.; Nikonov, S.S.; Philp, N.; Mears, A.J.; Swaroop, A.; Williams, D.S.; Pugh, E.N., Jr. Cone-like morphological, molecular, and electrophysiological features of the photoreceptors of the Nrl knockout mouse. *Investig. Ophthalmol. Vis. Sci.* **2005**, *46*, 2156–2167. [CrossRef]
18. Nikonov, S.S.; Daniele, L.L.; Zhu, X.; Craft, C.M.; Swaroop, A.; Pugh, E.N., Jr. Photoreceptors of $Nrl^{-/-}$ mice coexpress functional S- and M-cone opsins having distinct inactivation mechanisms. *J. Gen. Physiol.* **2005**, *125*, 287–304. [CrossRef]
19. Hao, H.; Kim, D.S.; Klocke, B.; Johnson, K.R.; Cui, K.; Gotoh, N.; Zang, C.; Gregorski, J.; Gieser, L.; Peng, W.; et al. Transcriptional regulation of rod photoreceptor homeostasis revealed by in vivo NRL targetome analysis. *PLoS Genet.* **2012**, *8*, e1002649. [CrossRef]
20. Kim, J.W.; Yang, H.J.; Brooks, M.J.; Zelinger, L.; Karakulah, G.; Gotoh, N.; Boleda, A.; Gieser, L.; Giuste, F.; Whitaker, D.T.; et al. NRL-Regulated Transcriptome Dynamics of Developing Rod Photoreceptors. *Cell Rep.* **2016**, *17*, 2460–2473. [CrossRef]
21. Liang, X.; Brooks, M.J.; Swaroop, A. Developmental genome-wide occupancy analysis of bZIP transcription factor NRL uncovers the role of c-Jun in early differentiation of rod photoreceptors in the mammalian retina. *Hum. Mol. Genet.* **2022**, *31*, 3914–3933. [CrossRef] [PubMed]
22. Berglund, K.; Schleich, W.; Krieger, P.; Loo, L.S.; Wang, D.; Cant, N.B.; Feng, G.; Augustine, G.J.; Kuner, T. Imaging synaptic inhibition in transgenic mice expressing the chloride indicator, Clomeleon. *Brain Cell Biol.* **2006**, *35*, 207–228. [CrossRef] [PubMed]
23. Kautzmann, M.A.; Kim, D.S.; Felder-Schmittbuhl, M.P.; Swaroop, A. Combinatorial regulation of photoreceptor differentiation factor, neural retina leucine zipper gene NRL, revealed by in vivo promoter analysis. *J. Biol. Chem.* **2011**, *286*, 28247–28255. [CrossRef]
24. Rompani, S.B.; Cepko, C.L. Retinal progenitor cells can produce restricted subsets of horizontal cells. *Proc. Natl. Acad. Sci. USA* **2008**, *105*, 192–197. [CrossRef]
25. Matsuda, T.; Cepko, C.L. Electroporation and RNA interference in the rodent retina in vivo and in vitro. *Proc. Natl. Acad. Sci. USA* **2004**, *101*, 16–22. [CrossRef] [PubMed]
26. Matsuda, T.; Cepko, C.L. Analysis of gene function in the retina. *Methods Mol. Biol.* **2008**, *423*, 259–278. [CrossRef] [PubMed]
27. Kim, S.Y.; Assawachananont, J. A New Method to Visualize the Intact Subretina From Retinal Pigment Epithelium to Retinal Tissue in Whole Mount of Pigmented Mouse Eyes. *Transl. Vis. Sci. Technol.* **2016**, *5*, 6. [CrossRef]
28. Veleri, S.; Nellissery, J.; Mishra, B.; Manjunath, S.H.; Brooks, M.J.; Dong, L.; Nagashima, K.; Qian, H.; Gao, C.; Sergeev, Y.V.; et al. REEP6 mediates trafficking of a subset of Clathrin-coated vesicles and is critical for rod photoreceptor function and survival. *Hum. Mol. Genet.* **2017**, *26*, 2218–2230. [CrossRef]
29. Bolger, A.M.; Lohse, M.; Usadel, B. Trimmomatic: A flexible trimmer for Illumina sequence data. *Bioinformatics* **2014**, *30*, 2114–2120. [CrossRef]
30. Kim, D.; Langmead, B.; Salzberg, S.L. HISAT: A fast spliced aligner with low memory requirements. *Nat. Methods* **2015**, *12*, 357–360. [CrossRef]
31. Li, H.; Handsaker, B.; Wysoker, A.; Fennell, T.; Ruan, J.; Homer, N.; Marth, G.; Abecasis, G.; Durbin, R.; 1000 Genome Project Data Processing Subgroup. The Sequence Alignment/Map format and SAMtools. *Bioinformatics* **2009**, *25*, 2078–2079. [CrossRef] [PubMed]
32. Love, M.I.; Huber, W.; Anders, S. Moderated estimation of fold change and dispersion for RNA-seq data with DESeq2. *Genome Biol.* **2014**, *15*, 550. [CrossRef] [PubMed]
33. Feng, J.; Liu, T.; Zhang, Y. Using MACS to identify peaks from ChIP-Seq data. *Curr. Protoc. Bioinform.* **2011**, *34*, 2.14.1–2.14.14. [CrossRef] [PubMed]
34. Ramirez, F.; Dundar, F.; Diehl, S.; Gruning, B.A.; Manke, T. deepTools: A flexible platform for exploring deep-sequencing data. *Nucleic Acids Res.* **2014**, *42*, W187–W191. [CrossRef]
35. Jalili, V.; Matteucci, M.; Masseroli, M.; Morelli, M.J. Using combined evidence from replicates to evaluate ChIP-seq peaks. *Bioinformatics* **2015**, *31*, 2761–2769. [CrossRef]
36. McLean, C.Y.; Bristor, D.; Hiller, M.; Clarke, S.L.; Schaar, B.T.; Lowe, C.B.; Wenger, A.M.; Bejerano, G. GREAT improves functional interpretation of cis-regulatory regions. *Nat. Biotechnol.* **2010**, *28*, 495–501. [CrossRef]
37. Koopmans, F.; van Nierop, P.; Andres-Alonso, M.; Byrnes, A.; Cijsouw, T.; Coba, M.P.; Cornelisse, L.N.; Farrell, R.J.; Goldschmidt, H.L.; Howrigan, D.P.; et al. SynGO: An Evidence-Based, Expert-Curated Knowledge Base for the Synapse. *Neuron* **2019**, *103*, 217–234.e4. [CrossRef]
38. Dennis, G., Jr.; Sherman, B.T.; Hosack, D.A.; Yang, J.; Gao, W.; Lane, H.C.; Lempicki, R.A. DAVID: Database for Annotation, Visualization, and Integrated Discovery. *Genome Biol.* **2003**, *4*, P3. [CrossRef]

39. Blanks, J.C.; Adinolfi, A.M.; Lolley, R.N. Photoreceptor degeneration and synaptogenesis in retinal-degenerative (rd) mice. *J. Comp. Neurol.* **1974**, *156*, 95–106. [CrossRef]
40. Sarin, S.; Zuniga-Sanchez, E.; Kurmangaliyev, Y.Z.; Cousins, H.; Patel, M.; Hernandez, J.; Zhang, K.X.; Samuel, M.A.; Morey, M.; Sanes, J.R.; et al. Role for Wnt Signaling in Retinal Neuropil Development: Analysis via RNA-Seq and In Vivo Somatic CRISPR Mutagenesis. *Neuron* **2018**, *98*, 109–126.e8. [CrossRef]
41. Carterdawson, L.D.; Lavail, M.M. Rods and Cones in the Mouse Retina.1. Structural-Analysis Using Light and Electron-Microscopy. *J. Comp. Neurol.* **1979**, *188*, 245–262. [CrossRef]
42. Dent, E.W.; Gupton, S.L.; Gertler, F.B. The growth cone cytoskeleton in axon outgrowth and guidance. *Cold Spring Harb. Perspect. Biol.* **2011**, *3*, a001800. [CrossRef]
43. Regus-Leidig, H.; Tom Dieck, S.; Specht, D.; Meyer, L.; Brandstatter, J.H. Early steps in the assembly of photoreceptor ribbon synapses in the mouse retina: The involvement of precursor spheres. *J. Comp. Neurol.* **2009**, *512*, 814–824. [CrossRef]
44. Anastassov, I.A.; Wang, W.; Dunn, F.A. Synaptogenesis and synaptic protein localization in the postnatal development of rod bipolar cell dendrites in mouse retina. *J. Comp. Neurol.* **2019**, *527*, 52–66. [CrossRef]
45. Breuninger, T.; Puller, C.; Haverkamp, S.; Euler, T. Chromatic bipolar cell pathways in the mouse retina. *J. Neurosci.* **2011**, *31*, 6504–6517. [CrossRef]
46. Ishikawa, M.; Hashimoto, Y.; Tonosaki, A.; Sakuragi, S. Preference of peanut agglutinin labeling for long-wavelength-sensitive cone photoreceptors in the dace retina. *Vis. Res.* **1997**, *37*, 383–387. [CrossRef] [PubMed]
47. Mack, A.F. Evidence for a columnar organization of cones, Muller cells, and neurons in the retina of a cichlid fish. *Neuroscience* **2007**, *144*, 1004–1014. [CrossRef]
48. Keeley, P.W.; Reese, B.E. Role of afferents in the differentiation of bipolar cells in the mouse retina. *J. Neurosci.* **2010**, *30*, 1677–1685. [CrossRef] [PubMed]
49. Whitaker, D.T.; Mondal, A.K.; Fann, H.; Hargrove, P.; Brooks, M.J.; Chaitankar, V.; Yu, W.; Wu, Z.; Kim, S.-Y.; Swaroop, A. NRL- and CRX-guided gene network modulates photoreceptor presynapse size and positioning during retinal development. *bioRxiv* **2019**, 753012. [CrossRef]
50. Cao, Y.; Masuho, I.; Okawa, H.; Xie, K.; Asami, J.; Kammermeier, P.J.; Maddox, D.M.; Furukawa, T.; Inoue, T.; Sampath, A.P.; et al. Retina-specific GTPase accelerator RGS11/G beta 5S/R9AP is a constitutive heterotrimer selectively targeted to mGluR6 in ON-bipolar neurons. *J. Neurosci. Off. J. Soc. Neurosci.* **2009**, *29*, 9301–9313. [CrossRef] [PubMed]
51. Martemyanov, K.A. G protein signaling in the retina and beyond: The Cogan lecture. *Investig. Ophthalmol. Vis. Sci.* **2014**, *55*, 8201–8207. [CrossRef] [PubMed]
52. Oh, E.C.; Cheng, H.; Hao, H.; Jia, L.; Khan, N.W.; Swaroop, A. Rod differentiation factor NRL activates the expression of nuclear receptor NR2E3 to suppress the development of cone photoreceptors. *Brain Res.* **2008**, *1236*, 16–29. [CrossRef] [PubMed]
53. Corbo, J.C.; Cepko, C.L. A hybrid photoreceptor expressing both rod and cone genes in a mouse model of enhanced S-cone syndrome. *PLoS Genet.* **2005**, *1*, e11. [CrossRef]
54. Assawachananont, J.; Kim, S.Y.; Kaya, K.D.; Fariss, R.; Roger, J.E.; Swaroop, A. Cone-rod homeobox CRX controls presynaptic active zone formation in photoreceptors of mammalian retina. *Hum. Mol. Genet.* **2018**, *27*, 3555–3567. [CrossRef] [PubMed]
55. Omori, Y.; Araki, F.; Chaya, T.; Kajimura, N.; Irie, S.; Terada, K.; Muranishi, Y.; Tsujii, T.; Ueno, S.; Koyasu, T.; et al. Presynaptic dystroglycan-pikachurin complex regulates the proper synaptic connection between retinal photoreceptor and bipolar cells. *J. Neurosci.* **2012**, *32*, 6126–6137. [CrossRef]
56. Koulen, P.; Fletcher, E.L.; Craven, S.E.; Bredt, D.S.; Wassle, H. Immunocytochemical localization of the postsynaptic density protein PSD-95 in the mammalian retina. *J. Neurosci.* **1998**, *18*, 10136–10149. [CrossRef]
57. Yang, X.R.; Kassam, F.; Innes, A.M. Rod-cone dystrophy in an adult with GNB1-related disorder: An expansion of the phenotype and natural history. *Am. J. Med. Genet. C Semin. Med. Genet.* **2023**, *193*, 183–187. [CrossRef]
58. Yu, J.; He, S.; Friedman, J.S.; Akimoto, M.; Ghosh, D.; Mears, A.J.; Hicks, D.; Swaroop, A. Altered expression of genes of the Bmp/Smad and Wnt/calcium signaling pathways in the cone-only Nrl$^{-/-}$ mouse retina, revealed by gene profiling using custom cDNA microarrays. *J. Biol. Chem.* **2004**, *279*, 42211–42220. [CrossRef]
59. Nikonov, S.S.; Lyubarsky, A.; Fina, M.E.; Nikonova, E.S.; Sengupta, A.; Chinniah, C.; Ding, X.Q.; Smith, R.G.; Pugh, E.N., Jr.; Vardi, N.; et al. Cones respond to light in the absence of transducin beta subunit. *J. Neurosci.* **2013**, *33*, 5182–5194. [CrossRef]
60. Pourhoseini, S.; Goswami-Sewell, D.; Zuniga-Sanchez, E. Neurofascin Is a Novel Component of Rod Photoreceptor Synapses in the Outer Retina. *Front. Neural Circuits* **2021**, *15*, 635849. [CrossRef]
61. Kohl, S.; Baumann, B.; Rosenberg, T.; Kellner, U.; Lorenz, B.; Vadala, M.; Jacobson, S.G.; Wissinger, B. Mutations in the cone photoreceptor G-protein alpha-subunit gene GNAT2 in patients with achromatopsia. *Am. J. Hum. Genet.* **2002**, *71*, 422–425. [CrossRef] [PubMed]
62. Chang, B.; Dacey, M.S.; Hawes, N.L.; Hitchcock, P.F.; Milam, A.H.; Atmaca-Sonmez, P.; Nusinowitz, S.; Heckenlively, J.R. Cone photoreceptor function loss-3, a novel mouse model of achromatopsia due to a mutation in Gnat2. *Investig. Ophthalmol. Vis. Sci.* **2006**, *47*, 5017–5021. [CrossRef] [PubMed]
63. Parsons, T.D.; Sterling, P. Synaptic ribbon. Conveyor belt or safety belt? *Neuron* **2003**, *37*, 379–382. [CrossRef] [PubMed]
64. Morgans, C.W.; Kensel-Hammes, P.; Hurley, J.B.; Burton, K.; Idzerda, R.; McKnight, G.S.; Bajjalieh, S.M. Loss of the Synaptic Vesicle Protein SV2B results in reduced neurotransmission and altered synaptic vesicle protein expression in the retina. *PLoS ONE* **2009**, *4*, e5230. [CrossRef]

65. Landgraf, I.; Muhlhans, J.; Dedek, K.; Reim, K.; Brandstatter, J.H.; Ammermuller, J. The absence of Complexin 3 and Complexin 4 differentially impacts the ON and OFF pathways in mouse retina. *Eur. J. Neurosci.* **2012**, *36*, 2470–2481. [CrossRef]
66. Masland, R.H. The neuronal organization of the retina. *Neuron* **2012**, *76*, 266–280. [CrossRef]
67. Hao, H.; Veleri, S.; Sun, B.; Kim, D.S.; Keeley, P.W.; Kim, J.W.; Yang, H.J.; Yadav, S.P.; Manjunath, S.H.; Sood, R.; et al. Regulation of a novel isoform of Receptor Expression Enhancing Protein REEP6 in rod photoreceptors by bZIP transcription factor NRL. *Hum. Mol. Genet.* **2014**, *23*, 4260–4271. [CrossRef]
68. Yoshida, S.; Mears, A.J.; Friedman, J.S.; Carter, T.; He, S.; Oh, E.; Jing, Y.; Farjo, R.; Fleury, G.; Barlow, C.; et al. Expression profiling of the developing and mature $Nrl^{-/-}$ mouse retina: Identification of retinal disease candidates and transcriptional regulatory targets of Nrl. *Hum. Mol. Genet.* **2004**, *13*, 1487–1503. [CrossRef]
69. Kim, J.W.; Yang, H.J.; Oel, A.P.; Brooks, M.J.; Jia, L.; Plachetzki, D.C.; Li, W.; Allison, W.T.; Swaroop, A. Recruitment of Rod Photoreceptors from Short-Wavelength-Sensitive Cones during the Evolution of Nocturnal Vision in Mammals. *Dev. Cell* **2016**, *37*, 520–532. [CrossRef]

life

MDPI

Article

Investigation of PACAP38 and PAC1 Receptor Expression in Human Retinoblastoma and the Effect of PACAP38 Administration on Human Y-79 Retinoblastoma Cells

Dénes Tóth [1,*], Eszter Fábián [2], Edina Szabó [2], Evelin Patkó [2], Viktória Vicena [2], Alexandra Váczy [2], Tamás Atlasz [2,3], Tamás Tornóczky [4] and Dóra Reglődi [2]

1 Department of Forensic Medicine, University of Pécs Medical School, Szigeti út 12, 7624 Pecs, Hungary
2 Department of Anatomy, HUN-REN-PTE PACAP Research Team, Centre for Neuroscience, University of Pécs Medical School, Szigeti út 12, 7624 Pecs, Hungary; eszter.fabian@aok.pte.hu (E.F.); szaboedina90@gmail.com (E.S.); evelin.patko@gmail.com (E.P.); vicena.viktoria@gmail.com (V.V.); vaczyalexandra@gmail.com (A.V.); attam@gamma.ttk.pte.hu (T.A.); dora.reglodi@aok.pte.hu (D.R.)
3 Department of Sportbiology, University of Pécs, Ifjúság út 6, 7624 Pecs, Hungary
4 Department of Pathology, University of Pécs Medical School and Clinical Center, 7624 Pecs, Hungary; tornoczki.tamas@pte.hu
* Correspondence: denes.toth@aok.pte.hu

Abstract: Retinoblastoma represents the most prevalent malignant neoplasm affecting the eyes in childhood. The clear-cut origin of retinoblastoma has not yet been determined; however, based on experiments, it has been suggested that *RB1* loss in cone photoreceptors causes retinoblastoma. Pituitary adenylate-cyclase activating polypeptide (PACAP) is a pleiotropic neuropeptide which has been shown to be affected in certain tumorous transformations, such as breast, lung, kidney, pancreatic, colon, and endocrine cancers. This study aimed to investigate potential changes in both PACAP38 and PAC1 receptor (PAC1R) expression in human retinoblastoma and the effect of PACAP38 administration on the survival of a human retinoblastoma cell line (Y-79). We analyzed human enucleation specimens removed because of retinoblastoma for PACAP38 and PAC1R immunostaining and the effect of PACAP38 on the survival of the Y-79 cell line. We described for the first time that human retinoblastoma cells from patients showed only perinuclear, dot-like immunopositivity for both PACAP38 and PAC1R, irrespective of laterality, genetic background, or histopathological features. Nanomolar (100 nM and 500 nM) PACAP38 concentrations had no effect on the viability of Y-79 cells, while micromolar (2 µM and 6 µM) PACAP38 significantly decreased tumor cell viability. These findings, along with general observations from animal studies showing that PACAP38 has strong anti-apoptotic, anti-inflammatory, and antioxidant effects on ocular tissues, together suggest that PACAP38 and its analogs are promising candidates in retinoblastoma therapy.

Keywords: PACAP; retinoblastoma; enucleation; immunohistochemistry; cell survival; cell viability

1. Introduction

Retinoblastoma, acknowledged as one of the leading childhood malignancies, is the most prevalent primary intraocular neoplasm that tends to manifest in early childhood [1]. The worldwide occurrence of retinoblastoma is approximately 1 in 16,000–18,000 live births [2]. However, there are variations in incidence among countries, regions, and ethnic groups, with almost similar incidence rates for males and females [3]. Developing countries with high birth rates have the highest mortality (40–70%) compared with Europe, the USA, and Canada (3–5%) [4]. The causes of these differences in mortality include delays in diagnosis, advanced stages of the disease, lack of access to health care systems, and other socioeconomic factors [5].

Biallelic loss-of-function mutations in the tumor suppressor gene RB1, located on chromosomal region 13q14, account for the initiation of retinoblastoma in 95% of cases [6].

Individuals with germline mutation (first hit) require only one additional hit (acquired somatic mutation). Therefore, heritable retinoblastoma occurs at a younger age and is usually multifocal or bilateral. In contrast, most sporadic cases are unilateral as a result of mutation in both alleles of *RB1* [7]. In rare cases, sporadic retinoblastoma could develop in the absence of *RB1* mutation as a consequence of the somatic amplification of the *MYCN* gene [8]. The exact cellular origin of retinoblastoma is not clear yet [9,10]. Based on live imaging data of early tumors from patients' eyes, the inner nuclear layer of the retina seemed to be the origin [11], but based on experiments, it has been suggested that human retinoblastoma arise from differentiating cones as it was found that retinoblastoma cells exhibit multiple elements of the cone precursor signaling circuitry and depend on this circuitry for their proliferation, survival [12], and *RB1* knockdown-induced human cone precursor proliferation [13]. Distinctions in molecular, clinical, and histopathological aspects among RB1−/− tumors reveal a progression marked by a loss of differentiation and a decline in the photoreceptor expression signature [14]. Human cone precursor maturation's unique cell-signaling circuitry makes them sensitive to *RB1* loss, leading to proliferation and lesion formation resembling retinoma and retinoblastoma [15]. A human Rb organoid model also identified maturing cone precursors as the origin of human retinoblastoma [16]. A multi-omics approach identified two retinoblastoma molecular subtypes expressing cone markers [17]. Interestingly, *RB1* loss was observed to induce proliferation and tumorigenesis in maturing cone precursors, while it was found that the somatic amplification of the *MYCN* gene induced proliferation in immature cone precursors [18].

Retinoblastoma is characterized primarily by leukocoria, with subsequent symptoms including strabismus, a red and painful eye, impaired visual tracking, and vision loss. Diagnosis is usually clinical (ophthalmoscopy) combined with additional examinations like ultrasonography, computer tomography, or magnetic resonance imaging [5]. Retinoblastoma is usually white and has a brain-like appearance with pale areas of calcification and yellowish necrotic zones. The histopathological characteristics of retinoblastoma encompass small hyperchromatic cells exhibiting a high nuclear-to-cytoplasmic ratio, accompanied by areas of variable-sized necrosis and dystrophic calcification [19]. The level of retinal differentiation in retinoblastoma varies and correlates negatively with the age of a child [20]. Differentiated tumors include fleurettes, representing advanced photoreceptor differentiation; Flexner–Wintersteiner rosettes, showing early retinal differentiation; and Homer Wright rosettes, exhibiting primitive neuroblastic differentiation [21,22].

The current therapy for retinoblastoma depends on the time of disease detection and socioeconomic factors. In low- and middle-income countries, therapy is aimed at protecting patients' lives due to enucleation, followed by salvage of the globe and vision. Unfortunately, enucleation is still the most frequent choice worldwide [23].

An evolutionarily conserved neuropeptide, pituitary adenylate-cyclase activating polypeptide (PACAP), is recognized for its multifunctionality and pleiotropy. Its established anti-apoptotic, anti-inflammatory, and antioxidant effects [24,25] are attributed to two functionally active isoforms: PACAP38 [26] and PACAP27 [27]. The latter represents 10% of the total PACAP in the body, while PACAP38 is the predominant form in mammalian tissues. PACAP acts on two nonspecific and one specific G-protein-coupled receptors. Highlighting its selective nature, the PAC1 receptor (PAC1R) responds exclusively to PACAP, in contrast to the shared VPAC1 and VPAC2 receptors that accommodate both PACAP and VIP. Across the central and peripheral nervous systems, as well as in peripheral organs, PACAP and PAC1R display broad and widespread expression [28].

The current understanding of the presence, distribution, and functional aspects of PACAP and its specific receptor in the human eye is limited. In a previous study, Olianas and coauthors [29] provided evidence showcasing that PACAP has the capacity to elevate cAMP levels in the retina. Additionally, from retinal homogenates obtained from human fetuses, they successfully identified the mRNA for both PACAP and its receptors. The first detailed exploration of the distribution of PACAP38 and PAC1R within the normal human eye was conducted in 2022 [30]. Corneal epithelial and endothelial cells, the iris (both muscle and stroma), the ciliary body, several retinal layers, and the optic nerve showed immunopositivity for both PACAP38 and PAC1R. Regarding the retina, the pigment epithelial layer—forming from the original outer eyecup layer—showed strong expression for PACAP38 and PAC1R. Weak or no immunostaining was observed in the outer nuclear and plexiform layers where rods and cones are located. In the majority of cases, the inner nuclear layer exhibited a markedly positive expression for both the peptide and PAC1R. The inner plexiform layer had strong PACAP38 and PAC1R immunoreactivity. Interestingly, a specific staining pattern was observed in the ganglion cell layer, where some ganglion cells showed very strong, and others showed negative, immunosignals [30]. PACAP takes part in a wide range of physiological [31–33] and pathological processes [34,35]. In vitro studies with human corneal endothelial cells and pigment epithelial cells proved that PACAP stimulates adenylate cyclase and various intracellular signaling pathways to protect the cells against various noxas, including hyperglycemia, oxidative stress, and growth factor deprivation [36–41]. In vivo studies showed the protective effect of PACAP in different types of retinopathies [40,42–44], glaucoma [45], and injuries [46,47]. PACAP also affects cellular differentiation [48], cell division, cell cycle, and cell death [49]. As PACAP regulates almost every aspect of stem cell physiology [50], it is not surprising that PACAP and its receptors were detected in numerous human cancer types [51–53]. Many different tumors show over- or under-expression of the PACAPergic system. The activation of PACAP receptors in specific neoplasms can lead to growth stimulation, whereas in others, it results in inhibitory effects. These effects depend on numerous factors, like the origin and type of the tumor, the stage of differentiation, and the tumoral environment [52,54]. As PACAP is shown to be involved in certain tumorous transformations [55] and changes in PACAP and PAC1R can be detected under pathological conditions [34,35], this study aimed to investigate potential alterations in the expression of PACAP38 and PAC1R within tumor tissue samples obtained from enucleation specimens of retinoblastoma patients and the effect of PACAP38 treatment on the survival of human Y-79 retinoblastoma cells.

2. Materials and Methods

2.1. Human Eyes

This study was conducted according to the ethical standards specified in the Declaration of Helsinki, along with due consideration of the corresponding regulations within Hungarian law. The collection of all samples strictly followed a protocol that received approval from the Institutional Ethics Committee at the University of Pecs (9188-PTE 2022; approval date: 10 June 2022). The identification of cases involved a comprehensive search through the pathological records of the Department of Pathology at the University of Pécs Medical School, covering the period from January 2001 to December 2017. This search specifically targeted enucleation specimens carrying a histopathological diagnosis of retinoblastoma. Patient medical records were reviewed for demographic information and clinical findings. The staging was conducted following the eighth edition of the Cancer Staging Manual of the American Joint Committee on Cancer (AJCC) [22]. In brief, pT1 denotes intraocular tumor(s) without local invasion, focal choroidal invasion, or involvement of the optic nerve head either pre- or intralaminarly. pT2 corresponds to intraocular tumor(s) with local invasion, pT3 indicates intraocular tumor(s) with substantial local invasion, and pT4 signifies the presence of extraocular tumor(s) [22]. The level of differentiation (grading) was classified as follows: G1 corresponds to a tumor displaying areas of retinocytoma, characterized by fleurettes or neuronal differentiation, accounting for more than half of the tumor. G2 signifies a tumor featuring numerous rosettes, including Flexner–Wintersteiner or Homer Wright rosettes, constituting more than half of the tumor. G3 denotes a tumor with occasional rosettes (Flexner–Wintersteiner or Homer Wright rosettes), accounting for less than half of the tumor. G4 represents a tumor with poorly differentiated cells lacking rosettes or displaying extensive areas of anaplasia [56].

Immunohistochemistry

Tissue samples underwent fixation in a 10% neutral-buffered formalin solution, followed by dehydration through a series of graded ethanol solutions and eventual embedding in paraffin. Subsequently, sections with a three-micrometer thickness were cut using a rotational microtome (Microm HM 325, Thermo Scientific, Ltd., Waltham, MA, USA) and affixed onto coated glass slides. Following deparaffinization and rehydration through graded ethanol, the samples underwent pretreatment using a heat-induced epitope retrieval method. This involved subjecting the samples to a microwave oven for 15 min at 750 W in a 1 mM (pH = 6.0) citrate buffer. Subsequent to cooling at room temperature, the samples were washed in a pH 7.6 TRIS-buffered saline solution (TBS). Samples were then incubated for 1 h at room temperature with anti-PACAP38 (Cat. Nr. T-4473, BMA Biomedicals, Ltd., Augst, Switzerland, 1:500) and anti-PAC1-R antibody (Cat. Nr. AVR-003, Alomone Labs, Ltd., Jerusalem, Israel, 1:125). After washing in TBS, the sections were exposed to an HISTOLS-AP-R anti-rabbit alkaline phosphatase labeled detection system (Cat. Nr. 30,011.R500A, Histopathology, Ltd., Pécs, Hungary) for a 30 min incubation at room temperature. Subsequently, they underwent another round of TBS washing, and the enzymatic reaction was initiated in a dark environment using an HISTOLS Resistant AP-Red Chromogen/substrate System (Cat. Nr. 30,019K, Histopathology, Ltd., Pécs, Hungary). Following a 10 min incubation with the chromogen/substrate working solution, staining intensity was controlled under a light microscope. This chromogen substance was chosen because its magenta staining was visible in the pigmented cells. Counterstaining was performed with hematoxylin, and tap water was used for bluing. After drying, samples were dehydrated in graded ethanol and cleared in xylene before being mounted with a permanent mounting medium. For negative control, the primary antibody was replaced with TBS, which resulted in no staining. The healthy parts of the eye served as an internal positive control. Using a Panoramic MIDI II automatic digital slide scanner (3DHISTECH Ltd., Budapest, Hungary), the slides were thoroughly scanned, and images were captured through CaseViewer 2.3 software (3DHISTECH Ltd., Budapest, Hungary).

2.2. Cell Culture

Y-79 human retinoblastoma cells, sourced from the American Type Culture Collection (ATCC, Manassas, VA, USA), were cultivated in RPMI-1640 medium supplemented with 10% fetal bovine serum, 100 U/mL penicillin, and 100 μg/mL streptomycin. The culture was maintained in a humidified incubator at 37 °C with 5% CO_2, and the culture medium was refreshed every second day. All cell culture reagents were procured from Sigma-Aldrich (St. Louis, MO, USA).

2.2.1. Cell Viability Assay

A colorimetric assay for evaluating cell viability, an MTT (3-(4,5-dimethylthiazol-2-yl)-2,5-diphenyltetrazolium bromide) assay from Sigma-Aldrich (St. Louis, MO, USA), was utilized to investigate the effect of PACAP38. Cells were seeded in 96-well plates (4×10^3 cells/well). Twelve wells served as control, and 100 μL of serum-free medium was added to these wells. Six wells in each column were treated with 10 μL 0.1 μM, 0.5 μM, 1 μM, 2 μM, and 6 μM PACAP38 (produced within the Department of Medical Chemistry at the University of Szeged, Szeged, Hungary) to observe the dose dependency. The wells were then filled with serum-free medium to a final volume of 100 μL. Following a 24 h incubation period, 10 μL of a 5 mg/mL MTT solution was added to each well, achieving a final concentration of 0.45 mg/mL. The plate underwent an additional 4 h incubation in a thermostat, after which the reduced formazan dye was dissolved using 100 μL DMSO (dimethyl-sulfoxide). After 30 min on a shaker, the absorbance was gauged at 630 nm using an ELISA reader (Dialab Ltd., Budapest, Hungary). The assay was performed in duplicate and repeated three times.

2.2.2. Statistical Analysis

Statistical analysis was conducted using one-way ANOVA followed by Dunnett's test with GraphPad Prism version 9.5.0 for Microsoft Windows (GraphPad Software LLC, San Diego, CA, USA). The presented data include means ± standard deviation (SD).

3. Results

3.1. Human Eyes

3.1.1. Clinical Data

Seven children (one girl and six boys) were included in our study who underwent primary enucleation because of retinoblastoma. The mean age at enucleation was 16.3 ± 10.5 months (median: 16.6 months; range: 32.2). Except for two cases, the right eye was affected, and there were no bilateral cases. In three cases, the tumor site involved the superotemporal quadrant after sectioning, while in one case, the lesion was near the optic disc. In the remaining cases, the tumor occupied the entire eye, and the precise point of origin could not be determined. In all cases, with one exception, a monofocal tumor was observed, and in the case of multifocal retinoblastoma, genetic involvement could not be confirmed. However, in the monofocal case, familial clustering and an RB1 mutation were identified. Histomorphologically, in three cases, poorly differentiated cells without rosettes were found (G4), while in one case, a tumor with occasional Homer Wright rosettes (G3) was observed. In the remaining cases, a tumor with many rosettes was detected (G2). Among the latter, Homer Wright rosettes were visible in two cases, and Flexner–Wintersteiner rosettes were observed in one case. In all cases, various degrees of necrosis were present within the tumors, and except for two cases, calcifications were also observed. The Mib-1 (Ki-67) labeling index was relatively high, exceeding 50% in all cases (excluding two instances where such an examination was not conducted for technical reasons). Table 1 provides an overview of the primary demographic and clinical features, while Table 2 summarizes the main pathological findings of the cases.

Table 1. Main clinical and demographic features of retinoblastoma cases.

Case	Sex	Age at Enucleation (Months)	Eye Involved	Tumor Site	Number of Tumor Foci	*RB1* Mutation
1	male	9.9	right	superotemporal	monofocal	no
2	female	16.6	right	superotemporal	multifocal	no
3	male	9.3	right	superotemporal	monofocal	no
4	male	4.1	right	adjacent to the optic disc	monofocal	yes
5	male	36.3	left	whole eye	monofocal	no
6	male	19.9	left	whole eye	monofocal	no
7	male	18.3	right	whole eye	monofocal	no

Table 2. Pathological findings of retinoblastoma cases (Mib-1: cell proliferation marker—percentage of immunoreactive tumor cells; n.d.: no data).

Case	Histomorphology	Necrosis	Calcification	Mib-1 (%)	Stage
1	tumor cells arranged in sheets, nests, and trabeculae (G4)	large	focal	80	pT2
2	tumor with many Homer Wright rosettes (G2)	small	focal	90	pT2
3	tumor with many Homer Wright rosettes (G2)	small	focal	90	pT2
4	tumor with many Flexner–Wintersteiner rosettes (G2)	large	no	50–60	pT2
5	tumor with occasional Homer Wight rosettes (G3)	large	focal	80	pT2
6	tumor cells arranged in sheets, nests, and trabeculae (G4)	large	focal	n.d.	pT2
7	tumor cells arranged in sheets, nests, and trabeculae (G4)	large	no	n.d.	pT2

3.1.2. Immunohistochemistry

In accordance with our earlier observations [30], we found PACAP38 and PAC1R immunopositivity in disease-free areas of the eye, including the cornea (both epithelial and endothelial cells), the iris (stroma and muscles), the ciliary body, and different retinal layers (pigment epithelial layer, inner plexiform layer, and ganglion cell layer). These findings served as a positive internal control. Figure 1 includes representative pictures from the disease-free section of the eyes, which served as an internal control in the study.

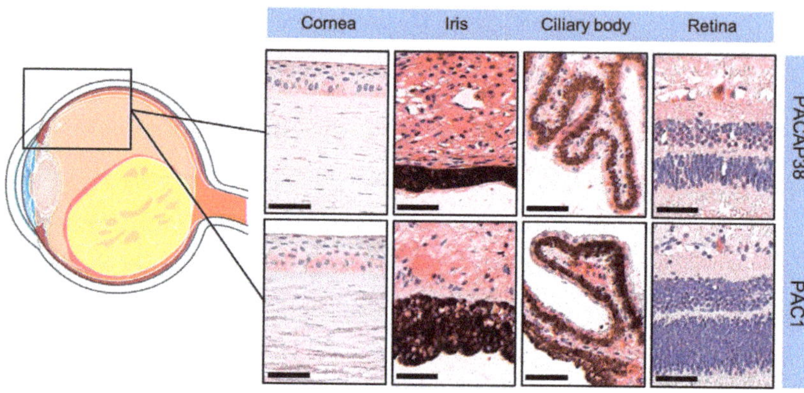

Figure 1. Representative pictures of PACAP38 and PAC1R immunopositivity in disease-free parts of the eyes (magnification 200×; scale bars: 50 μm). Certain parts of the figure were created using images from Servier Medical Art. Servier Medical Art by Servier is licensed under a Creative Commons Attribution 3.0 Unported License (https://creativecommons.org/licenses/by/3.0/, accessed on 1 November 2023).

Retinoblastoma cells showed only perinuclear, dot-like immunopositivity (black arrows) for both PACAP38 and PAC1R, irrespective of laterality and genetic background. In this immunopattern, there was no difference between the poorly differentiated samples or the different types of rosettes, regardless of the proportion of rosettes in the tumor (Figure 2).

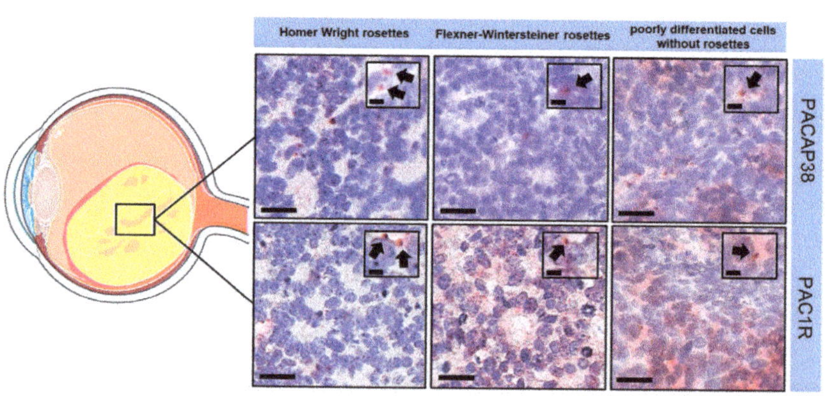

Figure 2. PACAP38 and PAC1R immunopositivity in the tumor samples (magnification: large pictures 400×, scale bars: 20 μm; index pictures 700×, scale bars: 5 μm). Black arrows indicated perinuclear, dot-like immunopositivity in retinoblastoma cells. The illustration includes components that were drawn utilizing images sourced from Servier Medical Art. Servier Medical Art by Servier is licensed under a Creative Commons Attribution 3.0 Unported License (https://creativecommons.org/licenses/by/3.0/, accessed on 1 November 2023).

3.2. Cell Culture

One-way analysis of variance indicated significant differences in cell survival among the treatment groups (F = 5.165, $p = 0.0047$). Dunnett's multiple comparisons test revealed statistically significant differences between the control group and both the 2 μM PACAP38- (mean diff: 16.5, 95% CI [1.471, 31.52], $p = 0.035$) and the 6 μM PACAP38 (mean diff: 20.38, 95% CI [8.488, 32.26], $p = 0.0053$)-treated groups. Although the mean difference for the 0.5 μM PACAP38-treated group was similar to that of the 2 μM PACAP38-treated group (17.09 vs. 16.5), there were no statistically significant differences observed when compared to the control group (95% CI [−1.290, 35.48], $p = 0.0653$) (Figure 3).

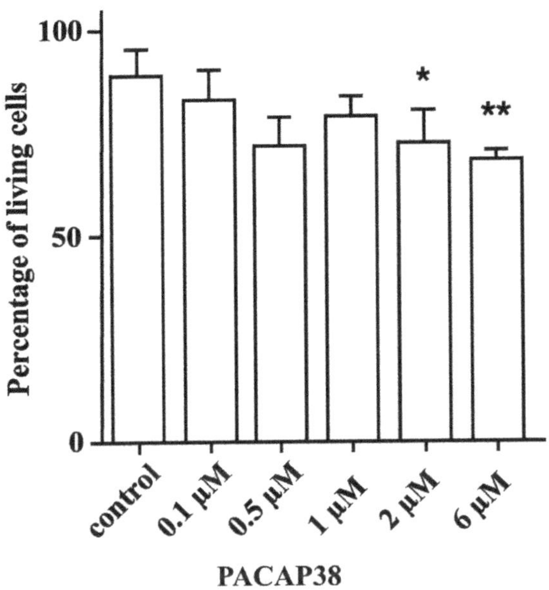

Figure 3. The percentage of living Y-79 cells after different concentrations of PACAP38 administration. (* $p < 0.05$; ** $p < 0.01$ compared to the control group. The data are presented as mean ± SD).

4. Discussion

In the first part of our study, we analyzed human enucleation specimens removed because of retinoblastoma for PACAP38 and PAC1R immunostaining and described, for the first time, the distribution of PACAP38 and PAC1R expression in human retinoblastoma. We found PACAP38 and PAC1R immunopositivity in the tumor-free area of the eyes consistent with the results of the first description of the distribution of PACAP38 and PAC1R in the human eye [30] which, therefore, served as a positive internal control. In retinoblastoma, we observed only focal, perinuclear dot-like immunopositivity for both PACAP38 and PAC1R. There were no differences in the immunopatterns between the different histological features, i.e., the presence of different types of rosettes.

A broad spectrum of human cancers has been observed to express PACAP38 and PAC1R. Furthermore, certain tumors show alteration of the PACAPergic system compared to the normal tissue. In papillary thyroid carcinoma, the overexpression of PACAP38-positive cells was detected compared to normal thyroid glands, while colloid showed weaker or no staining pattern. Regarding PAC1R, tumor cells showed only minimal or no expression compared to the normal glands, where strong granular expression was present [57]. Similar overexpression was observed in the case of invasive ductal adenocarcinoma of the breast both for PACAP38 and PAC1R [58,59]. Lower PACAP38 and PAC1R immunosignals were detected in the case of pancreatic ductal adenocarcinoma and insulinoma compared to those in healthy pancreatic tissues [60,61]. In the case of non-small cell lung cancer, colon adenocarcinoma, and kidney tumor samples, a significantly lower level of PACAP38-like immunoreactivity was detected by RIA compared with that in normal healthy tissues [62,63]. Prostatic adenocarcinoma showed essentially preserved PACAP38 immunopatterns compared to those in normal prostatic glands [64], and no significant alterations in PACAP38-like immunoreactivity detected by RIA were found in cases of urinary bladder tumor samples, prostatic adenocarcinoma samples, or various types of testicular malignancies, like seminoma, embryonal carcinoma, yolk sac tumor, and teratoma [63]. Regarding the tumors mentioned above in the immunohistochemical studies, both the peptide and its specific receptor showed widespread expression, including membrane and intracytoplasmic PAC1R expression, and in none of these cases were they able to detect a pattern where PACAP38 and PAC1R were expressed only in a perinuclear dot-like pattern, as observed in our study.

Retinoblastoma differs from these types of tumors as the exact cellular origin of retinoblastoma is still controversial. For this reason, in our experiment, we were not able to compare the PACAP38 and PAC1R immunoprofiles of 'normal' and tumor tissues. We attempted to compare the immunopatterns observed in disease-free parts of the eyes within the inner nuclear layer, outer nuclear layer (housing the nuclei of photoreceptor cells), and the layer of rods and cones with the immunopattern observed in retinoblastoma. No or only faint PACAP38 and PAC1R immunosignals were identified in the inner nuclear layer, depending on the cell types located in this layer. Moreover, the same type of cells showed individually variable patterns. In this retinal layer, the faint immunopatterns of PACAP38 and PAC1R were morphologically (perinuclear dot-like positivity) similar to that observed in retinoblastoma. The same observations were made in the case of the outer nuclear layer, as well as in the layer of rods and cones. Since the originating cell type of retinoblastoma is unknown, and the cells potentially involved in this context also exhibited individual expression variations, we could not establish a parallel with the pattern observed in retinoblastoma.

Regarding human retinoblastoma cell lines, literature data on PACAP38 and PAC1R expression are only available for the Y-79 cell line. Olianas and coworkers noted that around 60% of Y-79 cells express membrane-bound PAC1R, and PACAP induces a concentration-dependent increase in adenylyl cyclase activity, with PACAP38 being six-fold more potent than PACAP27 [65]. In contrast, we observed the absence of membrane PAC1R in retinoblastoma, highlighting a significant distinction between the in vitro model and the clinical manifestation. Thus, the Y-79 cell line, which is mostly used for in vitro retinoblastoma studies, has the limitation of representing patients' retinoblastoma. In the future, we need further studies to understand the precise reason for this.

In the second part of our experiment, we observed that nanomolar (100 nM and 500 nM) and 1 μM PACAP38 concentrations had no effect on the viability of Y-79 human retinoblastoma cells. In their investigation, Wojcieszak and Zawilska extensively explored how PACAP influences the viability of Y-79 cells derived from human retinoblastoma. Nanomolar (0.1–100 nM) concentrations of PACAP38 did not affect the viability of Y-79 cells, while micromolar (1–5 μM) concentrations of PACAP38 induced a dose-dependent decrease in tumor cell viability. The administration of a PAC1R antagonist, PACAP6-38, did not terminate this cytotoxic effect of PACAP38; furthermore, PACAP6-38 alone, in the same micromolar concentration, also produced a dose-dependent decrease in tumor cell viability. Micromolar concentrations of PACAP27 (0.1–5 μM) and a high-affinity selective PAC1 receptor agonist, maxadilan (1–2 μM), did not significantly affect the viability of Y-79 human retinoblastoma cells. [Disc^{6}]PACAP38 and FITC-Ahx-PACAP11-38, two membrane-penetrating PACAP38 analogs which are inactive in PAC1, VPAC1, and VPAC2 receptors, also decreased the viability of Y-79 cells but with lower potency than PACAP38. This suggests that PACAP can exert its cytotoxic effects in a membrane receptor-independent way [66]. Our results are in accordance with the aforementioned study [66], showing that concentrations of PACAP38 up to 100 nM had no effect on the viability of Y-79 cells. PACAP38 proved to be cytotoxic only when used in micromolar (2 μM and 6 μM) concentrations. Despite this, it is important to emphasize that Y-79 is just one of numerous human retinoblastoma cell lines, necessitating future testing of PACAP38 effects across diverse in vitro models.

The activation of PACAP receptors in specific neoplasms can lead to growth stimulation, whereas in others, it results in inhibitory effects. These effects are influenced by various factors, such as the species' origin, tumor type and origin, stage of differentiation, or the tumor environment [52,54]. For instance, in glioblastoma cases, the administration of PACAP27 resulted in the increased proliferation of mouse C6 glioma cells, while applying both PACAP isoforms led to a significant decrease in proliferation for T98G human glioma cells. In other human glioblastoma cell lines (M059K and M59J), PACAP agonists reduced cancer cell migration without affecting their proliferation. Furthermore, it has been demonstrated that the impact of PACAP is also dependent on the conditions within the tumor microenvironment [67]. On the other hand, there were instances where both pro-survival and anti-survival effects of PACAP38 were absent, as observed in JAR cytotrophoblast cells exposed to methotrexate treatment [68], and likewise in hepatocellular carcinoma cells (HEP-G2) [69]. The multifaceted nature of PACAP's impact is highlighted, elucidating that opposite effects can emerge within the same cell line depending on factors such as exposure time, like in the case of LNCaP human prostatic tumor cells [70], or concentration, like in the case of Y-79 human retinoblastoma cells [66]. In human Y-79 retinoblastoma cells, the cytotoxic effects of PACAP38 were observed at concentrations equal to or exceeding 2 μM. However, the exact mechanism of PACAP38-induced Y-79 cell cytotoxicity is still unknown. It was observed that PACAP38 exerts its cytotoxic effects in a PAC1/VPAC1-2 membrane receptor-independent way in Y-79 cells without the activation of PKA, PKC, MEK1/2, p38, and JNK kinases [66]. Previous investigations have also indicated that PACAP38 acts as an intracrine factor, exhibiting the capacity to penetrate the internal cell compartment through direct translocation and endocytosis. This phenomenon results in a significant upsurge in the intracellular fraction, particularly at micromolar concentrations. Moreover, intracellular PACAP38 is not entirely degraded by intracellular enzymes and is able to activate intranuclear PAC1Rs [71,72]. The perinuclear PACAP38 and PAC1R positivity in human retinoblastoma detected by our first experiment could confirm these in vivo findings.

PACAP has demonstrated favorable outcomes in numerous pathological conditions which are primarily attributed to its cell-protective, antioxidant, and anti-inflammatory properties [34,35]. On the other hand, as indicated above, the PACAPergic system is affected by numerous malignant transformations as PACAP might exhibit both stimulatory and inhibitory impacts on tumor growth or cancer cell migration depending on the type of tumor [52,54]. This implies the possible exploration of developing selective agonists or antagonists for PAC1R, as well as analogs or antagonists for PACAP38, as valuable tools for diverse approaches to cancer treatment [52,73,74]. While in most diseases, PACAP or PAC1R agonist/antagonist treatment may pose significant limitations due to inadequate in vivo stability or limited penetration through the blood–brain barrier [75], in ophthalmic diseases, local application can overcome these factors. Previous in vivo studies have confirmed that PACAP38, when administered as eye drops, can permeate ocular barriers and exhibit retinoprotective effects [76]. Additionally, PACAP38 has been identified as a new candidate medication for Dry Eye Disease [77]. A study suggests that VIP and PACAP analogs, explicitly developed for therapeutic purposes, can modulate molecular and cellular processes relevant to treating high-risk neuroblastoma [78]. In our study, we proved that human retinoblastoma expresses PAC1R, and it is also known that in Y-79 human retinoblastoma cells, the micromolar concentration of PACAP38 exerts cytotoxic effects. These data highlight the therapeutic potential of PACAP38 and its analogs in retinoblastoma therapy. The results of Boisvilliers et al. [78] especially strengthen this potential in the case of retinoblastoma (even in its rare form, where retinoblastoma develops in the absence of an *RB1* mutation as a consequence of the somatic amplification of the *MYCN* gene), as they found that these peptide analogs are capable of inducing a sustained decrease in n-MYC expression and hold potential not only for neuroblastoma therapy but also for addressing other tumors. In our study, human retinoblastoma was found to express PACAP38 and PAC1R exclusively in a perinuclear dot-like pattern, with PACAP38 demonstrating cytotoxicity at concentrations of 2 μM and above in Y-79 retinoblastoma cells. Nevertheless, our studies also have limitations. To begin with, the limited number of cases prevents us from making a broad generalization that all retinoblastomas exhibit a uniform expression pattern. None of the cases in our study showcased fleurettes or neuronal differentiation which constituted more than half of the tumor, thereby hindering our ability to establish the immunoprofile for grade 1 tumors. Additionally, we cannot currently conclude whether PACAP signaling is implicated in the carcinogenesis of retinoblastoma. These aspects should be investigated more thoroughly in future studies. Furthermore, it is crucial to emphasize the need for exploring PACAP38 and its analogs in diverse in vitro models, establishing the groundwork for potential in vivo studies.

5. Conclusions

Elucidating the cellular origin of human cancers, including retinoblastoma, and understanding how cellular/subcellular context influences the probability of cancer initiation and progression would help to develop novel preventive and early-intervention therapies or improve existing ones. PACAP is involved in dozens of physiological and pathophysiological processes, and an increasing amount of evidence suggests that PACAP has diagnostic and therapeutic potential in certain diseases. In our study, we described, for the first time, the distribution of PACAP38 and PAC1R immunoreactivity in human retinoblastoma and we confirmed the cytotoxic effect of micromolar PACAP38 concentrations in human retinoblastoma cells. The facts that (i) PACAP38 and PAC1R are present in human retinoblastoma and that (ii) PACAP38 and its analogs had a cytotoxic effect on retinoblastoma cells suggest the potential role of PACAP38 and its analogs in retinoblastoma therapy.

Author Contributions: Conceptualization, D.T., D.R. and T.A.; methodology, D.T., T.T., E.F., E.P. and A.V.; validation, E.F., T.T. and E.S.; formal analysis, E.F.; investigation, D.T., E.F., V.V., E.P. and T.T.; resources, D.T., D.R. and T.A.; data curation, V.V.; writing—original draft preparation, D.T., T.T., D.R., T.A., A.V., E.S. and E.F.; writing—review and editing, D.T., D.R., T.A. and A.V.; visualization, D.T., E.F., E.P. and V.V.; supervision, T.A., D.R. and T.T.; project administration, D.T. and D.R.; funding acquisition, D.R. and T.A. All authors have read and agreed to the published version of the manuscript.

Funding: This research was funded by the Hungarian Brain Research Program NAP 3.0, the Thematic Excellence Program 2021 TKP2021-EGA-16, HUN-REN-TKI-14016, and the Hungarian Scientific Research Fund (NKFIH K135457).

Institutional Review Board Statement: The study was conducted in accordance with the Declaration of Helsinki and approved by the Institutional Ethic Committee of the University of Pécs (9188-PTE 2022, approval date: 10 June 2022).

Informed Consent Statement: Not applicable. This was a retrospective study.

Data Availability Statement: The data presented in this study are available on request from the corresponding author. The data are not publicly available due to ethical restrictions.

Conflicts of Interest: The authors declare no conflicts of interest.

References

1. Yanoff, M.; Sassani, J.W. Retinoblastoma and simulating lesions. In *Ocular Pathology*, 8th ed.; Yanoff, M., Sassani, J.W., Eds.; Elsevier: London, UK, 2020; pp. 726–759.
2. Dimaras, H.; Corson, T.W.; Cobrinik, D.; White, A.; Zhao, J.; Munier, F.L.; Abramson, D.H.; Shields, C.L.; Chantada, G.L.; Njuguna, F.; et al. Retinoblastoma. *Nat. Rev. Dis. Primers* **2015**, *1*, 15021. [CrossRef]
3. Orjuela-Grimm, M.; Singh, N.; Bhatt-Carreño, S.; Singh, A.D. Retinoblastoma: Incidence and etiologic factors. In *Clinical Ophthalmic Oncology*; Berry, J.L., Kim, J.W., Damato, B.E., Singh, A.D., Eds.; Springer International Publishing: Cham, Switzerland, 2019; pp. 39–56.

4. Dimaras, H.; Kimani, K.; Dimba, E.A.; Gronsdahl, P.; White, A.; Chan, H.S.; Gallie, B.L. Retinoblastoma. *Lancet* **2012**, *379*, 1436–1446. [CrossRef] [PubMed]
5. Gupta, A.K.; Meena, J.P. A narrative review of retinoblastoma and recent advances in its management. *Pediatr. Med.* **2020**, *3*, 20. [CrossRef]
6. Knudson, A.G. Mutation and cancer: Statistical study of retinoblastoma. *Proc. Natl. Acad. Sci. USA* **1971**, *68*, 820–823. [CrossRef] [PubMed]
7. Wong, J.R.; Tucker, M.A.; Kleinerman, R.A.; Devesa, S.S. Retinoblastoma incidence patterns in the US surveillance, epidemiology, and end results program. *JAMA Ophthalmol.* **2014**, *132*, 478. [CrossRef] [PubMed]
8. Rushlow, D.E.; Mol, B.M.; Kennett, J.Y.; Yee, S.; Pajovic, S.; Thériault, B.L.; Prigoda-Lee, N.L.; Spencer, C.; Dimaras, H.; Corson, T.W.; et al. Characterisation of retinoblastomas without RB1 mutations: Genomic, gene expression, and clinical studies. *Lancet Oncol.* **2013**, *14*, 327–334. [CrossRef] [PubMed]
9. Bremner, R.; Sage, J. The origin of human retinoblastoma. *Nature* **2014**, *514*, 313. [CrossRef]
10. Bouchoucha, Y.; Matet, A.; Berger, A.; Carcaboso, A.M.; Gerrish, A.; Moll, A.; Jenkinson, H.; Ketteler, P.; Dorsman, J.C.; Chantada, G.; et al. Retinoblastoma: From genes to patient care. *Eur. J. Med. Genet.* **2023**, *66*, 104674. [CrossRef]
11. Rootman, D.B.; Gonzalez, E.; Mallipatna, A.; VandenHoven, C.; Hampton, L.; Dimaras, H.; Chan, H.S.L.; Gallie, B.L.; Heon, E. Hand-held high-resolution spectral domain optical coherence tomography in retinoblastoma: Clinical and morphologic considerations. *Br. J. Ophthalmol.* **2013**, *97*, 59–65. [CrossRef]
12. Xu, X.L.; Fang, Y.; Lee, T.C.; Forrest, D.; Gregory-Evans, C.; Almeida, D.; Liu, A.; Jhanwar, S.C.; Abramson, D.H.; Cobrinik, D. Retinoblastoma has properties of a cone precursor tumor and depends upon cone-specific MDM2 signaling. *Cell* **2009**, *137*, 1018–1031. [CrossRef]
13. Xu, X.L.; Singh, H.P.; Wang, L.; Qi, D.-L.; Poulos, B.K.; Abramson, D.H.; Jhanwar, S.C.; Cobrinik, D. Rb suppresses human cone-precursor-derived retinoblastoma tumours. *Nature* **2014**, *514*, 385–388. [CrossRef]
14. Kooi, I.E.; Mol, B.M.; Moll, A.C.; Van Der Valk, P.; De Jong, M.C.; De Graaf, P.; Van Mil, S.E.; Schouten-van Meeteren, A.Y.N.; Meijers-Heijboer, H.; Kaspers, G.L.; et al. Loss of photoreceptorness and gain of genomic alterations in retinoblastoma reveal tumor progression. *EBioMedicine* **2015**, *2*, 660–670. [CrossRef]
15. Singh, H.P.; Wang, S.; Stachelek, K.; Lee, S.; Reid, M.W.; Thornton, M.E.; Craft, C.M.; Grubbs, B.H.; Cobrinik, D. Developmental stage-specific proliferation and retinoblastoma genesis in RB-deficient human but not mouse cone precursors. *Proc. Natl. Acad. Sci. USA* **2018**, *115*, E9391–E9400. [CrossRef] [PubMed]
16. Liu, H.; Zhang, Y.; Zhang, Y.-Y.; Li, Y.-P.; Hua, Z.-Q.; Zhang, C.-J.; Wu, K.-C.; Yu, F.; Zhang, Y.; Su, J.; et al. Human embryonic stem cell-derived organoid retinoblastoma reveals a cancerous origin. *Proc. Natl. Acad. Sci. USA* **2020**, *117*, 33628–33638. [CrossRef] [PubMed]
17. Liu, J.; Ottaviani, D.; Sefta, M.; Desbrousses, C.; Chapeaublanc, E.; Aschero, R.; Sirab, N.; Lubieniecki, F.; Lamas, G.; Tonon, L.; et al. A high-risk retinoblastoma subtype with stemness features, dedifferentiated cone states and neuronal/ganglion cell gene expression. *Nat. Commun.* **2021**, *12*, 5578. [CrossRef]
18. Singh, H.P.; Shayler, D.W.H.; Fernandez, G.E.; Thornton, M.E.; Craft, C.M.; Grubbs, B.H.; Cobrinik, D. An immature, dedifferentiated, and lineage-deconstrained cone precursor origin of N-Myc–initiated retinoblastoma. *Proc. Natl. Acad. Sci. USA* **2022**, *119*, e2200721119. [CrossRef] [PubMed]
19. Singh, L.; Kashyap, S. Update on pathology of retinoblastoma. *Int. J. Ophthalmol.* **2018**, *11*, 2011–2016. [CrossRef]
20. Alsharif, H.; Helmi, H.; Maktabi, A. Histopathological characteristics and classification for prognostic indicators. In *Retinoblastoma—Past, Present and Future*; Manaa Alkatan, H., Ed.; IntechOpen: London, UK, 2019. [CrossRef]
21. Eagle, R.C. The pathology of ocular cancer. *Eye* **2013**, *27*, 128–136. [CrossRef]
22. Mallipatna, A.C.; Gallie, B.L.; Chévez-Barrios, P.; Rouic, L.L.-L.; Chantada, G.L.; Doz, F.; Brisse, H.J.; Munier, F.L.; Albert, D.M.; Català-Mora, J.; et al. Retinoblastoma. In *AJCC Cancer Staging Manual*, 8th ed.; Amin, M.B., Edge, S.B., Greene, F.L., Byrd, D.R., Brookland, R.K., Washington, M.K., Gershenwald, J.E., Compton, C.C., Hess, K.R., Sullivan, D.C., et al., Eds.; Springer International Publishing: Cham, Switzerland, 2017; pp. 827–839. [CrossRef]
23. Martínez-Sánchez, M.; Hernandez-Monge, J.; Rangel, M.; Olivares-Illana, V. Retinoblastoma: From discovery to clinical management. *FEBS J.* **2022**, *289*, 4371–4382. [CrossRef]
24. Manecka, D.-L.; Boukhzar, L.; Falluel-Morel, A.; Lihrmann, I.; Anouar, Y. PACAP signaling in neuroprotection. In *Pituitary Adenylate Cyclase Activating Polypeptide—PACAP*; Current Topics in Neurotoxicity; Reglodi, D., Tamas, A., Eds.; Springer International Publishing: Cham, Switzerland, 2016; Volume 11, pp. 549–561.
25. Toth, D.; Szabo, E.; Tamas, A.; Juhasz, T.; Horvath, G.; Fabian, E.; Opper, B.; Szabo, D.; Maugeri, G.; D'Amico, A.G.; et al. Protective effects of PACAP in peripheral organs. *Front. Endocrinol.* **2020**, *11*, 377. [CrossRef]
26. Miyata, A.; Arimura, A.; Dahl, R.R.; Minamino, N.; Uehara, A.; Jiang, L.; Culler, M.D.; Coy, D.H. Isolation of a novel 38 residue-hypothalamic polypeptide which stimulates adenylate cyclase in pituitary cells. *Biochem. Biophys. Res. Commun.* **1989**, *164*, 567–574. [CrossRef]
27. Miyata, A.; Jiang, L.; Dahl, R.D.; Kitada, C.; Kubo, K.; Fujino, M.; Minamino, N.; Arimura, A. Isolation of a neuropeptide corresponding to the n-terminal 27 residues of the pituitary adenylate cyclase activating polypeptide with 38 residues (PACAP38). *Biochem. Biophys. Res. Commun.* **1990**, *170*, 643–648. [CrossRef]

28. Vaudry, D.; Falluel-Morel, A.; Bourgault, S.; Basille, M.; Burel, D.; Wurtz, O.; Fournier, A.; Chow, B.K.C.; Hashimoto, H.; Galas, L.; et al. Pituitary adenylate cyclase-activating polypeptide and its receptors: 20 years after the discovery. *Pharmacol. Rev.* **2009**, *61*, 283–357. [CrossRef]
29. Olianas, M.C.; Ingianni, A.; Sogos, V.; Onali, P. Expression of pituitary adenylate cyclase-activating polypeptide (PACAP) receptors and PACAP in human fetal retina. *J. Neurochem.* **2002**, *69*, 1213–1218. [CrossRef]
30. Patko, E.; Szabo, E.; Toth, D.; Tornoczky, T.; Bosnyak, I.; Vaczy, A.; Atlasz, T.; Reglodi, D. Distribution of PACAP and PAC1 Receptor in the Human Eye. *J. Mol. Neurosci.* **2022**, *72*, 2176–2187. [CrossRef]
31. Reglodi, D.; Tamas, A.; Koppan, M.; Szogyi, D.; Welke, L. Role of PACAP in female fertility and reproduction at gonadal level—Recent advances. *Front. Endocrinol.* **2012**, *3*, 155. [CrossRef] [PubMed]
32. Watanabe, J.; Seki, T.; Shioda, S. PACAP and neural development. In *Pituitary Adenylate Cyclase Activating Polypeptide—PACAP*; Current Topics in Neurotoxicity; Reglodi, D., Tamas, A., Eds.; Springer International Publishing: Cham, Switzerland, 2016; Volume 11, pp. 65–82.
33. Reglodi, D.; Atlasz, T.; Szabo, E.; Jungling, A.; Tamas, A.; Juhasz, T.; Fulop, B.D.; Bardosi, A. PACAP deficiency as a model of aging. *GeroScience* **2018**, *40*, 437–452. [CrossRef]
34. Reglodi, D.; Helyes, Z.; Nemeth, J.; Vass, R.A.; Tamas, A. PACAP as a potential biomarker: Alterations of PACAP levels in human physiological and pathological conditions. In *Pituitary Adenylate Cyclase Activating Polypeptide—PACAP*; Current Topics in Neurotoxicity; Reglodi, D., Tamas, A., Eds.; Springer International Publishing: Cham, Switzerland, 2016; Volume 11, pp. 815–832.
35. Toth, D.; Reglodi, D.; Schwieters, L.; Tamas, A. Role of endocrine PACAP in age-related diseases. *Front. Endocrinol.* **2023**, *14*, 1118927. [CrossRef] [PubMed]
36. Fabian, E.; Reglodi, D.; Mester, L.; Szabo, A.; Szabadfi, K.; Tamas, A.; Toth, G.; Kovacs, K. Effects of PACAP on intracellular signaling pathways in human retinal pigment epithelial cells exposed to oxidative stress. *J. Mol. Neurosci.* **2012**, *48*, 493–500. [CrossRef] [PubMed]
37. Maugeri, G.; D'Amico, A.G.; Saccone, S.; Federico, C.; Cavallaro, S.; D'Agata, V. PACAP and VIP inhibit HIF-1α-mediated VEGF expression in a model of diabetic macular edema. *J. Cell. Physiol.* **2017**, *232*, 1209–1215. [CrossRef] [PubMed]
38. Maugeri, G.; D'Amico, A.G.; Castrogiovanni, P.; Saccone, S.; Federico, C.; Reibaldi, M.; Russo, A.; Bonfiglio, V.; Avitabile, T.; Longo, A.; et al. PACAP through EGFR transactivation preserves human corneal endothelial integrity. *J. Cell. Biochem.* **2019**, *120*, 10097–10105. [CrossRef] [PubMed]
39. Fabian, E.; Reglodi, D.; Horvath, G.; Opper, B.; Toth, G.; Fazakas, C.; Vegh, A.G.; Wilhelm, I.; Krizbai, I.A. Pituitary adenylate cyclase activating polypeptide acts against neovascularization in retinal pigment epithelial cells. *Ann. N. Y. Acad. Sci.* **2019**, *1455*, 160–172. [CrossRef] [PubMed]
40. Maugeri, G.; D'Amico, A.G.; Bucolo, C.; D'Agata, V. Protective effect of PACAP-38 on retinal pigmented epithelium in an in vitro and in vivo model of diabetic retinopathy through EGFR-dependent mechanism. *Peptides* **2019**, *119*, 170108. [CrossRef] [PubMed]
41. Maugeri, G.; Longo, A.; D'Amico, A.G.; Rasà, D.M.; Reibaldi, M.; Russo, A.; Bonfiglio, V.; Avitabile, T.; D'Agata, V. Trophic effect of PACAP on human corneal endothelium. *Peptides* **2018**, *99*, 20–26. [CrossRef] [PubMed]
42. Atlasz, T.; Szabadfi, K.; Kiss, P.; Tamas, A.; Toth, G.; Reglodi, D.; Gabriel, R. Evaluation of the protective effects of PACAP with cell-specific markers in ischemia-induced retinal degeneration. *Brain Res. Bull.* **2010**, *81*, 497–504. [CrossRef]
43. Vaczy, A.; Kovari, P.; Kovacs, K.; Farkas, K.; Szabo, E.; Kvarik, T.; Kocsis, B.; Fulop, B.; Atlasz, T.; Reglodi, D. Protective role of endogenous PACAP in inflammation-induced retinal degeneration. *Curr. Pharm. Des.* **2018**, *24*, 3534–3542. [CrossRef] [PubMed]
44. Kvarik, T.; Reglodi, D.; Werling, D.; Vaczy, A.; Kovari, P.; Szabo, E.; Kovacs, K.; Hashimoto, H.; Ertl, T.; Gyarmati, J.; et al. The protective effects of endogenous PACAP in oxygen-induced retinopathy. *J. Mol. Neurosci.* **2021**, *71*, 2546–2557. [CrossRef]
45. Szabo, E.; Patko, E.; Vaczy, A.; Molitor, D.; Csutak, A.; Toth, G.; Reglodi, D.; Atlasz, T. Retinoprotective effects of PACAP eye drops in microbead-induced glaucoma model in rats. *Int. J. Mol. Sci.* **2021**, *22*, 8825. [CrossRef]
46. Atlasz, T.; Szabadfi, K.; Reglődi, D.; Kiss, P.; Tamás, A.; Tóth, G.; Molnár, A.; Szabó, K.; Gábriel, R. Effects of pituitary adenylate cyclase activating polypeptide and its fragments on retinal degeneration induced by neonatal monosodium glutamate treatment. *Ann. N. Y. Acad. Sci.* **2009**, *1163*, 348–352. [CrossRef]
47. Seki, T.; Itoh, H.; Nakamachi, T.; Shioda, S. Suppression of ganglion cell death by PACAP following optic nerve transection in the rat. *J. Mol. Neurosci.* **2008**, *36*, 57–60. [CrossRef]
48. Lindholm, D.; Mäkelä, J.; Korhonen, L. PACAP and neural progenitor cells. In *Pituitary Adenylate Cyclase Activating Polypeptide—PACAP*; Current Topics in Neurotoxicity; Reglodi, D., Tamas, A., Eds.; Springer International Publishing: Cham, Switzerland, 2016; Volume 11, pp. 53–63.
49. Horvath, G.; Reglodi, D.; Fabian, E.; Opper, B. Effects of pituitary adenylate cyclase activating polypeptide on cell death. *Int. J. Mol. Sci.* **2022**, *23*, 4953. [CrossRef]
50. Denes, V.; Geck, P.; Mester, A.; Gabriel, R. Pituitary adenylate cyclase-activating polypeptide: 30 years in research spotlight and 600 million years in service. *J. Clin. Med.* **2019**, *8*, 1488. [CrossRef]
51. Moody, T.W.; Jensen, R.T. PACAP and cancer. In *Pituitary Adenylate Cyclase Activating Polypeptide—PACAP*; Current Topics in Neurotoxicity; Reglodi, D., Tamas, A., Eds.; Springer International Publishing: Cham, Switzerland, 2016; Volume 11, pp. 795–814.
52. Moody, T.W.; Nuche-Berenguer, B.; Jensen, R.T. Vasoactive intestinal peptide/pituitary adenylate cyclase activating polypeptide, and their receptors and cancer. *Curr. Opin. Endocrinol. Diabetes Obes.* **2016**, *23*, 38–47. [CrossRef] [PubMed]

53. Moody, T.W.; Jensen, R.T. Pituitary adenylate cyclase-activating polypeptide/vasoactive intestinal peptide [Part 1]: Biology, pharmacology, and new insights into their cellular basis of action/signaling which are providing new therapeutic targets. *Curr. Opin. Endocrinol. Diabetes Obes.* **2021**, *28*, 198–205. [CrossRef]
54. Zibara, K.; Zeidan, A.; Mallah, K.; Kassem, N.; Awad, A.; Mazurier, F.; Badran, B.; El-Zein, N. Signaling pathways activated by PACAP in MCF-7 breast cancer cells. *Cell. Signal.* **2018**, *50*, 37–47. [CrossRef] [PubMed]
55. Maugeri, G.; D'Amico, A.G.; Saccone, S.; Federico, C.; Rasà, D.M.; Caltabiano, R.; Broggi, G.; Giunta, S.; Musumeci, G.; D'Agata, V. Effect of PACAP on hypoxia-induced angiogenesis and epithelial—Mesenchymal transition in glioblastoma. *Biomedicines* **2021**, *9*, 965. [CrossRef]
56. Lochner, R.; Couce, M. Retinoblastoma. Available online: https://www.pathologyoutlines.com/topic/eyeretinaretinoblastoma.html (accessed on 15 November 2023).
57. Bardosi, S.; Bardosi, A.; Nagy, Z.; Reglodi, D. Expression of PACAP and PAC1 receptor in normal human thyroid gland and in thyroid papillary carcinoma. *J. Mol. Neurosci.* **2016**, *60*, 171–178. [CrossRef]
58. García-Fernández, M.O.; Bodega, G.; Ruíz-Villaespesa, A.; Cortés, J.; Prieto, J.C.; Carmena, M.J. PACAP expression and distribution in human breast cancer and healthy tissue. *Cancer Lett.* **2004**, *205*, 189–195. [CrossRef]
59. García-Fernández, M.O.; Collado, B.; Bodega, G.; Cortés, J.; Ruíz-Villaespesa, A.; Carmena, M.J.; Prieto, J.C. Pituitary adenylate cyclase-activating peptide/vasoactive intestinal peptide receptors in human normal mammary gland and breast cancer tissue. *Gynecol. Endocrinol.* **2005**, *20*, 327–333. [CrossRef]
60. Ferencz, S.; Reglodi, D.; Kaszas, B.; Bardosi, A.; Toth, D.; Vekony, Z.; Vicena, V.; Karadi, O.; Kelemen, D. PACAP and PAC1 receptor expression in pancreatic ductal carcinoma. *Oncol. Lett.* **2019**, *18*, 5725–5730. [CrossRef] [PubMed]
61. Ferencz, S.; Toth, D.; Kaszas, B.; Bardosi, S.; Vicena, V.; Karadi, O.; Reglodi, D.; Kelemen, D. PACAP and PAC1 receptor expression in human insulinomas. *Int. J. Pept. Res. Ther.* **2021**, *27*, 1719–1728. [CrossRef]
62. Szanto, Z.; Sarszegi, Z.; Reglodi, D.; Nemeth, J.; Szabadfi, K.; Kiss, P.; Varga, A.; Banki, E.; Csanaky, K.; Gaszner, B.; et al. PACAP immunoreactivity in human malignant tumor samples and cardiac diseases. *J. Mol. Neurosci.* **2012**, *48*, 667–673. [CrossRef]
63. Tamas, A.; Javorhazy, A.; Reglodi, D.; Sarlos, D.P.; Banyai, D.; Semjen, D.; Nemeth, J.; Lelesz, B.; Fulop, D.B.; Szanto, Z. Examination of PACAP-like immunoreactivity in urogenital tumor samples. *J. Mol. Neurosci.* **2016**, *59*, 177–183. [CrossRef] [PubMed]
64. García-Fernández, M.O.; Bodega, G.; Solano, R.M.; Ruíz-Villaespesa, A.; Sánchez-Chapado, M.; Carmena, M.J.; Prieto, J.C. Expression and distribution of pituitary adenylate cyclase-activating peptide in human prostate and prostate cancer tissues. *Regul. Pept.* **2002**, *110*, 9–15. [CrossRef] [PubMed]
65. Olianas, M.C.; Ennas, M.G.; Lampis, G.; Onali, P. Presence of pituitary adenylate cyclase-activating polypeptide receptors in Y-79 human retinoblastoma cells. *J. Neurochem.* **2002**, *67*, 1293–1300. [CrossRef] [PubMed]
66. Wojcieszak, J.; Zawilska, J.B. PACAP38 and PACAP6-38 exert cytotoxic activity against human retinoblastoma y79 cells. *J. Mol. Neurosci.* **2014**, *54*, 463–468. [CrossRef] [PubMed]
67. D'Amico, A.G.; Maugeri, G.; Vanella, L.; Pittalà, V.; Reglodi, D.; D'Agata, V. Multimodal role of PACAP in glioblastoma. *Brain Sci.* **2021**, *11*, 994. [CrossRef] [PubMed]
68. Brubel, R.; Boronkai, A.; Reglodi, D.; Racz, B.; Nemeth, J.; Kiss, P.; Lubics, A.; Toth, G.; Horvath, G.; Varga, T.; et al. Changes in the expression of pituitary adenylate cyclase-activating polypeptide in the human placenta during pregnancy and its effects on the survival of JAR choriocarcinoma cells. *J. Mol. Neurosci.* **2010**, *42*, 450–458. [CrossRef] [PubMed]
69. Horvath, G.; Brubel, R.; Kovacs, K.; Reglodi, D.; Opper, B.; Ferencz, A.; Szakaly, P.; Laszlo, E.; Hau, L.; Kiss, P.; et al. Effects of PACAP on oxidative stress-induced cell death in rat kidney and human hepatocyte cells. *J. Mol. Neurosci.* **2011**, *43*, 67–75. [CrossRef] [PubMed]
70. Farini, D.; Puglianiello, A.; Mammi, C.; Siracusa, G.; Moretti, C. Dual effect of pituitary adenylate cyclase activating polypeptide on prostate tumor LNCaP cells: Short- and long-term exposure affect proliferation and neuroendocrine differentiation. *Endocrinology* **2003**, *144*, 1631–1643. [CrossRef]
71. Doan, N.-D.; Chatenet, D.; Létourneau, M.; Vaudry, H.; Vaudry, D.; Fournier, A. Receptor-independent cellular uptake of pituitary adenylate cyclase-activating polypeptide. *Biochim. Biophys. Acta* **2012**, *1823*, 940–949. [CrossRef]
72. Yu, R.; Zhong, J.; Li, M.; Guo, X.; Zhang, H.; Chen, J. PACAP induces the dimerization of PAC1 on the nucleus associated with the cAMP increase in the nucleus. *Neurosci. Lett.* **2013**, *549*, 92–96. [CrossRef]
73. Reubi, J.C.; Läderach, U.; Waser, B.; Gebbers, J.O.; Robberecht, P.; Laissue, J.A. Vasoactive intestinal peptide/pituitary adenylate cyclase-activating peptide receptor subtypes in human tumors and their tissues of origin. *Cancer Res.* **2000**, *60*, 3105–3112.
74. Bourgault, S.; Vaudry, D.; Botia, B.; Couvineau, A.; Laburthe, M.; Vaudry, H.; Fournier, A. Novel stable PACAP analogs with potent activity towards the PAC1 receptor. *Peptides* **2008**, *29*, 919–932. [CrossRef] [PubMed]
75. Apostol, C.R.; Bernard, K.; Tanguturi, P.; Molnar, G.; Bartlett, M.J.; Szabò, L.; Liu, C.; Ortiz, J.B.; Saber, M.; Giordano, K.R.; et al. Design and synthesis of brain penetrant glycopeptide analogues of PACAP with neuroprotective potential for traumatic brain injury and parkinsonism. *Front. Drug Discov.* **2022**, *1*, 818003. [CrossRef] [PubMed]
76. Werling, D.; Banks, W.; Salameh, T.; Kvarik, T.; Kovacs, L.; Vaczy, A.; Szabo, E.; Mayer, F.; Varga, R.; Tamas, A.; et al. Passage through the ocular barriers and beneficial effects in retinal ischemia of topical application of PACAP1-38 in rodents. *Int. J. Mol. Sci.* **2017**, *18*, 675. [CrossRef]

77. Hirabayashi, T.; Shibato, J.; Kimura, A.; Yamashita, M.; Takenoya, F.; Shioda, S. Potential therapeutic role of pituitary adenylate cyclase-activating polypeptide for dry eye disease. *Int. J. Mol. Sci.* **2022**, *23*, 664. [CrossRef] [PubMed]
78. Boisvilliers, M.D.; Perrin, F.; Hebache, S.; Balandre, A.-C.; Bensalma, S.; Garnier, A.; Vaudry, D.; Fournier, A.; Festy, F.; Muller, J.-M.; et al. VIP and PACAP analogs regulate therapeutic targets in high-risk neuroblastoma cells. *Peptides* **2016**, *78*, 30–41. [CrossRef]

life MDPI

Article

Optical Coherence Tomography Biomarkers Predict the Long-Term Restorative Effect of Early Anti-VEGF Treatment on Diabetic Macular Edema

Süleyman Okudan *, Sule Acar Duyan, Abdullah Erdem, Ayse Bozkurt Oflaz and Banu Turgut Ozturk

Faculty of Medicine, Department of Ophthalmology, Selçuk University, 42130 Konya, Türkiye; dr.sulenuracar@gmail.com (S.A.D.); erdemabd@gmail.com (A.E.); draysebozkurtoflaz@yahoo.com (A.B.O.); ozturkbanuturgut@yahoo.com (B.T.O.)

* Correspondence: suleymanokudan@selcuk.edu.tr; Tel.: +90-3322415000

Abstract: Background/Purpose: This study compared the effects of three induction doses of anti-vascular endothelial growth factor (anti-VEGF) on diabetic macular edema (DME) with that of long-term treatment using biomarkers to find out the predictability potential of early response to anti-VEGF treatment for the long-term restorative effect. Methods: We retrospectively reviewed the clinical and optical coherence tomography (OCT) data of 71 DME eyes treated with three monthly anti-VEGF doses and followed for 1 year. BCVA, central subfield thickness (CST), subretinal fluid (SRF), intraretinal cysts, hyperreflective foci (HF), disorganization of inner retinal layers (DRILs), ellipsoid zone/external limiting membrane (EZ/ELM) integrity, and vitreoretinal relationships were assessed at baseline, 3, 6, and 12 months. Results: Patients (50.7% male) had a mean follow-up of 12 months. After three anti-VEGF doses, 19 eyes required no additional injections, 25 continued anti-VEGF, 20 switched to dexamethasone implants, and seven received combination therapy. Best corrected visual acuity (BCVA) improved from 0.52 to 0.40 logMAR at 3 months, 0.30 at 6 months, and stabilized at 0.40 at 12 months. CST decreased from 406 μm to 317 μm at 3 months and 307 μm at 12 months. Significant early improvements in BCVA, CST, SRF, and intraretinal cysts were sustained in the long-term follow-up. HF reduction became significant after 6 months, while DRIL and EZ/ELM integrity remained unchanged. Conclusions: The improvement of OCT biomarkers in DME patients supported that intravitreal anti-VEGF significantly restored the retinal microstructure, which was already evident at 3 months in the control after anti-VEGF induction.

Keywords: anti-VEGF therapy; diabetic macular edema; diabetic retinopathy; optical coherence tomography biomarkers

1. Introduction

Diabetic macular edema (DME) affects approximately 21 million people worldwide and is the main cause of vision loss in diabetic patients [1]. The etiology and pathogenesis of macular edema are multifactorial. Through various inflammatory and vasogenic mediators, including vascular endothelial growth factor (VEGF) upregulation and inflammatory cytokines and chemokines, pathological changes in the vascular endothelium are induced and the blood–retinal barrier is disrupted. Therefore, fluid extravasation occurs in the extracellular space, manifesting clinically as macular edema and causing vision loss [2,3]. VEGFs have a pivotal role in the complex pathogenesis of DME, and anti-VEGF agents are currently the standard therapy [4]. Current DME treatment commence with three monthly initiation doses of intravitreal anti-VEGF injections and continues with anti-VEGF or steroids. Following the initial dose, the number of additional injections and the interval between injections are determined according to the response of the individual patient and the ophthalmologist's preferred treatment regimen like PRN, treat and extend, etc. [5].

The response to treatment is quite variable in DME, and optical coherence tomography (OCT) has become the standard imaging modality used in the diagnosis and treatment follow-up of DME. It is able to quantitatively evaluate retinal thickness, volume, and all other morphological changes in retinal anatomy [6]. There are some biomarkers in OCT that are reported to be valuable in predicting prognosis and treatment response [7]. These include central subfield thickness (CST), subretinal fluid (SRF), intraretinal cysts, disorganization of the inner retinal layers (DRIL), hyperreflective foci (HF), ellipsoid zone (EZ), and outer limiting membrane (ELM) status [7,8].

Diabetic retinopathy is a leading cause of visual impairment worldwide, and the DRCR (Diabetic Retinopathy Clinical Research) Retina Network has been instrumental in developing treatment protocols, particularly through anti-VEGF therapies. These therapies have become a cornerstone in the treatment of diabetic macular edema (DME) and proliferative diabetic retinopathy (PDR) [9]. According to the DRCR retina network anti-VEGF treatment algorithm, the mean number of injections per year for a DME patient is greatest in year 1. It progressively decreases until year 5 according to the DRCR re-treatment algorithm [10]. Protocol I is a clinical trial conducted by the DRCR Retina Network and is an important milestone in the treatment of DME. It is one of the first large studies to evaluate the efficacy and safety of anti-VEGF therapy in diabetic macular edema. Protocol I reported the mean number of injections in DME patients as 8.1 in the first year, as 2.2 in the third year, and 1.9 in the fifth year. In the post hoc analysis of the Protocol I study, it was reported that the response after three anti-VEGF injections was preserved at 1 year and even at 3 years. It has been shown that deducing the long-term treatment response to anti-VEGF treatment is possible according to the outcome after three initial injections [11]. Though the number of injections is expected to lower with time, repeated injections usually become exhausting for DME patients after a period and raises questions about their benefits. Herein, we aimed to find out the effect of intravitreal treatment to regain vision and restore macular anatomy, and whether it is possible to predict the ultimate effect of anti-VEGF injections already following the three-monthly initiation dose. To reach this goal, the early responses of OCT biomarkers in DME patients are compared with those in the long-term response.

2. Materials and Methods

2.1. Study Design and Patients

This retrospective study included naive DME patients who received three initiation doses of monthly intravitreal anti-VEGF injections and were followed up for a minimum duration of 12 months. Patient files fulfilling this criterion were enrolled by scanning the registries of the Retina Unit between January 2018 and November 2023. The exclusion criteria were patients who underwent retinal laser photocoagulation for the treatment of diabetic retinopathy and those who underwent any other eye surgery procedure during 12 months. Additionally, patients with macular edema secondary to diseases other than diabetes and decreased visual acuity related to other causes like glaucoma, retinal detachment, macular hole, uveitis, optic neuropathy, or age-related macular degeneration were also excluded. The study protocol was approved by a clinical research ethics committee (Selçuk University Medical Faculty, 2024/650) and followed the tenets of the 1964 Helsinki Declaration.

Study parameters including age, gender, duration of DM, injections within 12 months, and best-corrected visual acuity (BCVA) (logMAR) were recorded from patients' files. The OCT (DRI-OCT Triton; Topcon Corporation, Tokyo, Japan) images at baseline, after 3 monthly anti-VEGF doses, at the 6th month, and at the 12th month were graded for the parameters.

The switch to steroids has been decided upon the residual inflammatory OCT biomarkers after the initial 3 anti-VEGF treatments. Lens status and history of glaucoma have been considered as contraindications for dexamethasone implant application. Anti-VEGF molecule preference has been decided regarding the reimbursement regulations of the national social security department. Accordingly, it is mandatory to start with the 3 initial doses of bevacizumab. If necessary, further doses can be scheduled as aflibercept and ranibizumab if the patient presents a poor response.

2.2. OCT Measurements

Among the OCT parameters, central subfield thickness (CST), intraretinal cysts, subretinal fluid (SRF), hyperreflective foci (HF), disorganization of the inner retinal layers (DRIL), EZ/ELM status, and vitreomacular relationship were examined.

The size of the intraretinal cysts was graded using numbers as a reference, taking into account the size of the largest intraretinal cyst that could be identified on the scan. Cysts were classified as absent, mild, moderate, and severe [12].

All scans were divided into two groups (high HF/low HF) according to the mean number of HF using 30 as the cut-off value [12].

The absence of the EZ and/or the ELM was considered a complete loss of foveal reflectivity at this level, identified as the first and the second hyperreflective bands of the four outermost layers on OCT, respectively. These layers were defined as disrupted if they were not perfectly discernible but still partially visible in the fovea; if EZ was non-gradable due to the presence of SRF, ELM only was considered [12].

2.3. Statistical Analysis

The study data were analyzed in a computer environment with SPSS (Statistical Package for Social Sciences) 18.0 package program. In descriptive analyses, frequency data were shown as number (n) and percentage (%), and numerical data were shown using mean ± standard deviation or median value (minimum–maximum) according to the conformity of data to a normal distribution with the Kolmogorov–Smirnov test. The distribution of normally distributed numerical data in two independent groups was evaluated with the independent samples t-test, and the distribution of non-parametric distributed numerical data was evaluated with the Mann–Whitney U test. Mc Nemar and Mc Nemar Bowker tests were used to evaluate categorical data measured at baseline and T follow-up. The statistical significance level was accepted as $p < 0.05$ for all tests.

3. Results

The study included the 71 eyes of 71 DME patients. The demographic data and diabetes duration of the patients are shown in Table 1.

Table 1. Age, diabetes duration, and gender distribution of patients.

	Years
Mean age	63.80 ± 7.96
Mean duration of diabetes	14.12 ± 6.44
Gender	***n* (%)**
Male	36 (50.7)
Female	35 (49.3)

The evaluation of patient files revealed a baseline BCVA of 0.52 (0.22–0.82) logMAR and CST of 406.0 μm (324–478). Grading of the initial OCT images showed IRF in 98.6% of eyes, of which 54.9% were severe, 26.8% were moderate, and 16.9% were mild cysts. Only 1.4% of the eyes had no cysts. SRF was evident in 26.6% of enrolled eyes. Those with HF less than 30 constituted 64.8% of the eyes, while those with HF more than 30 constituted 35.2% of the eyes. DRIL was present in 67.6% of eyes. At baseline, EZ/ELM was intact in 50.7%, disrupted in 35.2%, and absent in 14.1% of eyes.

The treatment results after three doses of monthly anti-VEGF therapy were analyzed in terms of treatment response according to the definition of refractory DME described as an increase in visual acuity (VA) $\leq$ 5 letters or a decrease in central subfield thickness (CST) $\leq$ 20% [13]. Accordingly, 46 eyes (64.8%) were labeled as responder and 25 eyes (35.2%) were labeled as non-responder to treatment and were recognized as refractory. After the third monthly visit, 19 eyes (26.8%) did not require additional injections, 25 eyes (35.2%) received additional anti-VEGF, 20 eyes (28.2%) received additional dexamethasone implants, and seven eyes (9.9%) received both anti-VEGF and dexamethasone implant treatment during 12 months follow-up. The initial three doses of intravitreal bevacizumab injection were performed. Of the eyes that received additional anti-VEGF injections, 15 received aflibercept, five received ranibizumab, and five continued with bevacizumab treatment. Among the eyes that received dexamethasone and anti-VEGF, three eyes received aflibercept and dexamethasone, three eyes received bevacizumab and dexamethasone, and one eye received ranibizumab and dexamethasone. The mean number of anti-VEGF injections per patient was 3.77 and 0.39 for dexamethasone during 12 months. The retreatment criteria were a loss of more than five letters or CST more than 250 μm. The BCVA improved after three doses of treatment compared to baseline (p = 0.001); this improvement continued at the sixth month (p < 0.001). At the final visit on month 12, it decreased slightly compared to the sixth month, but it was still significantly better than the baseline visit (p = 0.001) (Table 2, Figure 1a).

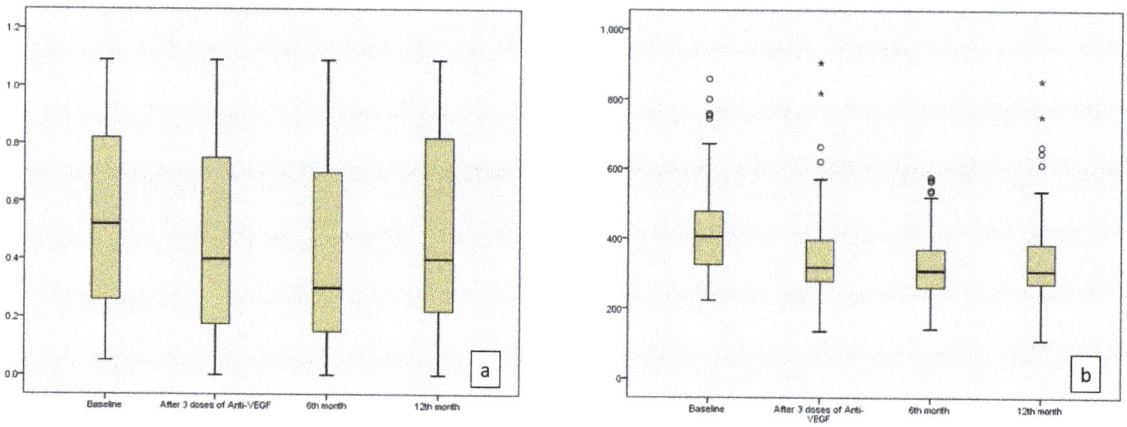

Figure 1. BCVA (logMAR) values (**a**) and CST (**b**) values demonstrated a statistically significant decrease at 3 months, 6 months, and 12 months compared to baseline values (p = 0.001, p < 0.001, p = 0.001 for BCVA and p < 0.001 for CST at the 3rd, 6th, and 12th months, respectively). (°; represents mild outliers, i.e., data points outside 1.5 to 3 times the interquartile range (IQR), but not extreme values. *; indicates extreme outliers, i.e., data points outside 3 times the IQR.)

The second most frequently used parameter for DME treatment is CST in daily routine. The median value of CST was 406 μm at the baseline visit, which decreased to 317 μm after three doses of anti-VEGF. It demonstrated a statistically significant decrease to 308 µm at the sixth month and 307 µm at the 12th month ($p < 0.001$) (Table 2, Figure 1b).

Table 2. Comparison of visual acuity and CST at baseline and after 3, 6, and 12 months.

	Baseline (n = 71) Median (1st–3rd Quartiles)	**After 3 Doses of Anti-VEGF (n = 71) Median (1st–3rd Quartiles)**	***p* Value**	**6th Month (n = 71) Median (1st–3rd Quartiles)**	***p* Value**	**12th Month (n = 71) Median (1st–3rd Quartiles)**	***p* Value**
BCVA (logMAR)	0.52 (0.22–0.82)	0.40 (0.15–0.80)	0.001	0.30 (0.15–0.70)	<0.001	0.40 (0.22–0.82)	0.001
CST (µ)	406.00 (324.00–478.00)	317.00 (275.00–398.00)	<0.001	308.00 (259.00–372.00)	<0.001	307.00 (269.00–384.00)	<0.001

The evaluation of OCT biomarkers during the follow-up period showed that the number of eyes with severe intraretinal cysts decreased significantly after three doses of treatment and at the sixth and 12th months compared to baseline, while the number of eyes with mild cysts increased ($p < 0.001$). At the end of the 12th month, there was still IRF in 88.7% of eyes, but 44.4% of that was mild (Figure 2a). After three doses of treatment, SRF resolved in 42.9% of eyes, which was found to be statistically significant compared to the baseline percentage of 26.6% ($p = 0.022$). While SRF resolved in two more eyes at the sixth month, no further improvement was observed at the 12th month (Figure 2b) (Table 3).

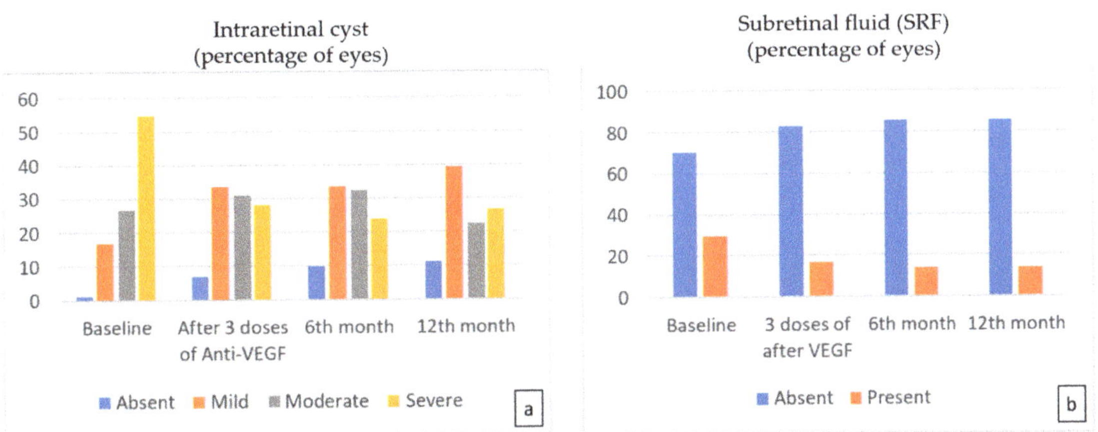

Figure 2. The percentage of eyes with moderate and severe intraretinal cysts (**a**) showed statistically significant improvement with time compared to baseline ($p < 0.001$ for the 3rd-, 6th-, and 12th-month visits, respectively). The percentage of eyes with SRF (**b**) presented a steady and statistically significant decrease during 12 monthly treatments ($p = 0.022$, $p = 0.007$, $p = 0.007$ for 3rd, 6th, and 12th months, respectively).

The change in the percentage of HF was slow during follow-up. Although there was a slight decrease in the number of eyes with more than 30 HF after three doses of anti-VEGF treatment, it was not statistically significant ($p = 0.227$); however, at the sixth-month and 12th-month visits, OCT images revealed a statistically significant decrease in the percentage of more than 30 HF compared to baseline ($p = 0.002$, $p < 0.001$) (Figure 3a).

Table 3. Comparison of OCT biomarkers at baseline and after 3 doses of monthly anti-VEGF at 6th month and 12th month.

	Baseline (n = 71) n (%)	After 3 Doses of Anti-VEGF (n = 71) n (%)	*p* Value	6th Month (n = 71) n (%)	*p* Value	12th Month (n = 71) n (%)	*p* Value
Intraretinal cysts							
Absent	1 (1.4)	5 (7.0)		7 (9.9)		8 (11.3)	
Mild	12 (16.9) *	24 (33.8)		24 (33.8)		28 (39.4)	
Moderate	19 (26.8)	22 (31.0)	<0.001	23 (32.4)	<0.001	16 (22.5)	<0.001
Severe	39 (54.9) *	20 (28.2)		17 (23.9)		19 (26.8)	
SRF							
Absent	50 (70.4)	59 (83.1)	0.022	61 (85.9)	0.007	61 (85.9)	0.007
Present	21 (29.6)	12 (16.9) *		10 (14.1) *		10 (14.1) *	
Hyperreflective foci							
Less than 30 in number	46 (64.8)	51 (71.8)	0.227	59 (83.1)	0.002	60 (84.5)	<0.001
More than 30 in number	25 (35.2)	20 (28.2)		12 (16.9) *		11 (15.5) *	
DRIL							
Absent	23 (32.4)	26 (36.6)	0.375	25 (35.2)	0.625	22 (31.0)	0.999
Present	48 (67.6)	45 (63.4)		46 (64.8)		49 (69.0)	
EZ/ELM status							
Intact	36 (50.7)	36 (50.7)		35 (49.3)		31 (43.7)	
Disrupted	25 (35.2)	27 (38.0)	0.506	25 (35.2)	0.842	29 (40.8)	0.076
Absent	10 (14.1)	8 (11.3)		11 (15.5)		11 (15.5)	
Vitreoretinal relationship							
Absent	26 (36.6)	18 (25.4)		17 (23.9)		16 (22.5)	
IVD	27 (38.0) *	24 (33.8)		22 (31.0)		18 (25.3)	
PVD	8 (11.3)	14 (19.7)	0.023	16 (22.5)	0.007	14 (19.7)	0.001
ERM	10 (14.1)	15 (21.1)		16 (22.5)		23 (32.4)	

* refers to the group from which the difference originates. Third month, sixth month, and twelfth month compared with the baseline.

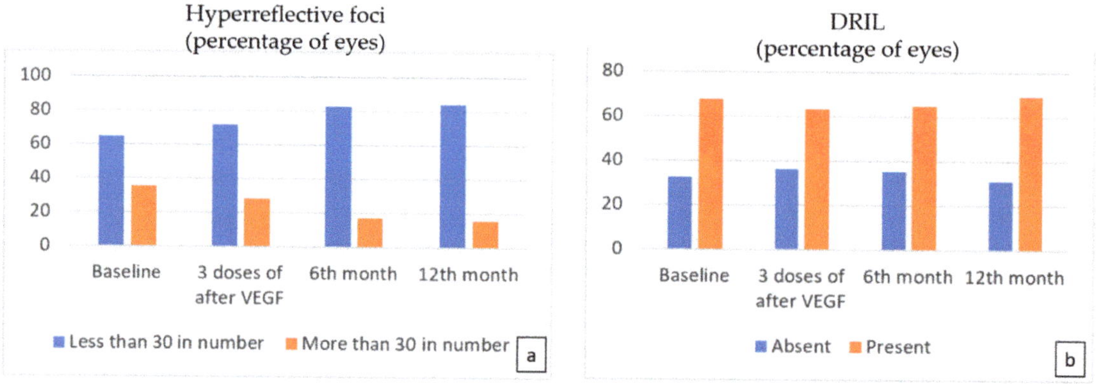

Figure 3. The number of eyes with more than 30 hyperreflective foci (**a**) decreased at each follow-up visit, but a statistically significant decrease was noted at the 6th and 12th months ($p = 0.002$, $p < 0.001$ for the 6th and 12th months, respectively). No significant difference in the number of eyes with DRIL (**b**) was found following treatment.

There was no significant difference in the status of DRIL at the third, sixth, and twelfth months (Figure 3b).

At months 3, 6, and 12, there was no significant difference in the status of EZ/ELM (Figure 4a). The number of eyes with IVD was found to be lower at the sixth month compared to the baseline ($p = 0.007$) (Table 3, Figure 4b).

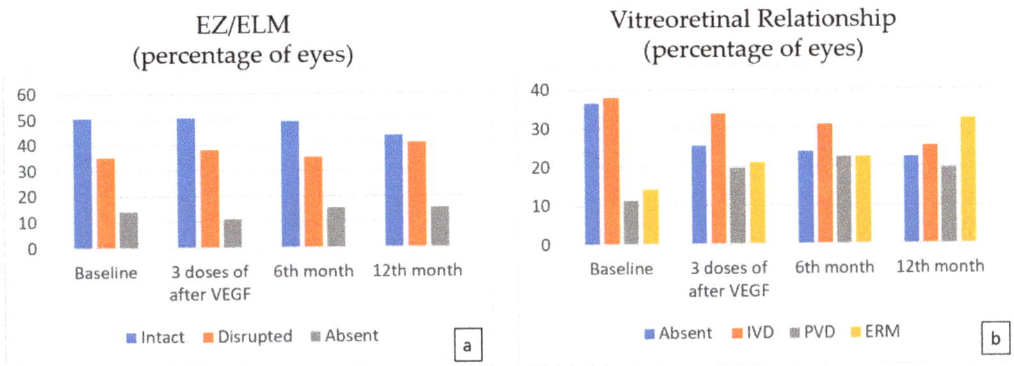

Figure 4. The EZ/ELM status (**a**) showed no significant difference following treatment. While a statistically significant decrease is observed in the number of eyes with IVD (**b**), an increase in the number of eyes with PVD was noted compared to the baseline ($p = 0.023$, $p = 0.007$, $p = 0.001$ at 3, 6, and 12 months, respectively).

A comparison of BCVA at baseline and the third-month visits between patients with SRF and without SRF showed worse BCVA in the SRF group ($p = 0.012$, $p = 0.039$, respectively) (Table 4, Figure 5).

Table 4. Comparison of BCVA according to the presence of SRF.

	Group Without SRF (n = 50) Median (1st–3rd Quartiles)	Group with SRF (n = 21) Median (1st–3rd Quartiles)	*p* Value
BCVA (Baseline)	0.40 (0.15–0.80)	0.80 (0.46–1.00)	0.012
BCVA (3rd month)	0.30 (0.13–0.70)	0.40 (0.30–0.91)	0.039
BCVA (6th month)	0.26 (0.13–0.70)	0.40 (0.21–0.81)	0.128
BVCA (12th month)	0.30 (0.21–0.82)	0.52 (0.30–0.91)	0.403

BCVA in the sixth month was found to be significantly worse in the group with a higher number of HFs at baseline compared to the group with a lower number of HFs ($p = 0.015$). When patients were divided into two groups in terms of DRIL existence, the mean BCVA was found to be worse in the group with DRIL at baseline, the third month, and sixth month ($p = 0.047$, $p = 0.002$, $p < 0.001$) (Table 5).

Table 5. Comparison of VA according to the number of HFs and the presence of DRIL.

	Less than 30 HF (n = 46) Median (1st–3rd Quartiles)	More than 30 HF (n = 25) Median (1st–3rd Quartiles)	*p* Value	Without DRIL (n = 23) Mean ± SD	With DRIL (n = 48) Mean ± SD	*p* Value
BCVA (Baseline)	0.40 (0.20–0.82)	0.70 (0.35–0.96)	0.110	0.44 ± 0.36	0.61 ± 0.32	0.047
BCVA (3rd month)	0.30 (0.13–0.56)	0.70 (0.18–0.87)	0.058	0.27 ± 0.26	0.54 ± 0.34	0.002
BCVA (6th month)	0.22 (0.13–0.52)	0.70 (0.17–0.87)	0.015	0.21 ± 0.17	0.51 ± 0.34	<0.001
BCVA (12th month)	0.35 (0.20–0.70)	0.60 (0.30–1.00)	0.114	0.42 ± 0.32	0.54 ± 0.35	0.186

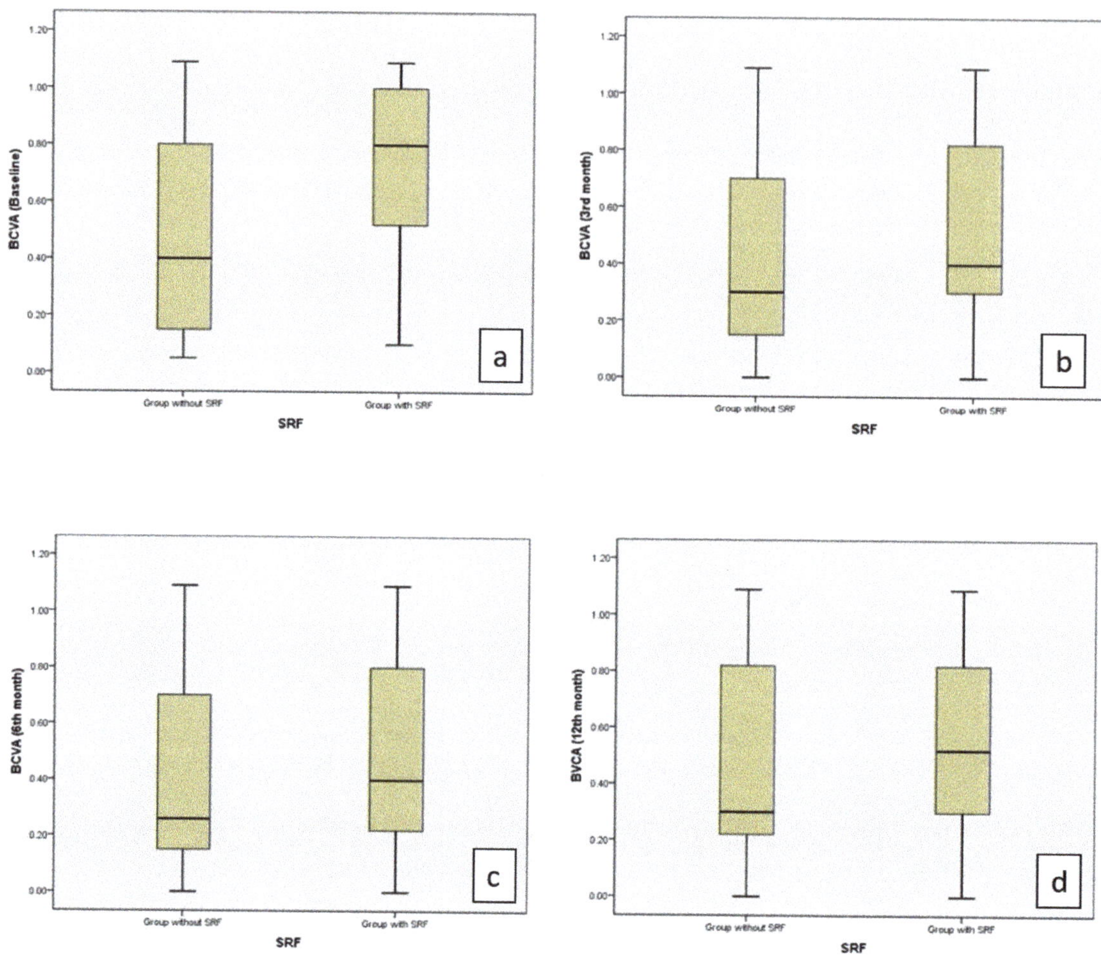

Figure 5. Values of BCVA (logMAR) were significantly higher in the group with SRF than in the group without SRF at the baseline (**a**) and at the 3rd month (**b**) ($p = 0.012$, $p = 0.039$, respectively). No significant difference was observed between the two groups in the 6th (**c**) and 12th months (**d**).

4. Discussion

The main objective of this retrospective observational study was to evaluate the response after three doses of monthly intravitreal anti-VEGF treatment and long-term response with OCT biomarkers in patients with DME.

The analysis of the data after three doses of monthly anti-VEGF treatment showed significant improvement in BCVA, CST, IRF, and SRF, while a significant decrease in HF occurred later at 6 months. The number of HF, intraretinal cysts, CST, and SRF was significantly reduced after intravitreal treatment, which suggested that the retinal microstructural integrity was pretty much restored by intravitreal agents.

In concordance with other previous studies, the improvement in VA was mostly observed in the third and sixth months of treatment, while VA remained more stable in the long term [4,14,15]. A similar study by Kriechbaum et al. [16] also reported that eyes treated with anti-VEGF therapy for DME showed a rapid improvement in VA in the first months of treatment, whereas VA improvement was slower in the long term.

Coscas et al. [17] first reported HFs as small punctiform hyper-reflective elements on SD-OCT. They suggested that HFs may represent aggregates of activated microglia cells. Vujosevic et al. [18] described HFs in the early stages of diabetic retinopathy (DR) and diabetic patients without DR. They also emphasized that these spots are mostly located in the inner retinal layers in the early stage and migrate toward the outer retinal layers as DR progresses. Previous studies have shown that HF, which is thought to have an inflammatory origin in DME, is significantly reduced after intravitreal treatment [19–21]. Vujosevic et al. [20] reported an early improvement in HF from the first month after anti-VEGF treatment. Schreur et al. [22] also showed a decrease in the number of HF after three monthly doses of anti-VEGF treatment. In our study, it was observed that there was no improvement in HF after three doses of anti-VEGF treatment, but there was a significant decrease at 6 months. While other OCT biomarkers such as CST, SRF, and intraretinal cysts responded earlier to treatment, HF responded later. In a study by Yoshitake et al. [23], it was highlighted that there was no improvement in the number of HFs in the first months after anti-VEGF treatment in eyes with DME, while there was a decrease in the number of HFs in the late period, as in our study. This late response might be explained by the longer duration of HF, but more research is needed to study the etiology and clinical significance of HF through histologic and epidemiologic studies.

CST is an important parameter for DME diagnosis and treatment decisions. The majority of published studies have shown that a reduction in CST is associated with an improvement in BCVA, but there are some studies reporting no direct correlation between BCVA and CST [8,21,24]. In our study, there was a statistically significant decrease in CST, as expected after three doses of anti-VEGF treatment and after six and twelve months.

Regarding SRF, the frequency of this biomarker in DME varies between 15% and 30% in different reports [25,26]. In this study, our results are consistent with the literature, as SRF was seen in 26.6% of the eyes at baseline. While the presence of SRF has been shown as a positive biomarker for better functional outcomes in some studies, there are also studies reporting the opposite. Vujosevic et al. [26] stated that the presence of SRF leads to ELM disruption and decreased retinal sensitivity. Seo et al. [27] reported that photoreceptors were more damaged in the presence of SRF and are therefore associated with poor visual outcomes. Moreover, they reported that DME eyes with SRF tended to respond poorly to anti-VEGF therapy. In contrast, another study reported that in DME patients, those with SRF had a better functional and anatomical response to anti-VEGF therapy than those without SRF [28].

In our study, the mean BCVA at baseline was worse in eyes with SRF. Although visual improvement was better in the group with SRF at the third and sixth month, it was not statistically significant.

Intraretinal cysts were another valuable OCT parameter investigated in our study. The pathophysiology of intraretinal cyst development is a combination of increased vascular permeability, leukostasis, inflammatory cytokines, VEGF, and Müller cell dysfunction [29]. In this study, there was a significant reduction in the number of patients with severe cysts in the early phase of treatment. Our observations that intraretinal cysts decreased during treatment with intravitreally injected anti-VEGF agents in patients with DME support the positive effect of anti-VEGF therapy on cystoid macular edema seen in previous studies [30].

DRIL is thought to indicate the disorganization or destruction of cells in the inner retinal layers and disruption of the visual pathways from photoreceptors to ganglion cells [31]. The International Retina Group demonstrated by OCT a reduction in DRIL extension after treatment of DME with dexamethasone, which they attributed to a favorable architectural effect on Muller cells due to a reduction in inflammation [32]. Vujosevic et al. [33] reported a significant decrease in DRIL extension with both dexamethasone and anti-VEGF treatment. In this study, no significant improvement was found in DRIL with anti-VEGF treatment. The mean baseline VA was worse in eyes with DRIL compared to eyes without DRIL, and no significant difference was observed in the BCVA of the DRIL-positive group with anti-VEGF treatment. The EZ/ELM status was also unresponsive to treatment; no significant improvement was observed with treatment. DRIL and EZ/ELM damage were mentioned as worse prognosis parameters, demonstrating limited improvement with treatment. Our results for these parameters are in accordance with the current literature [34].

In this study, patients who showed a poor response after three doses of anti-VEGF treatment were switched to anti-VEGF treatment and were switched to ranibizumab or aflibercept agents. In patients who showed a poor response and especially in those with prominent inflammatory biomarkers such as HF and SRF on OCT, treatment was switched to a dexamethasone implant. In patients who showed a poor response to anti-VEGF treatment, an early switch is important to prevent the permanent loss of retinal cells due to chronic edema [35].

The major limitation of this study was its retrospective nature and relatively small number of patients. Furthermore, the decision and timing of reinjection are variable. However, our results provide evidence from a real-world environment.

5. Conclusions

Our results revealed that intravitreal anti-VEGF treatment in DME resulted in a significant improvement of BCVA and OCT biomarkers including CST, SRF, and intraretinal cysts in the early phase. Hyperreflective foci known to be an inflammatory biomarker demonstrated a late response starting at 6 months. This improvement demonstrated that intravitreal treatment largely restored retinal microstructural integrity in DME starting up to three monthly initiation doses and provided the resolution of inflammatory findings in the long term.

Author Contributions: Conceptualization, S.O., S.A.D. and B.T.O.; methodology, S.A.D. and B.T.O.; software, S.A.D. and B.T.O.; validation, S.A.D., B.T.O. and A.B.O.; formal analysis, S.A.D. and B.T.O.; investigation, S.O., S.A.D. and B.T.O.; resources, A.E. and A.B.O.; data curation, A.E. and S.A.D.; writing—original draft preparation, S.O., S.A.D., A.B.O. and B.T.O.; writing—review and editing, S.O., S.A.D. and B.T.O.; visualization, S.O., S.A.D., A.E., A.B.O. and B.T.O.; supervision, S.O. and B.T.O.; project administration, S.O. and B.T.O.; funding acquisition, S.O. All authors have read and agreed to the published version of the manuscript.

Funding: This research received no external funding.

Institutional Review Board Statement: The study was conducted in accordance with the Declaration of Helsinki and approved by the Clinical Research Ethics Committee of Selcuk University Faculty of Medicine (2024/650).

Informed Consent Statement: Not applicable.

Data Availability Statement: Data are contained within the article.

Conflicts of Interest: The authors declare no conflicts of interest.

References

1. Arroba, A.I.; Valverde, A.M. Modulation of microglia in the retina: New insights into diabetic retinopathy. *Acta Diabetol.* **2017**, *54*, 527–533. [CrossRef] [PubMed]
2. Miller, K.; Fortun, J.A. Diabetic Macular Edema: Current Understanding, Pharmacologic Treatment Options, and Developing Therapies. *Asia Pac. J. Ophthalmol.* **2018**, *7*, 28–35. [CrossRef]
3. Romero-Aroca, P.; Baget-Bernaldiz, M.; Pareja-Rios, A.; Lopez-Galvez, M.; Navarro-Gil, R.; Verges, R. Diabetic macular edema pathophysiology: Vasogenic versus inflammatory. *J. Diabetes Res.* **2016**, 2156273. [CrossRef]

4. Elman, M.J.; Aiello, L.P.; Beck, R.W.; Bressler, N.M.; Bressler, S.B.; Edwards, A.R.; Ferris, F.L.; Friedman, S.M.; Glassman, A.R.; Miller, K.M.; et al. Randomized trial evaluating ranibizumab plus prompt or deferred laser or triamcinolone plus prompt laser for diabetic macular edema. *Ophthalmology* **2010**, *117*, 1064–1077. [CrossRef]
5. Hirano, T.; Toriyama, Y.; Takamura, Y.; Sugimoto, M.; Nagaoka, T.; Sugiura, Y.; Okamoto, F.; Saito, M.; Noda, K.; Yoshida, S.; et al. Outcomes of a 2-year treat-and-extend regimen with aflibercept for diabetic macular edema. *Sci. Rep.* **2021**, *11*, 4488. [CrossRef]
6. Schmidt-Erfurth, U.; Garcia-Arumi, J.; Bandello, F.; Berg, K.; Chakravarthy, U.; Gerendas, B.S.; Jonas, J.; Larsen, M.; Tadayoni, R.; Loewenstein, A. Guidelines for the management of diabetic macular edema by the European Society of Retina Specialists (EURETINA). *Ophthalmologica* **2017**, *237*, 185–222. [CrossRef] [PubMed]
7. Chang, Y.C.; Huang, Y.T.; Hsu, A.Y.; Meng, P.P.; Lin, C.H.; Lai, C.T.; Hsia, N.Y.; Chen, H.S.; Tien, P.T.; Lin, J.M.; et al. Optical Coherence Tomography Biomarkers in Predicting Treatment Outcomes of Diabetic Macular Edema after Ranibizumab Injections. *Medicina* **2023**, *59*, 629. [CrossRef] [PubMed]
8. Visioli, G.; Alisi, L.; Mastrogiuseppe, E.; Albanese, G.M.; Romano, E.; Iannetti, L.; Armentano, M.; Giovannetti, F.; Gharbiya, M. OCT biomarkers as predictors of visual improvement in diabetic macular edema eyes receiving dexamethasone implants. *Int. J. Retin. Vitr.* **2023**, *9*, 35. [CrossRef] [PubMed]
9. Sun, J.; Jampol, L. The Diabetic Retinopathy Clinical Research Network (DRCR.net) and Its Contributions to the Treatment of Diabetic Retinopathy. *Ophthalmic Res.* **2019**, *62*, 225–230. [CrossRef]
10. Elman, M.J.; Ayala, A.; Bressler, N.M.; Browning, D.; Flaxel, C.J.; Glassman, A.R.; Jampol, L.M.; Stone, T.W. Intravitreal Ranibizumab for diabetic macular edema with prompt versus deferred laser treatment: 5-year randomized trial results. *Ophthalmology* **2015**, *122*, 375–381. [CrossRef]
11. Bressler, S.B.; Odia, I.; Glassman, A.R.; Danis, R.P.; Grover, S.; Hampton, G.R.; Jampol, L.M.; Maureen, M.G.; Melia, M. Changes in Diabetic Retinopathy Severity when Treating Diabetic Macular Edema with Ranibizumab: DRCR.net Protocol I 5-Year Report. *Retina* **2018**, *38*, 1896–1904. [CrossRef]
12. Panozzo, G.; Cicinelli, M.V.; Augustin, A.J.; Parodi, M.B.; Cunha-Vaz, J.; Guarnaccia, G.; Kodjikian, L.; Jampol, L.M.; Jünemann, A. An optical coherence tomography-based grading of diabetic maculopathy proposed by an international expert panel: The European School for Advanced Studies in Ophthalmology classification. *Eur. J. Ophthalmol.* **2020**, *30*, 8–18. [CrossRef]
13. Busch, C.; Zur, D.; Fraser-Bell, S.; Lains, I.; Santos, A.R.; Lupidi, M.; Cagini, C.; Gabrielle, P.H.; Couturier, A.; Mane-Tauty, V.; et al. Shall we stay, or shall we switch? Continued anti-VEGF therapy versus early switch to dexamethasone implant in refractory diabetic macular edema. *Acta Diabetol.* **2018**, *55*, 789–796. [CrossRef]
14. Rajendram, R.; Fraser-Bell, S.; Kaines, A.; Michaelides, M.; Hamilton, R.D.; Esposti, S.D.; Peto, T.; Egan, C.; Bunce, C.; Leslie, R.D.; et al. A 2-year prospective randomized controlled trial of intravitreal bevacizumab or laser therapy (BOLT) in the management of diabetic macular edema: 24-month data: Report 3. *Arch. Ophthalmol.* **2012**, *130*, 972–979. [CrossRef] [PubMed]
15. Nguyen, Q.D.; Brown, D.M.; Marcus, D.M.; Boyer, D.S.; Patel, S.; Feiner, L.; Gibson, A.; Sy, J.; Rundle, A.C.; Hopkins, J.J.; et al. Ranibizumab for diabetic macular edema: Results from 2 phase III randomized trials: RISE and RIDE. *Ophthalmology* **2012**, *119*, 789–801. [CrossRef]
16. Kriechbaum, K.; Prager, S.; Mylonas, G.; Scholda, C.; Rainer, G.; Funk, M.; Kundi, M.; Schmidt-Erfurth, U. Intravitreal bevacizumab (Avastin) versus triamcinolone (Volon A) for treatment of diabetic macular edema: One-year results. *Eye* **2014**, *28*, 9–16. [CrossRef] [PubMed]
17. Coscas, G.; De Benedetto, U.; Coscas, F.; Calzi, C.I.L.; Vismara, S.; Roudot-Thoraval, F.; Bandello, F.; Souied, E. Hyperreflective dots: A new spectral-domain optical coherence tomography entity for follow-up and prognosis in exudative age-related macular degeneration. *Ophthalmologica* **2012**, *229*, 32–37. [CrossRef]
18. Vujosevic, S.; Bini, S.; Midena, G.; Berton, M.; Pilotto, E.; Miden, E. Hyperreflective intraretinal spots in diabetics without and with nonproliferative diabetic retinopathy: An in vivo study using spectral domain OCT. *J. Diabetes Res.* **2013**, *1*, 491835. [CrossRef] [PubMed]
19. Framme, C.; Schweizer, P.; Imesch, M.; Wolf, S.; Wolf-Schnurrbusch, U. Behavior of SD-OCT–detected hyperreflective foci in the retina of anti-VEGF–treated patients with diabetic macular edema. *Investig. Ophthalmol. Vis. Sci.* **2012**, *53*, 5814–5818. [CrossRef]
20. Vujosevic, S.; Berton, M.; Bini, S.; Casciano, M.; Cavarzeran, F.; Midena, E. Hyperreflective retinal spots and visual function after anti-vascular endothelial growth factor treatment in center-involving diabetic macular edema. *Retina* **2016**, *36*, 1298–1308. [CrossRef] [PubMed]
21. Vujosevic, S.; Torresin, T.; Bini, S.; Convento, E.; Pilotto, E.; Parrozzani, R.; Midena, E. Imaging retinal inflammatory biomarkers after intravitreal steroid and anti-VEGF treatment in diabetic macular oedema. *Acta Ophthalmol.* **2017**, *95*, 464–471. [CrossRef]
22. Schreur, V.; Altay, L.; van Asten, F.; Groenewoud, J.M.M.; Fauser, S.; Klevering, B.J.; Hoyng, C.B.; de Jong, E.K. Hyperreflective foci on optical coherence tomography associate with treatment outcome for anti-VEGF in patients with diabetic macular edema. *PLoS ONE* **2018**, *13*, e0206482. [CrossRef] [PubMed]
23. Yoshitake, T.; Murakami, T.; Suzuma, K.; Dodo, Y.; Fujimoto, M.; Tsujikawa, A. Hyperreflective Foci in the Outer Retinal Layers as a Predictor of the Functional Efficacy of Ranibizumab for Diabetic Macular Edema. *Sci. Rep.* **2020**, *10*, 873. [CrossRef] [PubMed]

24. Santo, A.R.; Gomes, S.C.; Figueira, J.; Nunes, S.; Lobo, C.L.; Cunha-Vaz, J.G. Degree of decrease in central retinal thickness predicts visual acuity response to intravitreal ranibizumab in diabetic macular edema. *Ophthalmologica* **2014**, *231*, 16–22. [CrossRef] [PubMed]
25. Vujosevic, S.; Torresin, T.; Berton, M.; Bini, S.; Convento, E.; Midena, E. Diabetic Macular Edema with and Without Subfoveal Neuroretinal Detachment: Two Different Morphologic and Functional Entities. *Am. J. Ophthalmol.* **2017**, *181*, 149–155. [CrossRef]
26. Reznicek, L.; Cserhati, S.; Seidensticker, F.; Liegl, R.; Kampik, A.; Ulbig, M.; Neubauer, A.S.; Kernt, M. Functional and morphological changes in diabetic macular edema over the course of anti-vascular endothelial growth factor treatment. *Acta Ophthalmol.* **2013**, *91*, 529–536. [CrossRef]
27. Seo, K.H.; Yu, S.Y.; Kim, M.; Kwak, H.W. Visual and Morphologic Outcomes of Intravitreal Ranibizumab for Diabetic Macular Edema Based on Optical Coherence Tomography Patterns. *Retina* **2016**, *36*, 588–595. [CrossRef] [PubMed]
28. Bonfiglio, V.; Reibaldi, M.; Pizzo, A.; Russo, A.; Macchi, I.; Faro, G.; Avitabile, T.; Longo, A. Dexamethasone for unresponsive diabetic macular oedema: Optical coherence tomography biomarkers. *Acta Ophthalmol.* **2019**, *97*, 540–544. [CrossRef]
29. Chung, Y.R.; Kim, Y.H.; Lee, S.Y.; Byeon, H.E.; Lee, K. Insights into the pathogenesis of cystoid macular edema: Leukostasis and related cytokines. *Int. J. Ophthalmol.* **2019**, *12*, 1202–1208. [CrossRef] [PubMed]
30. Wu, P.C.; Lai, C.H.; Chen, C.L.; Kuo, C.N. Optical coherence tomographic patterns in diabetic macula edema can predict the effects of intravitreal bevacizumab injection as primary treatment. *J. Ocul. Pharmacol. Ther.* **2012**, *28*, 59–64. [CrossRef]
31. Sun, J.K.; Radwan, S.H.; Soliman, A.Z.; Lammer, J.; Lin, M.M.; Prager, S.G.; Silva, P.S.; Aiello, L.B.; Aiello, L.P. Neural retinal disorganization as a robust marker of visual acuity in current and resolved diabetic macular edema. *Diabetes* **2015**, *64*, 2560–2570. [CrossRef] [PubMed]
32. Zur, D.; Iglicki, M.; Sala-Puigdollers, A.; Chhablani, J.; Lupidi, M.; Fraser-Bell, S.; Mendes, T.S.; Chaikitmongkol, V.; Cebeci, Z.; Dollberg, D.; et al. Disorganization of retinal inner layers as a biomarker in patients with diabetic macular oedema treated with dexamethasone implant. *Acta Ophthalmol.* **2019**, *98*, 217–223. [CrossRef] [PubMed]
33. Vujosevic, S.; Toma, C.; Villani, E.; Muraca, A.; Torti, E.; Florimbi, G.; Leporati, F.; Brambilla, M.; Nucci, P.; de Cilla, S. Diabetic macular edema with neuroretinal detachment: OCT and OCT-angiography biomarkers of treatment response to anti-VEGF and steroids. *Acta Diabetol.* **2020**, *57*, 287–296. [CrossRef] [PubMed]
34. Wirth, M.A.; Wons, J.; Freiberg, F.J.; Becker, M.D.; Michels, S. Impact of long-term intravitreal anti–vascular endothelial growth factor on preexisting microstructural alterations in diabetic macular edema. *Retina* **2018**, *38*, 1824–1829. [CrossRef] [PubMed]
35. Vitiello, L.; Salerno, G.; Coppola, A.; De Pascale, I.; Abbinante, G.; Gagliardi, V.; Lixi, F.; Pellegrino, A.; Giannaccare, G. Switching to an Intravitreal Dexamethasone Implant after Intravitreal Anti-VEGF Therapy for Diabetic Macular Edema: A Review. *Life* **2024**, *14*, 725. [CrossRef] [PubMed]

MDPI

Article

Use of VESsel GENeration with Optical Coherence Tomography Angiography and Fluorescein Angiography for Detection and Quantification of Vascular Changes in Mild and Moderate Diabetic Retinopathy

Mariana DuPont [1,2], Edmund Arthur [3], Yazen Shihab [4], Madelyn Kenny [3], Swetha Ravichandran [3], Patricia Parsons-Wingerter [5,†], Ruchi Vyas [6], Matthew C. Murray [7], Marina Predovic [7], Shiyin Lim [7], Nicole Jacobs [7], Sneha Ramesh [7], Amanda Vu [7], Srinivaas Sekaran [7], Kakarla V. Chalam [8], Ramana S. Moorthy [9], Jason Crosson [10], John Mason [10] and Maria B. Grant [1,*]

1 Department of Ophthalmology and Visual Sciences, Heersink School of Medicine, University of Alabama at Birmingham, Birmingham, AL 35294, USA; mdupont@uab.edu
2 Department of Hematology-Oncology, Heersink School of Medicine, University of Alabama at Birmingham, Birmingham, AL 35294, USA
3 School of Optometry, University of Alabama at Birmingham, Birmingham, AL 35294, USA; earthur@uab.edu (E.A.); mrkenny@uab.edu (M.K.); swetha30@uab.edu (S.R.)
4 Birmingham School of Medicine, University of Alabama, Birmingham, AL 35233, USA; yazen@uab.edu
5 National Aeronautics and Space Administration (NASA), Washington, DC 20546, USA; patriciaparsonsw@gmail.com
6 Mori Associates, Ames Research Center, National Aeronautics and Space Administration (NASA), Moffett Field, CA 94035, USA; ruchivyas@gmail.com
7 Blue Marble Space Institute of Science, Ames Research Center, NASA, Moffett Field, CA 94035, USA; matthewc.murray@hotmail.com (M.C.M.); predovic.marina@gmail.com (M.P.); shiyin.lim.000@gmail.com (S.L.); nicolejacobs@ucla.edu (N.J.); 97.sneha@gmail.com (S.R.); amandavu94@gmail.com (A.V.); sekaransrinivaas@gmail.com (S.S.)
8 Department of Ophthalmology, Loma Linda University Health Care, Loma Linda, CA 92354, USA; kchalam@llu.edu
9 Associated Vitreoretinal and Uveitis Consultants, Indianapolis, IN 46290, USA; rsmoorthy46032@yahoo.com
10 Retina Consultants of Alabama Birmingham, Birmingham, AL 35233, USA; jcrosson@uab.edu (J.C.); johnmason@uabmc.edu (J.M.)
* Correspondence: mariagrant@uabmc.edu; Tel.: +1-3528714313
† Retired.

Abstract: (1) Background: Previously, VESsel GENeration (VESGEN) software was used to map and quantify vascular changes observed on fluorescein angiography (FA) in subjects (n = 15 eyes) with retinal pathology ranging from mild non-proliferative diabetic retinopathy (NPDR) to proliferative diabetic retinopathy (PDR). In the current study, we used VESGEN for the assessment of individuals with early-stage NPDR imaged by FA (Cohort 1) and by optical coherence tomography angiography (OCTA; Cohort 2). (2) Methods: Cohort 1 included type 2 diabetics (T2D), represented 21 eyes (ranging from no DR to moderate DR), and also included nondiabetic controls (NDC; n = 15 eyes). Cohort 2 consisted of 23 eyes from T2D subjects (including no DR subjects and moderate DR subjects) and NDC (n = 18 eyes). (3) Results: In the FA-VESGEN study, total tortuosity (Tv) of microvessels ($G \geq 6$) increased in T2D with mild DR compared to the controls. In contrast, the VESGEN analysis of OCTA images showed that vessel length (characterized as density) was lower in T2D subjects before the diagnosis of DR and following the diagnosis of DR when compared to the controls. Additionally, T2D showed a significant decrease in vessel area (density). (4) Conclusions: FA elucidated the vessel morphology of small-generation microvessels to a greater degree than OCTA; however, OCTA identified changes in vessel density better than FA. VESGEN analysis can be used with both standard FA and OCTA to facilitate our understanding of early events in DR, including before the clinical diagnosis of DR.

Keywords: retinal imaging; VESGEN; image processing; fluorescein angiography; optical coherence tomography angiography; vascular segmentation

1. Introduction

Diabetic retinopathy (DR) represents the most common complication associated with diabetes mellitus (DM) and is recognized as the primary cause of blindness in diabetic adults. Thus, early detection, intervention, and treatment are crucial. By 2050, DR is projected to affect 16 million individuals, with 3.4 million experiencing vision-threatening problems or blindness [1,2].

Retinal changes are predictors of systemic vascular pathology, including hypertension [3], diabetes [2,4], renal disease [5], and sickle cell disease [6]. Yet it remains challenging to evaluate subtle changes in retinal blood vessels due to the complexity of the branching arterial and venous trees in the retina [7–10]; thus, current DR clinical examinations focus on secondary vascular changes such as microaneurysms and hemorrhages [11–13].

In a previous study conducted by Parsons-Wingerter et al. [14], vascular changes in patients with mild non-proliferative diabetic retinopathy (NPDR) to proliferative diabetic retinopathy (PDR) were quantified. Vessel Generation Analysis (VESGEN) [15], an automated microvascular mapping and quantification software, was used to map and quantify vascular changes in the previous study and indicated that vascular changes occurred in an oscillatory fashion, with vessel length and vessel number increasing in the moderate and PDR stages while being less in the mild and severe DR stages. However, this study had limitations, as it did not include diabetic individuals without clinical evidence of DR or age-matched nondiabetic controls. Also, the FA images were taken with a 55° lens camera which did not provide detail to the macular region, which is better visualized by a 35° lens camera [14].

The current study sought to overcome the constraints of the prior study by comparing the DR subjects to diabetics without any evidence of clinical DR and to age-matched nondiabetic controls. We also asked whether examining optical coherence tomography angiography (OCTA) images could provide advantages to FA.

2. Materials and Methods

2.1. Study Participants

Informed consent was obtained from all subjects involved in the studies. Following informed consent, subjects (control, DM alone and DM with confirmed DR) [16] were recruited from Indiana University (IRB: 1402550709), the University of Alabama at Birmingham (IRB-300009129), and the University of Florida (IRB#535-2011). Grading of NPDR was carried out by 2 masked board-certified retina specialists using standard early treatment diabetic retinopathy study (ETDRS) guidelines [17]. Individuals with mild NPDR had ETDRS levels between 20 and 35, whereas those with moderate NPDR were classified as ETDRS levels 43–47. No subjects had diabetic macular edema. Normal subjects had an ETDRS grading of 10. Exclusion criteria for controls and diabetics were ages less than 18 years, significant lens changes or media opacities that prevented imaging, chronic, or acute infections (HIV, hepatitis, or tuberculosis), any known history of other retinal diseases or diabetic complications other than DR, pregnancy, presence of malignancy, and prisoners. Inclusion criteria for diabetic subjects include diabetes defined as hemoglobin (Hb)A1c $\geq$ 6.5% with or without DR. Cohort 1: FA images from 21 diabetic eyes no DR (n = 5), mild DR (n = 8), and moderate DR (n = 8) and NDC (n = 15 eyes) (Table 1). Cohort 2: OCTA images from 23 diabetic eyes with no DR (n = 17) and moderate DR (n = 6), which were compared to the controls (18 eyes). One diabetic subject (1 eye) was removed from the study due to a failed screening, resulting in a total of thirteen diabetics (23 eyes). Baseline demographics of the entire cohort were comparable in all groups and are shown in Table 2.

Table 1. Baseline characteristics of fluorescein angiography for controls and diabetics.

		FA	
		Control	**Type 2 Diabetics**
N		8	12
Total eyes		15	21
Sex *			
	Male	4	9
	Female	11	5
Age (range) **		25–52	32–69
Race ***			
	White	7	5
	Black	2	7
	Asian	4	0
	Other	2	0
Retinopathy type			
	None	-	5
	Mild	-	8
	Moderate	-	8
HbA1c% (Mean ± SD) ****		-	8.3 ± 1.8

Age and HbA1c% were determined based on total subjects, while sex, race, and retinopathy were based on total eye count. * 7 patients with unknown sex were not included in the count; ** 4 patients with unknown ages were not included in the count; *** 9 patients with unknown race were not included in the count; **** 6 patients with unknown A1C were not included in count; FA = Fluorescein Angiography, HbA1c% = Hemoglobin A1C, SD = Standard deviation.

Table 2. Baseline characteristics of optical coherence tomography angiography for controls and diabetics.

		OCTA	
		Control	**Type 2 Diabetics**
N		17	13
Total eyes		18	23
Sex			
	Male	5	1
	Female	13	22
Age (range)		46–74	43–65
Race			
	White	15	2
	Black	3	21
	Asian	0	0
	Other	0	0
Retinopathy type			
	None	-	17
	Mild	-	0
	Moderate	-	6
HbA1c% (Mean ± SD) *		-	7.54 ± 1.80 *

Age and HbA1c% were determined based on total subjects, while sex, race, and retinopathy were based on total eye count. * Two individuals had a reading of <8 and were listed as 8 during avg. calculations; OCTA = Optical Coherence Tomography Angiography, HbA1c% = Hemoglobin A1C, SD = Standard deviation.

2.2. FA Image Acquisition

Retinal imaging was conducted on all subjects using a 35° lens field. FA was performed using an intravenous injection of a 5 mL 20% sodium fluorescein. Retinal imaging was then conducted on all subjects with a SPECTRALIS imaging system using a 35° lens field. The photographers were intentionally blinded to the severity of the retinal condition of each participant, ensuring an unbiased data collection. High-quality, mid-phase images were obtained after complete arteriovenous filling of the posterior pole. Images with the maximum capillary plexus detail devoid of/with limited dye leakage were selected for each eye for further analysis.

2.3. OCTA Image Acquisition

Macular-centered OCTA images, that were 20 × 20°, of the superficial vascular plexus (SVP) consisting of 512 B-scans, 512 A-scans per B-scan, 12-micron spacing between the B-scans, and 5 frames averaged per each B-scan location (Spectralis HRA + OCT; Eye Explorer version 1.10.4.0; Heidelberg Engineering Heidelberg, Germany) were obtained for 30 participants (13 DR patients and 17 controls) for a total of 41 eyes. The superficial vascular plexus (SVP) was defined as the composite retinal vasculature from the inner limiting membrane (ILM) to the inner plexiform layer (IPL)/inner nuclear layer (INL) boundary [18,19]. The signal quality values of all the OCTA images from the vendor software ranged from 30 to 40, with 40 being the highest possible value that can be obtained.

2.4. Image Processing

Each grayscale FA and OCTA image was processed manually in Photoshop (Adobe, Inc., San Jose, CA, USA) to a binary (black/white) image together with a region of interest (ROI) image, subject to inspection by at least one expert reviewer. Vascular branching in the binary images was separated into arterial and venous vascular patterns (FA only) using color fundus images for FA images based on previously established characteristics [7]. Each binary image of arterial or venous pattern, together with its ROI, served as the input images for the VESGEN analysis described below. Due to poor image quality, one control (FA) and two diabetic (one FA and one OCTA) retinal images were removed from the study. FA images of the controls were previously used for a prior study on retinal vessel changes in pulmonary arterial hypertension [15]. Total vessels were analyzed for OCTA imaging due to the inability of distinguishing arteries from veins.

2.5. Vascular Quantification

VESGEN analysis (Bethesda, MD, USA), a JAVA-based software for interactive use, was employed to analyze the vascular features of the FA and OCTA images [14,15,20]. This software works as a plug-in to the ImageJ software from the U.S. National Institutes of Health, Bethesda, MD, USA, and is available upon request from the U.S. National Aeronautics and Space Administration (NASA, https://software.nasa.gov/software/ARC-17621-1 (accessed on 15 December 2023; recently reviewed). VESGEN has several analysis options [21].

The results in VESGEN 2D are provided in pixel units. The software uses a mix of standard ImageJ processing algorithms and custom algorithms unique to the software. For example, it offers automatic methods for detecting regions of interest, including a method to identify the midpoint boundary between a selected vessel region and other non-selected vessels. VESGEN 2D utilizes specialized algorithms to determine branch boundaries and assign branches to their respective branching generations (G1, . . ., Gx) for the analysis of vascular trees.

Users can interact with VESGEN 2D to optimize the output analysis. After analyzing and mapping a vascular tree into distinct generations, the vessels can be combined into groups such as small, medium, and large branching generations (see below). This helps to facilitate further analysis. The output results include vascular maps illustrating the generational assignments for vascular trees and distance maps displaying local vessel diameter, as well as the ROI used for calculating the vessel density parameters.

VESGEN was employed in this study to examine all the FA and OCTA images. The analysis considered the following vascular parameters: vessel tortuosity (Tv), vessel length density, and vessel area density. The VESGEN software allows for the automated classification of blood vessels into generation(G)s 1–3 (large) and G4–5 (medium) according to user decisions. Vessels of generation 6 and smaller are considered microvascular vessels ($G \geq 6$). The VESGEN vascular tree option was used to analyze images displaying branching, asymmetric, nonhomogeneous structures with tapering vessels, such as those seen in 35° FA. Additionally, the study utilized the VESGEN vascular tree network composite option for the OCTA images displaying a closed vascular network geometry.

Vessel area density (Av) and vessel length density (Lv) are key metrics for quantifying vascular structures within a region of interest (ROI). Vessel area density (Av) is defined as the density of the total vascular area within the ROI, calculated by dividing the vascular area by the ROI. Vessel length density (Lv) measures the density of the total vessel length. Additionally, vessel tortuosity is assessed using the distance method, which involves dividing the length of the vessel by the distance between its two endpoints. The ROI plays a crucial role in these measurements, as it defines the specific area within which the density functions, such as vessel area density (Av) and vessel length density (Lv), are computed.

2.6. Statistical Analyses

Following normalizing of data, the Shapiro–Wilk test of normality was performed to assess the Gaussian distribution of the data, where $p < 0.05$ indicates non-normal/nonparametric data. Grouped data determined to be normally distributed were assessed for statistical significance by an ordinary one-way parametric ANOVA with Holm–Šídák's multiple comparisons test. Data determined to be non-normally distributed were assessed for statistical significance by a nonparametric Kruskal–Wallis test with a Dunn multiple comparison test. Tukey's multiple comparison tests with a single pooled variance were also used. An ordinary two-way ANOVA was used when applicable in addition to Tukey's multiple comparison test with a single pooled variance. Due to the small sample size, the outliers were not removed. Statistical analysis was performed using GraphPad software. A p value < 0.05 assessed statistical significance.

3. Results

3.1. Clinical Characteristics of FA Subjects

The clinical characteristics of Cohort 1 are outlined in Table 1. FA images (Figure 1) with a 35° field were processed for the VESGEN input. As depicted in Figure 1, the FA images in the first column were processed and prepared for binarization to be input into VESGEN in the second column.

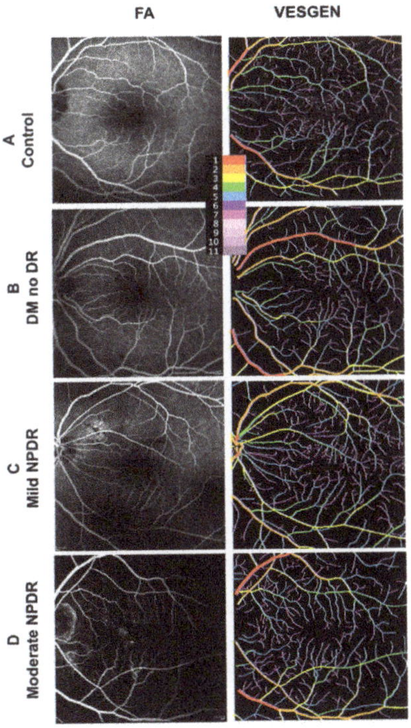

Figure 1. VESGEN characterization of FA images from controls and diabetic individuals. Vascular images and VESGEN maps from the right retina of (**A**) control, (**B**) T2D without DR, (**C**) T2D with mild NPDR, and (**D**) T2D with moderate NPDR. Each row represents increasing severity starting with control (**A**), followed by a T2D individuals with no DR (none; **B**), mild NPDR (**C**), and moderate NPDR (**D**). Legend identifies branching generation (center top). FA = fluorescein angiography.

3.2. FA Images Demonstrate Changes in Tortuosity

Our primary intention of the FA study was to determine if vascular changes were detectable by VESGEN before the onset of clinically diagnosed DR. For this study, individuals were separated based on status as (i) control, (ii) T2D with no history of DR (DM), and (iii) T2D with a history of DR (DM w/DR, Figure 2). When using FA studies, no differences were detected in total vessel length (measured as density) (Figure 2A) and total vessel density (Figure 2B) between the three cohorts. This lack of differences in overall changes in vessel length and densities was confirmed by comparisons of the results for all generations examined, G1–3, G4–5, and G $\geq$ 6. In contrast, a significant increase in tortuosity (Tv) was detected when comparing the controls to diabetic subjects with DR ($p = 0.0054$) (Figure 2C). FA showed the peripheral retina in more detail and demonstrated increased tortuosity in the T2D subjects with DR compared to the controls.

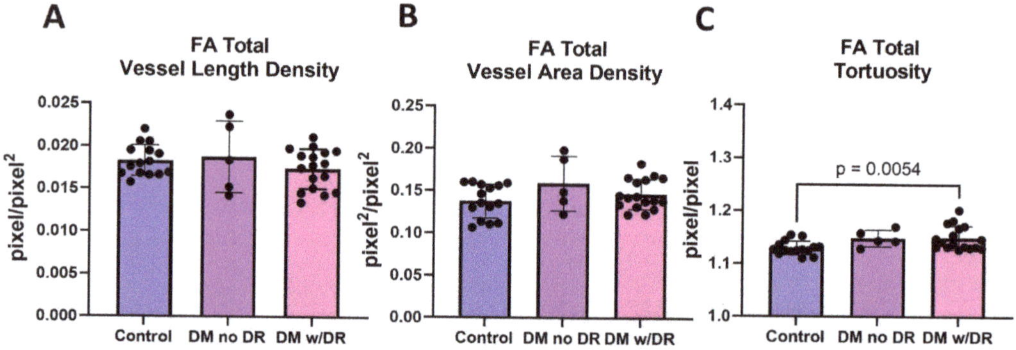

Figure 2. VESGEN analysis of FA diabetics with and without diabetic retinopathy. (**A**) Total vessel length (measured as density) was compared in control, individuals with DM without DR, and individuals with DM with DR. The three cohorts were compared using parametric one-way ANOVA, but no significant differences were found. (**B**) Total vessel area density was compared in the cohorts using one-way nonparametric ANOVA, and no significant differences were found. (**C**) Total tortuosity showed significant differences between controls and DM subjects with DR ($p = 0.054$) when using one-way nonparametric ANOVA. DM = diabetes mellitus; DR = diabetic retinopathy; FA = fluorescein angiography.

3.3. FA Demonstrates That Arterial but Not Venous Tortuosity Was Increased in Diabetics with DR

Because FA detected an increase in total tortuosity, we further explored this by separating the arteries and veins. Diabetics with NPDR had more tortuous arteries but not veins compared to the controls (Figure 3A, $p = 0.0041$). When arteries were further separated into diabetic with mild NPDR and moderate NPDR, arterial tortuosity showed a significant increase between the control and mild NPDR (Figure 3C, $p = 0.0219$), suggesting that tortuosity is an early event in the arteries. No significant differences were observed in venous tortuosity in any of the cohorts (Figure 3B,D).

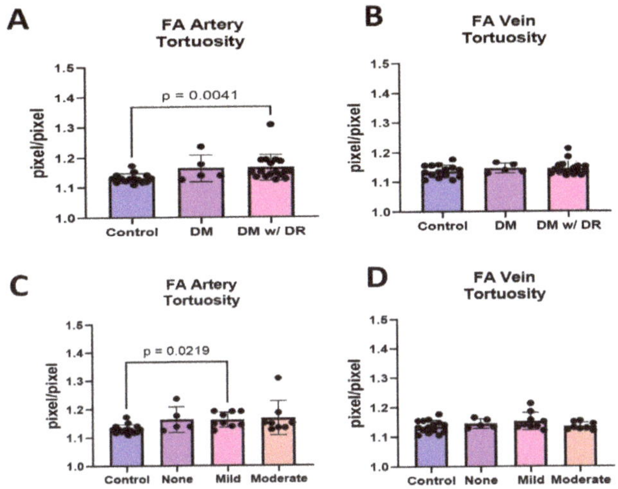

Figure 3. Arterial and venous VESGEN analysis in diabetics with and without DR. (**A**) Arterial tortuosity was compared in the three cohorts. Significant differences were found in the arteries of individuals with DM and DR when compared to controls ($p = 0.0041$). (**B**) Venous tortuosity showed no differences between any of the cohorts. (**C**) Individuals with mild NPDR showed increased arterial tortuosity compared to controls, suggesting that tortuosity may represent a feature of early NPDR ($p = 0.0219$). (**D**) Venous tortuosity did not show a difference between mild NPDR and control. All measurements were performed using a one-way nonparametric ANOVA. DM = diabetes mellitus; DR = diabetic retinopathy; FA = fluorescein angiography.

3.4. FA Distinguishes Total Tortuosity by Vessel Generation

We next sought to determine whether the size of the vessels impacted tortuosity. Large (G1–3), medium (G4–5), and small ($G \geq 6$) total vessels (arteries + veins) were assessed. The tortuosity of the microvessels ($G \geq 6$) showed a significant increase ($p = 0.0281$) when comparing the controls and diabetics with mild NPDR but not moderate NPDR (Figure 4).

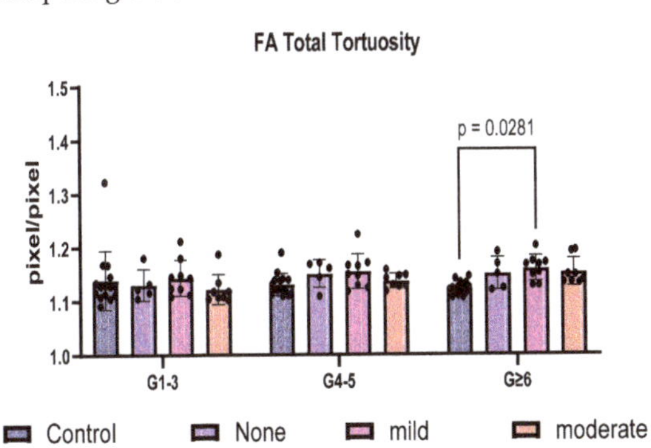

Figure 4. VESGEN analysis between diabetics with and without diabetic retinopathy based on vessel size in FA imaging. Total tortuosity based on vessel generational size was compared between controls and individuals with mild and moderate NPDR. Total tortuosity of FA imaging was analyzed using two-way ANOVA and showed a significant increase in microvessel $G \geq 6$ of mild retinopathy ($p = 0.0281$) when compared to controls. DM = diabetes mellitus; DR = diabetic retinopathy; FA = fluorescein angiography.

3.5. Clinical Characteristics of OCTA Subjects

The clinical characteristics of the OCTA study subjects are summarized in Table 2. OCTA images (Figure 5; first column) were prepared for input into VESGEN (Figure 5; second column).

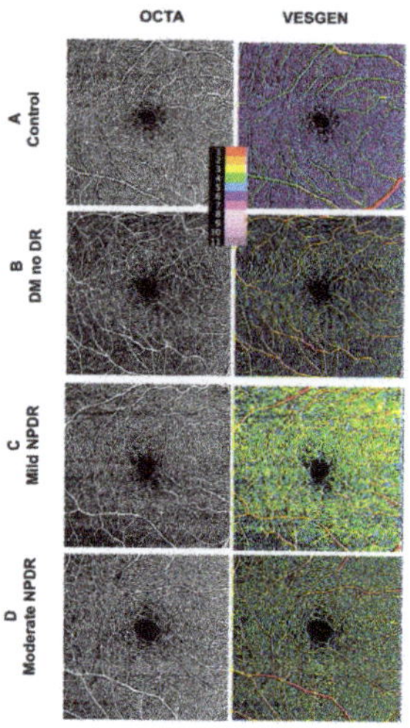

Figure 5. VESGEN characterization of OCTA from controls and diabetes individuals with and without NPDR. OCTA vascular images and VESGEN maps from (**A**) control, (**B**) T2D without DR, (**C**) T2D with mild NPDR, and (**D**) T2D with moderate NPDR (**D**). After the control row, each subsequent row represents increasing severity of DR. Legend identifies branching generation (center top).

3.6. OCTA Images Demonstrate Changes in Multiple Endpoints

The use of OCTA indicated a reduction in vessel length density, vessel area density, and tortuosity in diabetics with no retinopathy. Vessel length density showed a significant decrease in diabetics with no retinopathy ($p = 0.0013$) and a further decrease in diabetics with mild or moderate NPDR ($p < 0.0001$; Figure 6A). Vessel area density and tortuosity similarly revealed a significant decrease in diabetics with no retinopathy ($p = 0.0065$; Figure 6B and $p = 0.0153$; Figure 6C, respectively). OCTA focused more on the central retina and indicated a reduction in tortuosity or straightening of vessels before the onset of clinical retinopathy.

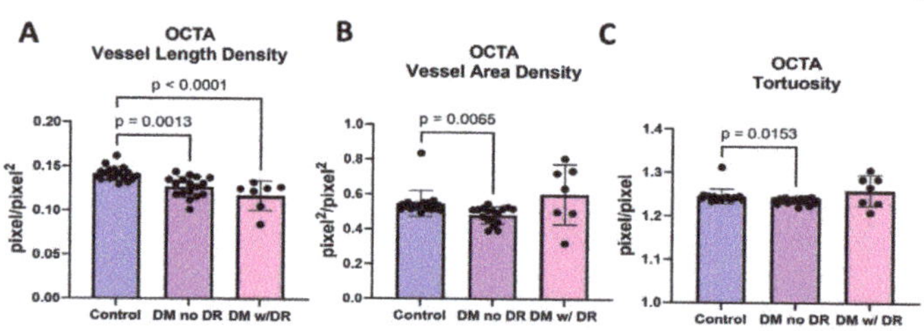

Figure 6. Total VESGEN analysis of OCTA between diabetics with and without diabetic retinopathy (**A**) Total vessel length density was analyzed using parametric one-way ANOVA and showed significant reduction in subjects with DM and no DR and controls ($p = 0.0013$) and a further reduction between DM subjects with DR and controls ($p < 0.0001$). (**B**) Total vessel area density was analyzed using nonparametric one-way ANOVA and showed a reduction in diabetics without DR ($p = 0.0065$). (**C**) Total tortuosity was analyzed using nonparametric one-way ANOVA and demonstrated a reduction in the subjects with diabetes without DR compared to control ($p = 0.0153$). DM = diabetes mellitus; DR = diabetic retinopathy; OCTA = optical coherence tomography angiography.

3.7. OCTA Distinguishes Total Tortuosity by Vessel Generation

Lastly, we sought to determine whether the generation of the vessels impacted tortuousity. As previously mentioned, large (G1–3), medium (G4–5), and small ($G \geq 6$) total vessels (arteries + veins) were compared. Using OCTA, G1–3 tortuosity was significantly increased in DR compared to the control ($p = 0.0024$) and DM without retinopathy ($p = 0.0001$; Figure 7). When using OCTA, however, the smaller DR microvessels (G4–5 and $G \geq 6$) did not appear to change in tortuosity.

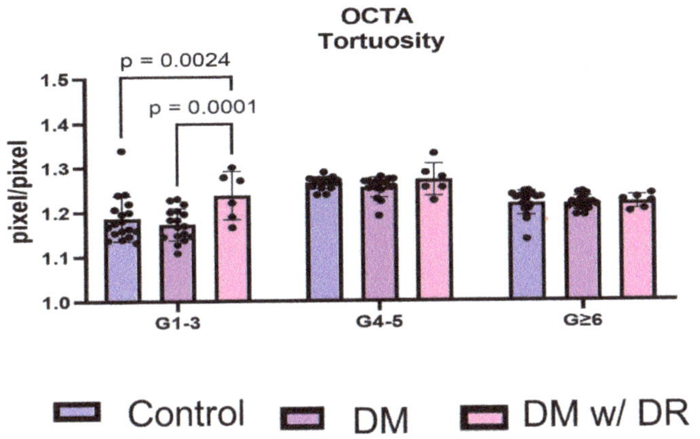

Figure 7. VESGEN analysis between diabetics with and without diabetic retinopathy based on vessel size in OCTA images. Total tortuosity was analyzed using two-way ANOVA and was compared between controls and diabetic individuals with and without DR. Less tortuosity was observed in the G1–3 vessels in controls compared to diabetic individuals with DR ($p = 0.0024$). In T2D individuals without clinical evidence of DR, tortuosity was decreased compared to T2D individuals with DR ($p = 0.0001$). DM = diabetes mellitus; DR = diabetic retinopathy; OCTA = optical coherence tomography angiography.

4. Discussion

In vessel pathology, degeneration and subsequent angiogenesis are hallmark features of DR. Current clinical evaluation of this condition relies on the assessment of microaneurysms and hemorrhages, without the assessment of vascular branching and other morphological changes. Previously, VESGEN was used to measure the "oscillation" of vessel density with the progression of DR in 50° FA [15]. The study found that in DR, there were alternating phases of revascularization and vascular dropout. Initially, changes were observed in the arteries, followed by modifications to the veins. These consistent, progressive arterial changes were found to be more predictive of future DR stages compared to traditional indicators such as microaneurysms, vascular leakage, and cotton wool spots. This suggests that VESGEN analysis could offer improved and more predictive diagnostic capabilities based on early vascular branching changes [15]. However, this initial study did not include a cohort of age-matched controls, nor did it include diabetic subjects without DR. The current study addresses these limitations and uses 35° FA imaging, which facilitates better visualization of smaller microvessels near the macula compared to the 50° FA images used in the initial study. The present study also adds critical information regarding the utility of using VESGEN for the evaluation of OCTA images.

Both FA and OCTA show that tortuosity changes are observed during the initial stages of DR. However, tortuosity measurements need to compare vessels of the same size rather than the specific generation, as our study supports. FA studies indicate increased tortuosity in the small vessels of generations > six in diabetics with mild DR. OCTA confirmed an increase in tortuosity of the larger microvessels (G1–3); however, these generations correspond to branching levels to the smallest vessels that are detected by FA ($G \geq 6$) [22].

Thus, we found that both FA and OCTA imaging showed tortuosity changes, suggesting that the visibility of early-stage DR changes is not preferentially limited to one imaging modality over another. It has been generally appreciated that the tortuosity of retinal vessels is present in advanced diabetes and hypertension [23]; this study is the first to detect differences in early DR.

Previously, Cui et al. [24] compared OCTA and color fundus photography (CFP) to FA for identifying DR lesions. They found that both methods were equally effective in identifying lesions. Additionally, they suggested that using a combination of OCTA and CFP could provide a less invasive alternative to FA for diagnosing DR [24]. Weinhaus et al. reported that FA does not image the deeper capillary plexus well; however, this study used nonhuman primate eyes [25]. They suggested that scattering from the deeper retinal layers obliterated the images of the capillaries in the deeper plexus.

OCTA separates and quantifies the retinal circulation into superficial and deep capillary plexus. The radial capillary network in inner and outer retinal enface zones can be readily imaged by OCTA and decreases in the perfusion of the retinal vasculature easily measured. Sambhav et al. showed that the deeper plexus is involved much earlier than the superficial plexus in DR [26].

Interestingly, in diabetics before the onset of clinically present DR, a reduction was seen in vessel length density, vessel area density, and tortuosity by OCTA. This suggests that OCTA was able to detect changes in vascularity prior to the clinical diagnosis of retinopathy. While examining arteries and veins separately, which can only be achieved using FA, we showed that arteries are tortuous but not veins, which is consistent with previous arteriovenous differences in tortuosity reported by Parsons-Wingerter et al. [14].

NASA has developed preliminary automated vessel extraction methods using artificial intelligence (AI), machine learning, and other computational techniques. This process involves extracting binary vascular patterns from grayscale images. The prototype computer code has been incorporated into the VESGEN software, which was launched in 2020. These AI and machine learning models need training and evaluation datasets to learn how to perform specific tasks and have the capability to generalize and perform effectively on previously unseen images. Although AI was not utilized in this study, the aim for clinical application would involve extensive AI training to facilitate the use of VESGEN in clinical practice.

Limitations of our study include that we did not analyze arteries and veins separately in the OCTA images due to our inability to manually separate them, which we were able to do with FA. However, others have used OCT axial reflectance profiles to validate the use of OCTA in guiding artery–vein (AV) classification [27]. Depth-resolved OCT profile analysis enables reliable AV classification, thereby facilitating the early diagnosis of retinal diseases and potentially enhancing deep learning-based AV classification algorithms [27]. Future studies may use these algorithms for the preparation of binary images of arteries and veins from OCTA prior to VESGEN analysis. Another limitation of our study is that the study subjects in the control cohorts were predominantly white, while the T2D subjects included more Black individuals. However, our study was not designed to assess the impact of race on retinal vasculature, and this question will require additional studies with larger cohorts.

5. Conclusions

VESGEN analysis software in combination with FA and OCTA imaging can detect differences in tortuosity, vessel area density, and vessel length density. Both 35° FA and OCTA images show similar vascular changes and support our previous findings made with the VESGEN analysis of 50° FA retina scans. Both FA and OCTA combined with VESGEN have the potential to assist in the early detection of subtle vascular changes. VESGEN analysis could distinguish even small improvements with pharmacological or other therapeutic interventions and thus serve as an excellent clinical research tool and can also be used to identify retinal vascular changes in other systemic diseases.

Author Contributions: Investigation, E.A., M.D., Y.S., R.V., A.V., M.K., M.C.M., M.P., N.J., S.L., S.R. (Swetha Ravichan-dran), S.R. (Sneha Ramesh) and S.S.; data curation M.D., J.C., R.S.M., K.V.C. and J.M.; writing—original draft preparation, M.D. and M.B.G.; writing—review and editing, M.D., M.B.G., E.A. and P.P.-W.; visualization, M.D.; funding acquisition, M.B.G. and E.A. All authors have read and agreed to the published version of the manuscript.

Funding: This research was funded by NEI R01EY012601, R01EY028858, R01EY028037, R01EY025383, R01EY032753, EY035539 Research to Prevent Blindness unrestricted grant awarded to Department of Ophthalmology and Visual Sciences at UAB (PI. MB Grant), UAB ICAR Pilot Grant (PI, E. Arthur), UAB DRC Pilot Grant (PI, E. Arthur), and ARVO Genentech Career Development Award (PI, E. Arthur).

Institutional Review Board Statement: This study was conducted according to the relevant guidelines and was approved by the Institutional Review Board of Indiana University (IRB: 1402550709), the University of Alabama at Birmingham (IRB-300009129), and the University of Florida (IRB#535-2011).

Informed Consent Statement: Informed consent was obtained from all subjects involved in the study.

Data Availability Statement: The data presented in this study are available on request from the corresponding author.

Acknowledgments: The authors acknowledge and thank the subjects that participated in this study and their families. The authors also wish to acknowledge Krishnan Radhakrishnan, Substance Abuse and Mental Health Administration, USA, for his helpful review of the statistical methods.

Conflicts of Interest: The authors declare no conflicts of interest.

References

1. Lin, K.-Y.; Hsih, W.-H.; Lin, Y.-B.; Wen, C.-Y.; Chang, T.-J. Update in the epidemiology, risk factors, screening, and treatment of diabetic retinopathy. *J. Diabetes Investig.* **2021**, *12*, 1322–1325. [CrossRef] [PubMed]
2. Shukla, U.V.; Tripathy, K. Diabetic Retinopathy. In *StatPearls*; StatPearls Publishing: Treasure Island, FL, USA, 2024.
3. Modi, P.; Arsiwalla, T. Hypertensive Retinopathy. In *StatPearls*; StatPearls Publishing: Treasure Island, FL, USA, 2024.
4. Ghamdi, A.H.A. Clinical predictors of diabetic retinopathy progression: A systematic review. *Curr. Diabetes Rev.* **2020**, *16*, 242–247. [PubMed]
5. Grunwald, J.E.; Alexander, J.; Ying, G.S.; Maguire, M.; Daniel, E.; Whittock-Martin, R.; Parker, C.; McWilliams, K.; Lo, J.C.; Go, A.; et al. Retinopathy and chronic kidney disease in the Chronic Renal Insufficiency Cohort (CRIC) study. *Arch. Ophthalmol.* **2012**, *130*, 1136–1144. [CrossRef] [PubMed]
6. Jin, J.; Kandula, V.; Miller, R.E. Monitoring retinal pathology and cerebral injury in sickle cell disease using spectral-domain optical coherence tomography in pediatric patients. *Pediatr. Blood Cancer* **2021**, *68*, e29028. [CrossRef] [PubMed]
7. Farrah, T.E.; Dhillon, B.; Keane, P.A.; Webb, D.J.; Dhaun, N. The eye, the kidney, and cardiovascular disease: Old concepts, better tools, and new horizons. *Kidney Int.* **2020**, *98*, 323–342. [CrossRef] [PubMed]
8. Cade, W.T. Diabetes-related microvascular and macrovascular diseases in the physical therapy setting. *Phys. Ther.* **2008**, *88*, 1322–1335. [CrossRef] [PubMed]
9. Alghadyan, A.A. Diabetic retinopathy—An update. *Saudi J. Ophthalmol.* **2011**, *25*, 99–111. [CrossRef] [PubMed]
10. Duh, E.J.; Sun, J.K.; Stitt, A.W. Diabetic retinopathy: Current understanding, mechanisms, and treatment strategies. *JCI Insight* **2017**, *2*, e93751. [CrossRef] [PubMed]
11. De Madrid, D.A.P.; Berástegui, I. Diabetic retinopathy. *An. Sist. Sanit. Navar.* **2008**, *31* (Suppl. S3), 23–34. [PubMed]
12. Moreno, A.; Lozano, M.; Salinas, P. Diabetic retinopathy. *Nutr. Hosp.* **2013**, *28* (Suppl. S2), 53–56. [PubMed]
13. Kanski, J.J. Diabetic retinopathy—A preventable cause of blindness. *Practitioner* **1985**, *229*, 343–348. [PubMed]
14. Parsons-Wingerter, P.; Radhakrishnan, K.; Vickerman, M.B.; Kaiser, P.K. Oscillation of angiogenesis with vascular dropout in diabetic retinopathy by VESsel GENeration analysis (VESGEN). *Investig. Ophthalmol. Vis. Sci.* **2010**, *51*, 498–507. [CrossRef] [PubMed]
15. Vickerman, M.B.; Keith, P.A.; McKay, T.L.; Gedeon, D.J.; Watanabe, M.; Montano, M.; Karunamuni, G.; Kaiser, P.K.; Sears, J.E.; Ebrahem, Q.; et al. VESGEN 2D: Automated, user-interactive software for quantification and mapping of angiogenic and lymphangiogenic trees and networks. *Anat. Rec.* **2009**, *292*, 320–332. [CrossRef] [PubMed]
16. Fukuda, M. Classification and treatment of diabetic retinopathy. *Diabetes Res. Clin. Pract.* **1994**, *24*, S171–S176. [CrossRef] [PubMed]
17. The effect of intensive diabetes treatment on the progression of diabetic retinopathy in insulin-dependent diabetes mellitus. The Diabetes Control and Complications Trial. *Arch. Ophthalmol.* **1995**, *113*, 36–51. [CrossRef] [PubMed]
18. Arthur, E.; Ravichandran, S.; Snyder, P.J.; Alber, J.; Strenger, J.; Bittner, A.K.; Khankan, R.; Adams, S.L.; Putnam, N.M.; Lypka, K.R.; et al. Retinal mid-peripheral capillary free zones are enlarged in cognitively unimpaired older adults at high risk for Alzheimer's disease. *Alzheimers Res. Ther.* **2023**, *15*, 172. [CrossRef] [PubMed]

19. Arthur, E.; Alber, J.; Thompson, L.I.; Sinoff, S.; Snyder, P.J. OCTA reveals remodeling of the peripheral capillary free zones in normal aging. *Sci. Rep.* **2021**, *11*, 15593. [CrossRef]
20. DuPont, M.; Lambert, S.; Rodriguez-Martin, A.; Hernandez, O.; Lagatuz, M.; Yilmaz, T.; Foderaro, A.; Baird, G.L.; Parsons-Wingerter, P.; Lahm, T.; et al. Retinal vessel changes in pulmonary arterial hypertension. *Pulm. Circ.* **2022**, *12*, e12035. [CrossRef] [PubMed]
21. Lagatuz, M.; Vyas, R.J.; Predovic, M.; Lim, S.; Jacobs, N.; Martinho, M.; Valizadegan, H.; Kao, D.; Oza, N.; Theriot, C.A.; et al. Vascular patterning as integrative readout of complex molecular and physiological signaling by vessel generation analysis. *J. Vasc. Res.* **2021**, *58*, 207–230. [CrossRef] [PubMed]
22. Pramil, V.; Levine, E.S.; Waheed, N.K. Macular vessel density in diabetic retinopathy patients: How can we accurately measure and what can it tell us? *Clin. Ophthalmol.* **2021**, *15*, 1517–1527. [CrossRef] [PubMed]
23. Han, H.-C. Twisted blood vessels: Symptoms, etiology, and biomechanical mechanisms. *J. Vasc. Res.* **2012**, *49*, 185–197. [CrossRef] [PubMed]
24. Cui, Y.; Zhu, Y.; Wang, J.C.; Lu, Y.; Zeng, R.; Katz, R.; Vingopoulos, F.; Le, R.; Laíns, I.; Wu, D.M.; et al. Comparison of widefield swept-source optical coherence tomography angiography with ultra-widefield colour fundus photography and fluorescein angiography for detection of lesions in diabetic retinopathy. *Br. J. Ophthalmol.* **2021**, *105*, 577–581. [CrossRef]
25. Weinhaus, R.S.; Burke, J.M.; Delori, F.C.; Snodderly, D.M. Comparison of fluorescein angiography with microvascular anatomy of macaque retinas. *Exp. Eye Res.* **1995**, *61*, 1–16. [CrossRef] [PubMed]
26. Sambhav, K.; Abu-Amero, K.K.; Chalam, K.V. Deep Capillary Macular Perfusion Indices Obtained with OCT Angiography Correlate with Degree of Nonproliferative Diabetic Retinopathy. *Eur. J. Ophthalmol.* **2017**, *27*, 716–729. [CrossRef] [PubMed]
27. Adejumo, T.; Kim, T.-H.; Le, D.; Son, T.; Ma, G.; Yao, X. Depth-resolved vascular profile features for artery-vein classification in OCT and OCT angiography of human retina. *Biomed. Opt. Express* **2022**, *13*, 1121–1130. [CrossRef] [PubMed]

life

MDPI

Article

Intravitreal Antiangiogenic Treatment for Diabetic Retinopathy: A Mexican Real-Life Scenario Experience

Sonia López-Letayf [1,2], Oscar Vivanco-Rojas [1,2], Valentina Londoño-Angarita [2], Fátima Sofía Magaña-Guerrero [2], Beatriz Buentello-Volante [2] and Yonathan Garfias [1,2,*]

[1] Department of Biochemistry, Faculty of Medicine, Universidad Nacional Autónoma de México, Av. Universidad 3000, Mexico City 04510, Mexico; sonyletayf@comunidad.unam.mx (S.L.-L.); oscarv@bq.unam.mx (O.V.-R.)

[2] Cell and Tissue Biology, Research Unit, Institute of Ophthalmology, Conde de Valenciana, Chimalpopoca 14, Mexico City 06800, Mexico; dra.vlondono@gmail.com (V.L.-A.); fatima.magana@institutodeoftalmologia.org (F.S.M.-G.); bbuentello@institutodeoftalmologia.org (B.B.-V.)

* Correspondence: ygarfias@bq.unam.mx

Abstract: The objective of this study was to analyze the effectiveness of two intravitreal antiangiogenic drugs, ranibizumab and aflibercept, in a Mexican population over a period of 5 years, evaluating the improvement in visual acuity (VA) and central retinal thickness (CRT) in a real-world scenario. This is a retrospective study with subjects diagnosed with diabetic retinopathy (DR), proliferative diabetic retinopathy (PDR), and diabetic macular edema (DME) receiving intravitreal injections of ranibizumab and/or aflibercept. In this study, we analyzed 588 eyes of 294 patients who received intravitreal antiangiogenic injections. The results showed an improvement regardless of antiangiogenic treatment or diagnosis in both VA and CRT. We found that both aflibercept and ranibizumab improved VA, while subjects with DME responded less to antiangiogenic treatment ($p < 0.05$), and that this difference did not correspond to the CRT measured by OCT. These results support evidence that intravitreal antiangiogenic medications are effective for ophthalmic complications of diabetes in our population; however, damage to visual structures is not reversed in most patients. And that the perception by the patient (VA) and that of the ophthalmologist (CRT) do not completely correlate in our study.

Keywords: diabetic retinopathy; anti-VEGF; ranibizumab; aflibercept; diabetic macular edema; Mexican population

1. Introduction

Diabetes mellitus is a disease characterized by chronically high blood glucose levels, with a global prevalence of 10.5% until 2021, with an estimated 12.2% of the world's population being affected by diabetes mellitus by 2048. Mexico is among the first 10 countries with a high rate of diabetes [1], with a prevalence of 18.3% [2]. A hyperglycemic state causes microvascular changes, causing complications such as neuropathy, retinopathy (DR), and diabetic nephropathy (DN) [3].

DR is a microvascular complication, with a prevalence in North America and the Caribbean of 33.3% in diabetics; this prevalence is projected to increase by 2030 [4,5]. Neural, inflammatory, and vascular dysfunctions of retinal tissue are the main causes of DR pathogenesis. The latter are associated with retinal blood vessel permeability, with vascular endothelial growth factor (VEGF) being one of the major molecules involved in this endothelial permeability.

This factor is responsible for maintaining endothelial cell survival, proliferation, and migration, as well as the growth of the newly formed vessels. It serves as the main therapeutic target for the treatment of this disease; however, it can only be used in advanced stages of DR and in diabetic macular edema (DME) [6]. Currently, the only means of preventing the early stages from advancing to a more severe stage is systemic risk factor control [5]. DME is a complication that can appear at any stage of DR, and it is defined as a thickening of the retina that involves the fovea due to abnormal liquid accumulation. It is the main reason for vision loss among patients with DR, with a prevalence of 5.47% globally among people with diabetes [7].

The VEGF-A isoform has the greatest angiogenic effect and is the main therapeutic target for the treatment of DR and DME. VEGF dysregulation is dependent on multiple factors, but damage also involves neuronal damage and injury to immune components [6]. Ranibizumab (Lucentis, Novartis Pharma Stein AG, Switzerland) is a humanized recombinant monoclonal antibody, and aflibercept (Wetlia, Bayer AG, Germany) is a recombinant fusion protein. The latter not only recognizes VEGF-A but also targets placental growth factor (PGF), which has been associated with the progression of DR [8]. While anti-VEGF biologicals are used to treat neovascularization, they also inhibit blood vessel permeability, which is another consequence of the vascular changes that are present in DR. DME is a complication of increased permeability, DR, which increases the risk of at any point presenting DME, and at the same time, having DME increases the risk of hypoxic events that lead to the progression of DR. Therefore, these two entities are mutually engaged. Pan-retinal photocoagulation (PRP) is usually used in conjunction with intravitreal anti-VEGF biologicals. How PRP works is not fully understood, but it is believed that the laser causes some cells to die in areas of high oxygen demand, diminishing hypoxia signals in areas of poor retinal perfusion. This exerts a decrease in signals that would lead to neovascularization, such as VEGF [9].

Although the use of intravitreal anti-VEGF biologicals to treat DM complications such as DR and DME has been accepted in the medical community, there are still controversies about their effectiveness. For example, Antoszyk, M., and colleagues showed that intravitreal aflibercept was equally effective as PRP in proliferative DR with vitreous hemorrhage in a 24-week follow-up [10]. Moreover, in a recent review, it is mentioned that PRP should be avoided when the anti-VEGF therapeutic strategy is well planned; nonetheless, in a real-world situation, anti-VEGF treatment adherence is challenging; therefore, PRP is eligible for those eyes with severe DR and for those patients for whom anti-VEGF therapy is not an option due to inadequate adherence [8]. Additionally, as anti-VEGF treatment has no permanent effects, a cost-effective study of each patient must be performed. Also, there is still active research comparing the effectiveness of the two most commonly used anti-VEGF biological drugs: aflibercept and ranibizumab. Similar results are found among them [11–13], however, ranibizumab is the more cost-effective treatment in DME in comparison to aflibercept [14]. In contrast, aflibercept is a more effective treatment than ranibizumab when initial visual acuity is moderate to severe [15,16]. Therefore, the aim of the present study is to analyze the effectiveness of aflibercept and ranibizumab in a Mexican population over a span of 5 years in a real-life scenario.

2. Material and Methods

2.1. Study Design

This is a retrospective real-scenario cross-sectional descriptive study of diabetic patients with intravitreal anti-angiogenic treatment (Ranibizumab [ranib] and/or Aflibercept [aflib]). The study included all medical records that fulfilled all the selected criteria between January 2019 and December 2023. The protocol was evaluated and approved by the Institutional Review Board (CEI-2022-09/05; date of approval: 19 September 2022).

2.2. Study Population

Inclusion criteria: Medical records were selected from all patients with type-2 diabetes mellitus (T2DM) with a diagnosis of non-proliferative diabetic retinopathy (DR), proliferative diabetic retinopathy (PDR), and diabetic macular edema (DME) who had intravitreal antiangiogenic injections. Also, visual acuity (VA) and/or optical coherence tomography (OCT) before and after antiangiogenic treatment. Routinely, the antiangiogenic treatment schedule is once a month for three months; thereafter, the patient is evaluated, treated, and extended as needed. *Exclusion criteria:* Diagnosis of systemic diseases concomitant to T2DM; inconclusive retinal disease diagnosis; incomplete medical record. *Elimination criteria:* Repeated patient file numbers; patients treated with another intravitreal treatment such as bevacizumab and dexamethasone implants.

2.3. Data Collection

To determine the effectiveness of antiangiogenic treatment, VA and OCT parameters were considered. Before antiangiogenic treatment and after 3 months of intravitreal treatment, VA was measured using Snellen fractions, and low visual acuity was assessed by counting fingers (CF), hand movement (HM), color and light perception (LPCP), light perception (LP), and no-light perception (NLP). For the database, the transformation to the *LogMAR* scale was performed using the formula published by Moussa et al., 2021 [17]. OCT scans were obtained using Spectralis OCT (Heidelberg Engineering, Heidelberg, Germany) and Eye Explorer software version 2 (HEYEX 2) (Heidelberg Engineering, Heidelberg, Germany). Central retinal thickness (CRT) (μm) was measured at the center of the fovea; this measure is an average of at least six measurements automatically calculated by the aforementioned device. For VA and OCT, it was measured before (PRE) intravitreal (IV) and after (POST) intravitreal treatment. Data acquisition were reviewed by at least 3 researchers (S.L.-L., O.V.-R. and V. L.-A.).

2.4. Statistical Analysis

The results are presented as means and standard deviation (SD), or frequencies. The Kolmogorov-Smirnov test was performed to check for normality. The Wilcoxon test was performed to compare the PRE and POST group treatments. The Mann–Whitney U test was performed to evaluate the differences between diagnoses (DX) and treatments (TX). Data were recorded using Microsoft Excel 365 and added to IBM SPSS Statistics 25 for analysis in addition to PRISM 10.

3. Results

3.1. Demographics

After reviewing all the medical records between 2019 and 2023, a total of 4323 patients received at least an intravitreal medication, eight hundred and thirty-five subjects had DR and/or PDR and/or DME, and a total of 294 patients fulfilled the inclusion criteria with a total of 588 eyes. The reduced continuity of patient treatment, the absence of data in the registries, and the comorbidities of the patients were the reasons for the exclusion of most of the patients. Women represented 60% (177 patients, 354 eyes) of the sample and men 40% (117 patients, 234 eyes); the mean age was 63.3 ± 9.3 years. Three hundred and eighty-one patients out of the 588 eyes received intravitreal injections; 169 received ranibizumab injections; 99 received aflibercept injections; and 113 received both ranibizumab and aflibercept. From the treated eyes, 195 had VA measurements, while 186 had OCT measurements. The mean number of injections after pre-measurement OCT was 2.91 ± 2.93, and the mean number of injections was for aflibercept 4.06 ± 2.36 and for ranibizumab 2.94 ± 3.05; the mean number of injections when considering patients with both treatments was 2.81 ± 3.16. The data are summarized in Table 1.

Table 1. Demographic characteristics of the involved subjects in the study.

		N = 294 (588 eyes)	
		n	%
Gender			
	Female	177 (354 eyes)	60%
	Male	117 (234 eyes)	40%

Table 1. *Cont.*

		N = 294 (588 eyes)	
		n	%
		Mean ± SD	**CI**
Age		63.39 ± 9.3	62.32–64.48
	Female	64.32 ± 9.4	62.98–65.78
	Male	61.91 ± 9.1	60.25–63.58
Diagnoses			%
	DR	242	41.2%
	DME	155	25.6%
	PDR	191	32.4%
Treatment			
	IV (Total)	381	64.7%
	Aflibercept	99	25.9%
	Ranibizumab	169	44.4%
	Aflib/ranib	113	29.7%
Measurements			
	VA	195	51.1%
	OCT	186	48.9%
Follow up OCT			
	Mean No. Injection	2.91 ± 2.93	
	Mode No. Injection	3	
Number of Injections			
	Ranibizumab	2.94 ± 3.05	
	Aflibercept	4.06 ± 2.36	
	Aflib/ranib	2.81 ± 3.16	

DR: diabetic retinopathy; DME: diabetic macular edema; PDR: proliferative diabetic retinopathy; VA: visual acuity; OCT: optical coherence tomography.

3.2. Evaluation of Improvement between VA and CRT

In the analysis of patients with anti-angiogenic treatment, it was found that the evaluation by VA and CRT maintains the trend towards improvement after intravitreal antiangiogenic treatment, considering treatments with aflibercept, ranibizumab, or their combination, regardless of the diagnoses (DR, PDR, and DME). Taken together, all pretreatment *LogMAR* values are shown as 1.1 ± 0.66. Interestingly, these values were 0.98 ± 0.60 after intravitreal antiangiogenic, which means a significant ($p < 0.05$) VA improvement. Similarly, behavior presented CRT measurements with OCT; measures in the pretreatment presented 372 ± 152 μm, whilst after treatment, this measurement significantly ($p < 0.05$) decreased around 50 μm (321 ± 171 μm), showing an improvement in the macular thickness (Figure 1).

To identify whether there were differences among treatments, patients were grouped according to the different antiangiogenic agents to which they were subjected: ranibizumab (ranib), aflibercept (aflib), or a combination of both (ranib/aflib). Antiangiogenic intravitreal injections were effective alone or in combination in both VA and CRT measurements. Although there were statistically significant differences in all the groups when VA was analyzed, aflibercept alone showed the most significant improvement in comparison with ranibizumab alone or ranibizumab in combination with aflibercept; moreover, aflibercept demonstrated to be slightly superior ($p = 0.028$) to ranibizumab when comparisons were made among both post-treatment groups (Figure 2A). Antiangiogenic intravitreal injections were also effective alone and in combination when CRT measurements were analyzed, showing that in the three groups there were statistically significant differences before and after treatments. Interestingly, in this case, ranibizumab intravitreal injection showed the most significant difference compared to the other two groups (aflibercept alone and the combination of both ranibizumab and aflibercept) (Figure 2B). On the other hand, the analysis of the populations according to the treatment used and the number of samples may be affected by factors such as the ophthalmologist's assessment, the development of the treated condition, and criteria associated with the evolution of the patient, in addition to the number of injections administered. However, in the assumption that the differential population would substantially affect the analyses between treatments, ranibizumab vs. the other conditions would expose the same event in CRT measurements, where it shows no significant differences among group analyses (rani vs. alflib, $p = 0.79$; rani vs. aflib/rani, $p = 0.24$.; aflib vs. aflib/rani, $p = 0.43$).

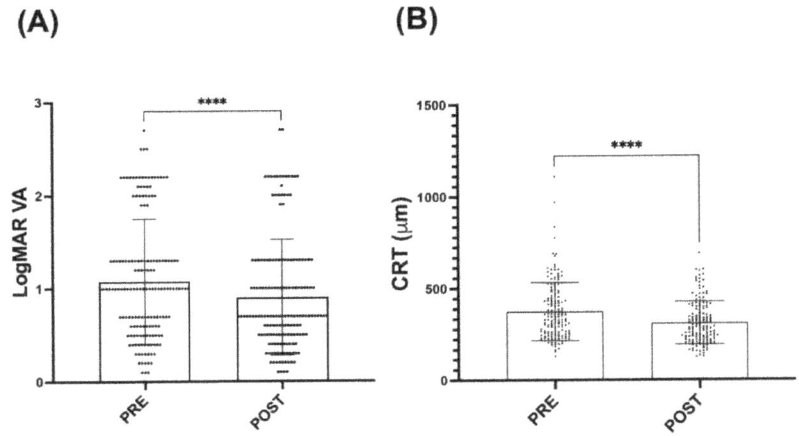

Figure 1. Effect of the antiangiogenic intravitreal injections on VA and CRT measurements. (**A**) Bars graph showing *LogMAR* values before (PRE) and after (POST) antiangiogenic intravitreal injections in all study subjects. (**B**) Bars graph of CRT measurements (µm) before and after antiangiogenic intravitreal injections in all study subjects (**B**). It is observed that there is an improvement in both measurements (VA and CRT) after treatment with antiangiogenic intravitreal injections. A Wilcoxon statistical test was performed. The bars symbolize the mean (±SD). (**** $p < 0.0001$).

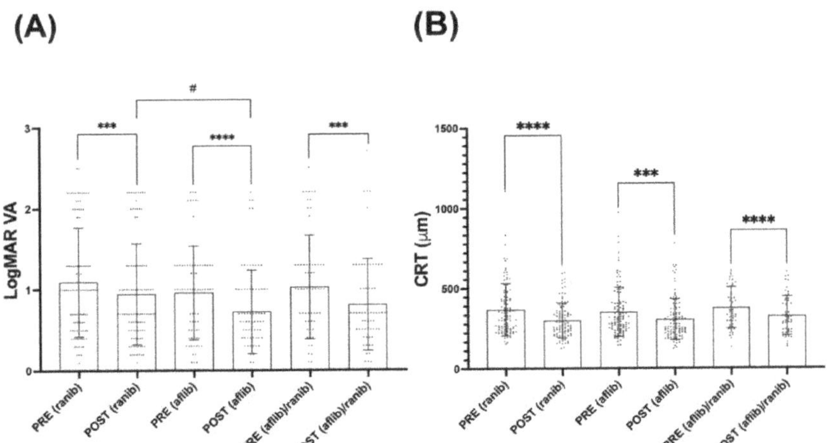

Figure 2. Analysis of the improvement in VA and CRT based on antiangiogenic therapy. Bars graphs representing the distribution of *LogMAR* AV (**A**) and CRT (**B**) values analyzing pre- and post-treatment: ranibizumab [PRE and PRO (ranib)], aflibercept [PRE and PRO (aflib)], ranibizumab/aflibercept [PRE and PRO (ranib/aflib)]. The analysis shows that the treatment of all aflibercept, ranibizumab, and the use of both significantly improved VA and CRT outcomes after antiangiogenic intravitreal injections. Moreover, it was demonstrated to be slightly superior to ranibizumab when VA was taken into account. The bars symbolize the mean (±SD). Wilcoxon statistical test **** $p < 0.0001$; *** $p < 0.001$; U Mann–Whitney test # $p < 0.05$.

Although VA significantly improved after intravitreal antiangiogenic injections (ranibizumab, aflibercept, and ranibizumab/aflibercept) in DR and PDR, antiangiogenic intravitreal treatment did not show any difference in VA in the DME group (Figure 3A). In this context, there is evidence demonstrating the efficacy of anti-VEGF injections on eyes with PDR in improving visual acuity [8]. Moreover, anti-VEGF treatment is believed to have neovascular regression effects in acute PDR subjects [18], and it has also been reported that aflibercept has the capacity to regress the neovascularization process in PDR [19]. Thus, retinal neovascularization regression exerted by anti-VEGF treatments might be a possible mechanism that explains an improvement in DR and PDR without DME.

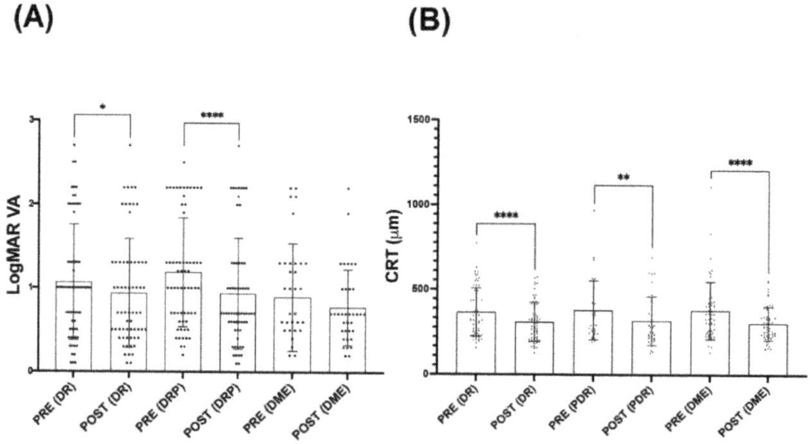

Figure 3. Effect of intravitreal antiangiogenic injections on VA and CRT measurements based on the diabetic retinopathy diagnosis. Bars graphs represent the *LogMAR* (**A**) and CRT (**B**) values based on the retinopathy diagnosis before (PRE) and after (POST) antiangiogenic intravitreal treatments. Diabetic retinopathy (DR), proliferative diabetic retinopathy (PDR), and diabetic macular edema (DME). VA ameliorates in both DR and PDR treated with antiangiogenic intravitreal injections, while this treatment apparently did not modify VA in DME. PDR responded better than DR when VA was measured. Antiangiogenic intravitreal injections ameliorated CRT measurements in all DR, PDR, and EMD retinopathy conditions. The bars symbolize the mean (±SD). Wilcoxon statistical test **** $p < 0.0001$; ** $p < 0.01$; * $p < 0.05$.

It is worth mentioning that PDR responded better than DR in terms of statistical values ($p < 0.0001$ vs. $p < 0.05$, respectively) when antiangiogenic intravitreal injections were applied and VA was evaluated. Intravitreal antiangiogenic injections were able to ameliorate CRT measurements in all DR, PDR, and DME conditions, showing that their use was equally effective in DR and DME (Figure 3).

In addition, the behavior of the number of eyes treated with the antiangiogenic agents rani, aflib, and rani/aflib was compared according to the type of condition analyzed, and it was observed that there is a tendency to use rani for DR, PDR, and DME; aflib is more used for DR; and for more complex conditions such as PDR and DME, the use of both antiangiogenic agents (rani/aflib) is chosen (Table 2).

Table 2. Number of eyes with different pathologies treated with antiangiogenic agents.

	DR	PDR	DME	Total
Rani	70	55	44	169
Aflib	43	27	29	99
Rani/aflib	37	36	40	113
Total	150	118	113	381

4. Discussion

In our study, we analyzed a Mexican population treated with the most commonly used antiangiogenic biologicals, aflibercept and ranibizumab, to treat DR, PDR, and DME. The evaluation of improvement was assessed using the techniques used by ophthalmological personnel to follow up by means of VA and CRT.

Our results indicate that the use of antiangiogenic intravitreal medications helps patients with DR, PDR, and DME by improving visual acuity, a subjective measure that indicates the patient's perception of vision. Also, the treatment was able to reduce the thickness of the retina by means of OCT, which is a patient-independent procedure and shows results more accurately. It is known that the main effect of antiangiogenic intravitreal treatment is to prevent the progression of the DR but not its reversal, since the reduction in macular thickness does not always correlate with an improvement in VA. It has been reported that even in different stages of the pathology, the relationship between VA and CRT is not entirely similar, as in glaucoma, where macular parameters of CRT were associated with VA in moderate-advanced glaucoma but did not present a difference in early glaucoma [20]. In addition, patients treated with bevacizumab have been evaluated and found that the decrease in central retinal thickness was better in the DME group compared to serous macular detachment but that there was no effect on visual prognosis [21]. Since VA is a clinically significant variable, it bears the most weight when evaluating a patient's quality of life, and a lack of improvement in VA means that, despite neovascularization and abnormal liquid accumulation leading to visual impairment, the neural damage cannot be reversed through antiangiogenic medications. Although a reduction in macular thickness has been linked clinically, it may not necessarily be a sign of improvement because in diabetic patients, it has been associated with diabetic peripheral neuropathy [22].

Successful treatment of DR and DME will come with a better understanding of the underlying pathogenesis. Many articles in the literature attempt to explain other causes of the pathogenesis and include immunological, neural, and genetic factors, as well as metabolic syndrome, as multifactorial culprits for the lack of success of anti-VEGF therapies and the overall microvascular complications of diabetes [7,23].

Antiangiogenic medications have been studied at length, and observations include a lack of progression of DR and less presentation of DME. Overall, ranibizumab has been shown to be a safe and good option for the treatment of advanced DR stages, high-risk DR, and DME; nevertheless, in our study, a lack of response to ranibizumab prompted a change to aflibercept. By analyzing which anti-VEGF medication was better, we found that in a meta-analysis by Virgili E., and colleagues., in which they analyzed three anti-VEGF medications and PRP, they established that aflibercept and ranibizumab had better outcomes than both PRP alone and bevacizumab, and between them, aflibercept had better outcomes [24]. Similarly, the CLARITY study showed that PDR eyes treated with aflibercept had better VA outcomes at 1 year compared with eyes treated with PRP, and that aflibercept was superior to PRP [25].

In our study, when assessing improvement, we found that it responds positively regardless of treatment, but when comparing the treatment groups, it was found that aflibercept has a greater improvement compared to ranibizumab or the use of both in the measurement of VA. Although more doses of aflibercept were administered on average, it is also important to consider that the population that is hardly able to pay for the treatment needs to focus on the treatment they can afford. In most cases, the most economical method is chosen. Many of the patients had undergone PRP treatments prior to treatment. The addition of antiangiogenic intravitreal use to PRP treatment has been reported to significantly increase its improvement, according to several meta-analyses [26,27]. However, there have been reported cases where the ophthalmologist decides to switch from PRP to use anti-VEGF mid-treatment. Studies have shown that delayed cross-treatment is not as promising as the results of rapid anti-VEGF treatment [28]. That is why studies are needed to support individual or combined use to avoid neovascularization, in addition to having a follow-up of patients, since in several cases the treatment is interrupted, changed from one to another due to a change in costs, or simply abandoned.

Most of the observed improvement was related only to the thickness of the macula, where the treatment has its main effect, but the VA results consider the patient's perception, which often differs because it depends on factors that the ophthalmologist does not control, such as day, time, and emotional state, among many others, which could sometimes cause a false positive, overestimating the VA measurement. Therefore, in some cases, the data are contradictory between the VA and the CRT measurement. In addition, it supports the idea that the progression and sequelae of DR depend not only on angiogenic status but also on the multifactorial nature of the disease, such as damage to the nervous system, the immune system, and the changes in the extracellular matrix caused by diabetes per se [6]. Even genetic factors, which are generally standardized, do not take into account population factors and the different risk factors implicit in the population [29]. Furthermore, it has been reported that eyes with PDR treated with anti-VEGF alone may experience marked disease progression with potentially serious visual consequences if treatment is discontinued for uncontrolled or conscious reasons [30]. However, an individualized treatment that depends on the patient's adherence, the economic burden, and the distances to reach the healthcare center are determining factors in obtaining successful results.

The major limitation of our study is that it is based on a real-world scenario. The three-dose scheme is dependent on the economic status of the patients or if they have an insurance coverage policy. These factors determine the adherence to the treatment; therefore, to our observation, this situation is the main reason patients do not follow the suggested intravitreal treatment in the present study. In this context, we are aware that in many countries, intravitreal treatment success is really dependent on insurance support.

One of the main problems lies in the fact that treatments require greater durability and efficacy. The treatment of DME and DR requires modifying the paradigms that are established in these conditions and proposing new pathways and molecules involved in their pathogenesis. Even new anti-VEGF therapies have been implemented, such as faricimab, which has been applied to patients with DME and has a favorable prognosis, but even in this case, the regimen is based on 12 weeks [31]. In this context, it is worth mentioning that faricimab is now registered in Mexico, and more studies need to be performed in our population to determine its efficacy. In the assessment of new drugs, they should be evaluated in an appropriate form for specific populations, since in different reports it has been suggested that conditions like DR have a racial factor. In this context, it has been described that the prevalence of DR in patients with T2DM is 46% higher in non-Hispanic blacks and 84% higher in Mexican Americans than in Caucasians [32]. The population ethnicity is important considering the great variability observed in different features such as incidence, progression, heritability, pathogenesis, and response to treatment [33].

The lack of improvement in a large percentage of patients is what prompts the search for other explanations for the pathogenesis of DR. Thus far, we know that patients with DR also have prior neural damage, sometimes to the appearance of DME or neovascularization, due to the nature of the pathogenesis of diabetes, which in our country is commonly found combined with systemic hypertension, which was found to worsen the damage to visual structures, and which, combined with the increased glucose levels, affects not just the cells but the extracellular matrix, making injuries worse, recovery slower or impossible, and dysregulation of the immune system. This is also because nerve damage is one of the main complications caused by diabetes [6,34–36].

The importance of this study is to understand the current state of anti-angiogenic treatments at an ophthalmologic reference center in Mexico. Patients with a diagnosis of DR, PDR, or DME, which are complications of T2DM, a condition with a high prevalence in Mexico (18%), were evaluated, in addition to encouraging other reference centers to evaluate the efficacy of anti-angiogenic agents in the Mexican population. Although the Mexican population has been evaluated in international or Latin American studies [37], most categorize them as Hispanic, grouping them with other ethnicities such as Brazil, which has a high incidence of T1DM, which makes it difficult to compare the real effect of anti-angiogenics. Hence, it is important to start promoting the effects of these treatments on the Mexican population.

Finally, we are convinced that visual acuity and central retinal thickness measurements are indirect indexes to evaluate the antiangiogenic effect of the intravitreal drugs. Although we have OCT-angiography (OCT-A) technology, this technique is not routinely achieved due to its high cost; therefore, OCT-A measurements were not included in this study.

5. Conclusions

This study, along with several others, predominantly demonstrates the importance of early diagnosis and control of diabetes because, although the treatments exist and are somewhat successful, they do not reverse the damage done to visual structures provoked by the disease. Despite the macula reducing its thickness, VA does not improve in at least half of the patients. This study is only a small view of the large problem facing a large portion of the diabetic population. To understand the real impact of this disease in countries such as Mexico, where diabetes presents a high prevalence and complications due to comorbidities are common, the lack of specific treatment for the Mexican population, the reduced integration between health institutions to maintain a consensus on treatment, and the lack of a central database with patients' medical records are hindrances to establishing the real impact of diseases like DR on the population and health services. The creation of a database focusing on diabetic patients would help lawmakers and healthcare institutions establish prevention programs and allocate resources to treat patients before they present the worst complications from diabetes.

Author Contributions: Conceptualization, Y.G.; data curation, S.L.-L. and O.V.-R.; formal analysis, S.L.-L. and O.V.-R.; investigation, V.L.-A.; methodology, S.L.-L., O.V.-R. and Y.G.; project administration, Y.G.; supervision, F.S.M.-G., B.B.-V. and Y.G.; validation, F.S.M.-G. and B.B.-V.; writing—original draft, S.L.-L. and O.V.-R.; writing—review and editing, F.S.M.-G., B.B.-V. and Y.G. All authors have read and agreed to the published version of the manuscript.

Funding: This article was funded by the Consejo Nacional de Humaninades, Ciencia y Tecnología (CONAHCYT) with number 319469; by the Universidad Nacional Autónoma de México (UNAM) with number DGAPA-PAPIIT-UNAM IN210224; SECTEI/159/2023; and the Institute of Ophthalmology, Conde de Valenciana Foundation.

Institutional Review Board Statement: The study was conducted in accordance with the Declaration of Helsinki, and approved by the Institutional Review Board of The Institute of Ophthalmology, Conde de Valenciana (CEI-2022-09/05; date of approval: 19 September 2022).

Data Availability Statement: The data presented in this study are available on request from the corresponding author (the dataset contains private information protected by law).

Acknowledgments: Authors would like to give special thanks to Mohamed Ali Pereyra Morales for his invaluable support in this manuscript.

Conflicts of Interest: The authors declare no conflicts of interest.

References

1. Ogurtsova, K.; da Rocha Fernandes, J.D.; Huang, Y.; Linnenkamp, U.; Guariguata, L.; Cho, N.H.; Cavan, D.; Shaw, J.E.; Makaroff, L.E. IDF Diabetes Atlas: Global estimates for the prevalence of diabetes for 2015 and 2040. *Diabetes Res. Clin. Pract.* **2017**, *128*, 40–50. [CrossRef] [PubMed]
2. Basto-Abreu, A.; López-Olmedo, N.; Rojas-Martínez, R.; Aguilar-Salinas, C.A.; Moreno-Banda, G.L.; Carnalla, M.; Rivera, J.A.; Romero-Martinez, M.; Barquera, S.; Barrientos-Gutiérrez, T. Prevalencia de prediabetes y diabetes en México: Ensanut 2022. *Salud Pública México* **2023**, *65*, s163–s168. [CrossRef] [PubMed]
3. Faselis, C.; Katsimardou, A.; Imprialos, K.; Deligkaris, P.; Kallistratos, M.; Dimitriadis, K. Microvascular Complications of Type 2 Diabetes Mellitus. *Curr. Vasc. Pharmacol.* **2020**, *18*, 117–124. [CrossRef]
4. Teo, Z.L.; Tham, Y.C.; Yu, M.; Chee, M.L.; Rim, T.H.; Cheung, N.; Bikbov, M.M.; Wang, Y.X.; Tang, Y.; Lu, Y.; et al. Global Prevalence of Diabetic Retinopathy and Projection of Burden through 2045: Systematic Review and Meta-analysis. *Ophthalmology* **2021**, *128*, 1580–1591. [CrossRef] [PubMed]

5. Tan, T.E.; Wong, T.Y. Diabetic retinopathy: Looking forward to 2030. *Front. Endocrinol.* **2022**, *13*, 1077669. [CrossRef] [PubMed]
6. Antonetti, D.A.; Silva, P.S.; Stitt, A.W. Current understanding of the molecular and cellular pathology of diabetic retinopathy. *Nat. Rev. Endocrinol.* **2021**, *17*, 195–206. [CrossRef]
7. Zhang, J.; Zhang, J.; Zhang, C.; Zhang, J.; Gu, L.; Luo, D.; Qiu, Q. Diabetic Macular Edema: Current Understanding, Molecular Mechanisms and Therapeutic Implications. *Cells* **2022**, *11*, 3362. [CrossRef]
8. Arrigo, A.; Aragona, E.; Bandello, F. VEGF-targeting drugs for the treatment of retinal neovascularization in diabetic retinopathy. *Ann. Med.* **2022**, *54*, 1089–1111. [CrossRef]
9. Everett, L.A.; Paulus, Y.M. Laser Therapy in the Treatment of Diabetic Retinopathy and Diabetic Macular Edema. *Curr. Diabetes Rep.* **2021**, *21*, 35. [CrossRef]
10. Antoszyk, A.N.; Glassman, A.R.; Beaulieu, W.T.; Jampol, L.M.; Jhaveri, C.D.; Punjabi, O.S.; Salehi-Had, H.; Wells, J.A.; Maguire, M.G.; Stockdale, C.R.; et al. Effect of Intravitreous Aflibercept vs Vitrectomy With Panretinal Photocoagulation on Visual Acuity in Patients With Vitreous Hemorrhage From Proliferative Diabetic Retinopathy: A Randomized Clinical Trial. *JAMA* **2020**, *324*, 2383–2395. [CrossRef]
11. Kucukevcilioglu, M.; Yeşiltaş, Y.S.; Durukan, A.H.; Unlu, N.; Onen, M.; Alp, M.N.; Kalayci, D.; Acar, M.A.; Sekeroglu, M.A.; Citirik, M.; et al. Real Life Multicenter Comparison of 24-Month Outcomes of Anti-VEGF Therapy in Diabetic Macular Edema in Turkey: Ranibizumab vs. Aflibercept vs. Ranibizumab-Aflibercept Switch. *Medicina* **2023**, *59*, 263. [CrossRef] [PubMed]
12. Kaya, M.; Öztürk, T.; Koçak, N.; Akbulut Yağcı, B.; Ataş, F.; Kaynak, S. Ranibizumab or Aflibercept Monotherapies in Treatment-Naive Eyes with Diabetic Macular Edema: A Head-to-Head Comparison in Real-Life Experience. *Turk. J. Ophthalmol.* **2023**, *53*, 30–36. [CrossRef] [PubMed]
13. Chatzirallis, A.; Theodossiadis, P.; Droutsas, K.; Koutsandrea, C.; Ladas, I.; Moschos, M.M. Ranibizumab versus aflibercept for diabetic macular edema: 18-month results of a comparative, prospective, randomized study and multivariate analysis of visual outcome predictors. *Cutan. Ocul. Toxicol.* **2020**, *39*, 317–322. [CrossRef]
14. Kiss, S.; Malangone-Monaco, E.; Wilson, K.; Varker, H.; Stetsovsky, D.; Smith, D.; Garmo, V. Real-World Injection Frequency and Cost of Ranibizumab and Aflibercept for the Treatment of Neovascular Age-Related Macular Degeneration and Diabetic Macular Edema. *J. Manag. Care Spec. Pharm.* **2020**, *26*, 253–266. [CrossRef] [PubMed]
15. Wells, J.A.; Glassman, A.R.; Ayala, A.R.; Jampol, L.M.; Aiello, L.P.; Antoszyk, A.N.; Arnold-Bush, B.; Baker, C.W.; Bressler, N.M.; Browning, D.J.; et al. Aflibercept, bevacizumab, or ranibizumab for diabetic macular edema. *N. Engl. J. Med.* **2015**, *372*, 1193–1203. [CrossRef] [PubMed]
16. Gabrielle, P.H.; Nguyen, V.; Creuzot-Garcher, C.; Arnold, J.J.; Mehta, H.; Duran, M.A.; Bougamha, W.; Carreño, E.; Viola, F.; Squirrell, D.; et al. Three-Year Treatment Outcomes of Aflibercept Versus Ranibizumab for Diabetic Macular Edema: Data from the Fight Retinal Blindness! Registry. *Retina* **2022**, *42*, 1085–1094. [CrossRef] [PubMed]
17. Moussa, G.; Bassilious, K.; Mathews, N. A novel excel sheet conversion tool from Snellen fraction to LogMAR including 'counting fingers', 'hand movement', 'light perception' and 'no light perception' and focused review of literature of low visual acuity reference values. *Acta Ophthalmol.* **2021**, *99*, e963–e965. [CrossRef] [PubMed]
18. Avery, R.L. Regression of retinal and iris neovascularization after intravitreal bevacizumab (Avastin) treatment. *Retina* **2006**, *26*, 352–354. [CrossRef] [PubMed]
19. Gross, J.G.; Glassman, A.R.; Jampol, L.M.; Inusah, S.; Aiello, L.P.; Antoszyk, A.N.; Baker, C.W.; Berger, B.B.; Bressler, N.M.; Browning, D.; et al. Panretinal Photocoagulation vs Intravitreous Ranibizumab for Proliferative Diabetic Retinopathy: A Randomized Clinical Trial. *JAMA* **2015**, *314*, 2137–2146. [CrossRef]
20. Wu, J.H.; Moghimi, S.; Nishida, T.; Mohammadzadeh, V.; Kamalipour, A.; Zangwill, L.M.; Weinreb, R.N. Association of macular OCT and OCTA parameters with visual acuity in glaucoma. *Br. J. Ophthalmol.* **2023**, *107*, 1652–1657. [CrossRef]
21. Koytak, A.; Altinisik, M.; Sogutlu Sari, E.; Artunay, O.; Umurhan Akkan, J.C.; Tuncer, K. Effect of a single intravitreal bevacizumab injection on different optical coherence tomographic patterns of diabetic macular oedema. *Eye* **2013**, *27*, 716–721. [CrossRef]
22. Liu, B.; Wang, W.; Zhou, R.; Zeng, X.; Zhu, Z. Retinal Neurodegeneration in Diabetic Peripheral Neuropathy by Optical Coherence Tomography: A Systematic Review and Meta-analysis. *Curr. Eye Res.* **2021**, *46*, 1201–1208. [CrossRef]
23. Rezaei, M.; Rabizadeh, S.; Mirahmad, M.; Hajmiri, M.S.; Nakhjavani, M.; Hemmatabadi, M.; Shirzad, N. The association between advanced glycation end products (AGEs) and ABC (hemoglobin A1C, blood pressure, and low-density lipoprotein cholesterol) control parameters among patients with type 2 diabetes mellitus. *Diabetol. Metab. Syndr.* **2022**, *14*, 122. [CrossRef] [PubMed]
24. Virgili, G.; Curran, K.; Lucenteforte, E.; Peto, T.; Parravano, M. Anti-vascular endothelial growth factor for diabetic macular oedema: A network meta-analysis. *Cochrane Database Syst. Rev.* **2023**, *10*, CD007419. [CrossRef]
25. Sivaprasad, S.; Prevost, A.T.; Vasconcelos, J.C.; Riddell, A.; Murphy, C.; Kelly, J.; Bainbridge, J.; Tudor-Edwards, R.; Hopkins, D.; Hykin, P.; et al. Clinical efficacy of intravitreal aflibercept versus panretinal photocoagulation for best corrected visual acuity in patients with proliferative diabetic retinopathy at 52 weeks (CLARITY): A multicentre, single-blinded, randomised, controlled, phase 2b, non-inferiority trial. *Lancet* **2017**, *389*, 2193–2203. [CrossRef]
26. Yin, H.; Zhong, S. Efficacy of ranibizumab combined with photocoagulation for diabetic retinopathy: A meta-analysis study. *Medicine* **2023**, *102*, e34170. [CrossRef] [PubMed]
27. Zhang, W.; Geng, J.; Sang, A. Effectiveness of Panretinal Photocoagulation Plus Intravitreal Anti-VEGF Treatment Against PRP Alone for Diabetic Retinopathy: A Systematic Review With Meta-Analysis. *Front. Endocrinol.* **2022**, *13*, 807687. [CrossRef] [PubMed]

28. Bressler, S.B.; Glassman, A.R.; Almukhtar, T.; Bressler, N.M.; Ferris, F.L.; Googe, J.M., Jr.; Gupta, S.K.; Jampol, L.M.; Melia, M.; Wells, J.A., 3rd. Five-Year Outcomes of Ranibizumab With Prompt or Deferred Laser Versus Laser or Triamcinolone Plus Deferred Ranibizumab for Diabetic Macular Edema. *Am. J. Ophthalmol.* **2016**, *164*, 57–68. [CrossRef]
29. Vivanco-Rojas, O.; López-Letayf, S.; Londoño-Angarita, V.; Magaña-Guerrero, F.S.; Buentello-Volante, B.; Garfias, Y. Risk Factors for Diabetic Retinopathy in Latin America (Mexico) and the World: A Systematic Review and Meta-Analysis. *J. Clin. Med.* **2023**, *12*, 6583. [CrossRef]
30. Wubben, T.J.; Johnson, M.W.; Sohn, E.H.; Peairs, J.J.; Kay, C.N.; Kim, S.J.; Gardner, T.W.; Paulus, Y.M.; Zacks, D.N.; Steinle, N.C.; et al. Anti-Vascular Endothelial Growth Factor Therapy for Diabetic Retinopathy: Consequences of Inadvertent Treatment Interruptions. *Am. J. Ophthalmol.* **2019**, *204*, 13–18. [CrossRef]
31. Gonzalez-Cortes, J.H.; Martinez-Pacheco, V.A.; Gonzalez-Cantu, J.E.; Bilgic, A.; de Ribot, F.M.; Sudhalkar, A.; Mohamed-Hamsho, J.; Kodjikian, L.; Mathis, T. Current Treatments and Innovations in Diabetic Retinopathy and Diabetic Macular Edema. *Pharmaceutics* **2023**, *15*, 122. [CrossRef] [PubMed]
32. Harris, M.I.; Klein, R.; Cowie, C.C.; Rowland, M.; Byrd-Holt, D.D. Is the Risk of Diabetic Retinopathy Greater in Non-Hispanic Blacks and Mexican Americans Than in Non-Hispanic Whites With Type 2 Diabetes? A U.S. population study. *Diabetes Care* **1998**, *21*, 1230–1235. [CrossRef] [PubMed]
33. Simó-Servat, O.; Hernández, C.; Simó, R. Genetics in Diabetic Retinopathy: Current Concepts and New Insights. *Curr. Genom.* **2013**, *14*, 289–299. [CrossRef]
34. Lachin, J.M.; Genuth, S.; Nathan, D.M.; Zinman, B.; Rutledge, B.N.; Group, D.E.R. Effect of glycemic exposure on the risk of microvascular complications in the diabetes control and complications trial--revisited. *Diabetes* **2008**, *57*, 995–1001. [CrossRef] [PubMed]
35. Lee, M.W.; Koo, H.M.; Lee, W.H.; Park, J.H.; Lee, Y.H.; Kim, J.Y. Impacts of Systemic Hypertension on the Macular Microvasculature in Diabetic Patients Without Clinical Diabetic Retinopathy. *Investig. Ophthalmol. Vis. Sci.* **2021**, *62*, 21. [CrossRef] [PubMed]
36. Mandava, N.; Tirado-Gonzalez, V.; Geiger, M.D.; Patnaik, J.L.; Frazer-Abel, A.; Lynch, A.M.; Palestine, A.G.; Holers, V.M.; Wagner, B.D.; Sanchez-Santos, I.; et al. Complement Activation in the Vitreous of Patients With Proliferative Diabetic Retinopathy. *Investig. Ophthalmol. Vis. Sci.* **2020**, *61*, 39. [CrossRef]
37. Arevalo, J.F.; Lasave, A.F.; Wu, L.; Maia, M.; Diaz-Llopis, M.; Alezzandrini, A.A.; Brito, M.; Pan-American Collaborative Retina Study Group. Intravitreal Bevacizumab for Proliferative Diabetic Retinopathy: Results From the Pan-American Collaborative Retina Study Group (PACORES) at 24 Months of Follow-up. *Retina* **2017**, *37*, 334–343. [CrossRef]

life

MDPI

Article

Interpreting Deep Neural Networks in Diabetic Retinopathy Grading: A Comparison with Human Decision Criteria

Sangeeta Biswas [1,*], Md. Ahanaf Arif Khan [1], Md. Hasnain Ali [1], Johan Rohdin [2], Subrata Pramanik [1], Md. Iqbal Aziz Khan [1], Sanjoy Kumar Chakravarty [1] and Bimal Kumar Pramanik [1]

1 Faculty of Engineering, University of Rajshahi, Rajshahi 6205, Bangladesh; sprmnk@ru.ac.bd (S.P.); iqbal_aziz_khan@ru.ac.bd (M.I.A.K.); sanjoy.cse@ru.ac.bd (S.K.C.); bkp@ru.ac.bd (B.K.P.)
2 Faculty of Information Technology, Brno University of Technology, 61200 Brno, Czech Republic; rohdin@fit.vutbr.cz
* Correspondence: sangeeta.cse@ru.ac.bd

Abstract

Diabetic retinopathy (DR) causes visual impairment and blindness in millions of diabetic patients globally. Fundus image-based Automatic Diabetic Retinopathy Classifiers (ADRCs) can ensure regular retina checkups for many diabetic patients and reduce the burden on the limited number of retina experts by referring only those patients who require their attention. Over the last decade, numerous deep neural network-based algorithms have been proposed for ADRCs to distinguish the severity levels of DR. However, it has not been investigated whether DNN-based ADRCs consider the same criteria as human retina professionals (HRPs), i.e., whether they follow the same grading scale when making decisions about the severity level of DR, which may put the reliability of ADRCs into question. In this study, we investigated this issue by experimenting on publicly available datasets using MobileNet-based ADRCs and analyzing the output of the ADRCs using two eXplainable artificial intelligence (XAI) techniques named Gradient-weighted Class Activation Map (Grad-CAM) and Integrated Gradients (IG).

Keywords: deep neural network; d iabetic retinopathy classification; Grad-CAM; fundus image; integrated gradients

1. Introduction

Diabetic Retinopathy (DR) is a microvascular complication of diabetes. In the retinal blood vessels of diabetic patients, high glucose levels damage the pericyte cells that wrap around the capillaries in the retina, as well as the endothelial cells that regulate exchanges between the bloodstream and the surrounding tissues. The destruction of essential cells causes retinal blood vessels to lose the ability of maintaining normal vascular tension, grow unstable, and be easily damaged by oxides, resulting in insufficient blood circulation, vascular leakage, retinal hemorrhages, and ultimately leading to DR [1–3]. It is one of the most serious and frequently occurring complications in diabetic patients. According to the International Diabetes Federation (IDF), approximately 103.12 million out of 537 million diabetic patients had DR in 2021 [4,5]. It is a leading cause of partial or full vision loss globally [5–7]. Besides vision impairments, there is evidence that DR is associated with other micro- and macrovascular complications of diabetes such as subclinical atherosclerosis [8–11], cardiovascular disease [12–16], cerebrovascular incident [17–21], cognitive impairment [22], dementia [23], and high risk of mortality [19,24,25].

DR is not curable; however, early treatment can prevent, delay, or reduce vision loss. Regular retina checkups can ensure that DR patients receive treatment at appropriate times. With a global rise in diabetic patients and the shortage of Human Retina Professionals (HRPs), many patients do not receive a proper diagnosis on a regular basis in most countries, especially in third-world countries. Capturing fundus images requires the physical presence of the patients at the eye centers. In third-world countries, the number of eye centers is inadequate and often unreachable for many patients. Moreover, manually detecting DR by examining fundus photographs is time-consuming. These obstacles reduce the possibility of regular checkups for many diabetic patients. Therefore, many diabetic patients remain at risk of vision loss without accessible and affordable solutions, which inspires researchers to develop digitized fundus image-based Automatic Diabetic Retinopathy Classifiers (ADRCs). By providing automated imaging, DR grading, and reporting within minutes during a diabetic patient's regular exam with minimal involvement of short-term trained operators, and by recommending patients having referable DR (such as moderate DR, severe DR, and proliferative DR) for further detailed examinations, ADRCs reduce the burden of retina specialists.

At the beginning, ADRCs were mainly based on basic image processing techniques [26,27]. Then, non-neural network-based machine learning techniques were proposed as ADRCs [28–31], and gradually neural network-based ADRCs drew the main attention. Even though neural network-based ADRC was proposed in 1996 [32], it gained popularity after 2016. Since then, Deep Neural Network (DNN)-based algorithms [33–55] have been dominating ADRC development.

A clear understanding of the criteria underlying DR grading is crucial for an HRP, such as a retina doctor, consultant, or specialist, to ensure reliable use of ADRCs in automatic DR grading and to prevent incorrect treatments. This is why the interpretability of DNN-based systems is an important issue. If it turns out that humans and machines use different criteria, they probably could complement each other, or in the longer perspective, machines can be improved by learning to use the same criteria humans use. If they already use the same criteria as humans, then the analysis of their domain robustness could be reduced to analyzing the domain robustness of the segmentation of individual lesions. Even though so many DNN-based algorithms are published for DR classification, very few works ([37,43,51]) have shown which parts of fundus images contribute to the final decision on DR grades. Moreover, none of these works determined whether the decision-making criteria of ADRCs are similar to those of HRPs. In this work, we investigate whether a DNN-based ADRC that classifies DR into different grades considers the same criteria as HRPs. We believe the findings from our investigation will help developers improve the accuracy of ADRCs while assisting HRPs to better understand how ADRCs operate.

We describe currently applicable DR severity classification schemes in Section 2 and previous works on interpretability issues in Section 3. In Section 4, we outline our approach for analyzing ADRC's decision. Our experimental setup is described in Section 5. In Section 6, we present our ADRCs' performance. In Section 7, we analyze our ADRCs' decisions, and finally in Section 8, we draw our conclusion.

2. DR Severity Classification Schemes for HRPs

In 1890, Hirschberg proposed the earliest known classification for DR [56], which then evolved as human understanding of disease pathophysiology was improved, methods of imaging were changed, methods of DR assessment were updated, and effective treatments were developed. After going through many modifications ([57–63]) for more than 130 years, the severity classification of DR has come to a stage where DR is mainly divided into two grades considering the absence or presence of new, tiny, and abnormal blood vessels:

(1) Non-Proliferative DR (NPDR) and (2) Proliferative DR (PDR). Various signs such as microaneurysms, hard and/or soft exudates, venous caliber abnormalities, venous sheathing, perivenous exudate, arteriolar abnormalities, intra-retinal microvascular abnormalities (IRMAs), and arteriovenous nicking are included in NPDR. Conversely, retinal or disc neovascularization, fibrous proliferation, retinal detachment, and preretinal and vitreous haemorrhage are included in PDR.

By subdividing NPDR in different ways, different DR grading schemes were proposed, such as Early Treatment of Diabetic Retinopathy Study (ETDRS) [63], International Clinical Diabetic Retinopathy (ICDR) [64], Scottish scheme [65], and National Health Service (NHS) England scheme [66], to accurately describe the progression of DR, quantify its severity, and predict the risk of progression. Nowadays, the ETDRS severity scale, proposed in 1991, is primarily utilized in research and intervention studies worldwide. However, it is not suitable for routine clinical use. The NHS and Scottish grading schemes have limited use outside England, Wales, and Scotland.

In 2002, the ICDR grading scheme was proposed by simplifying the ETDRS severity scale. Due to its convenience and ease of adoption, the ICDR grading scheme is by far the most common DR grading system in clinical use worldwide. The ICDR grading scheme defines the DR severity levels as follows: DR-0 (No DR) indicates no evidence of retinopathy or related abnormalities; the presence of only microaneurysms characterizes DR-1 (Mild DR); DR-2 (Moderate DR) includes one or more microaneurysms, retinal dot or blot hemorrhages, hard or soft exudates, without signs of severe retinopathy; DR-3 (Severe DR) requires more than 20 intraretinal hemorrhages in each of the four quadrants, definite venous beading in at least two quadrants, or prominent intraretinal microvascular abnormalities in at least one quadrant, but no signs of proliferative retinopathy; finally, DR-4 (PDR) is defined by the presence of neovascularization, or vitreous or preretinal hemorrhages. Many retina datasets based on the ICDR scale are publicly available for researching automatic DR classification, such as the Indian Diabetic Retinopathy Image Dataset (IDRiD), Kaggle EyePACS Diabetic Retinopathy Detection dataset (Kaggle EyePACS), and Messidor-2.

3. Previous Works on Interpretability of ADRCs

The accuracy of predictions, as well as the interpretability of the reasons behind them, is essential for ADRCs. Self-explanatory ADRCs enable HRPs to compare the information reported by an ADRC with their own knowledge, increasing the probability of an accurate diagnosis, which may significantly influence a patient's treatment. Even though many DNN-based ADRCs have been proposed since 2016, very few studies (e.g., [35–37,40,43,49,51,53]) have addressed the interpretability issue of DNN-based ADRCs. All these works generated heatmaps using eXplainable Artificial Intelligence (XAI) techniques and highlighted regions important for final decision-making by overlaying heatmaps on fundus images.

Gargeya et al. [35] generated heatmaps with the help of the Classification Activation Maps (CAMs) [67]. To generate heatmaps, they implanted a convolutional visualization layer at the end of their network. By overlaying the heatmap on the fundus image, the authors demonstrated that the features of PDR, such as retinal hemorrhage, hard exudates, and neovascularization, can be highlighted with the help of the visualization layer of their ADRC. Even though they provided evidence of their method's ability to highlight severely affected pathologic regions, they did not provide any evidence that their method is capable of highlighting regions affected by mild or moderate DR. Even in their example image (Figure 5 in [35]), many DR-affected regions are not highlighted.

Quellec et al. [36,49] modified CAM to produce heatmaps. Wang et al. [37] proposed Regression Activation Maps (RAMs), an updated version of CAM, to generate heatmaps. They provided evidence that their proposed RAM learned to highlight DR severity-specific signs, such as the narrowing of the retinal arteries for mild-conditioned DR patients and balloon-like structures for severe DR. Gao et al. [40] also utilized CAM to generate heatmaps. They provided evidence that their model focused on primary lesions during classification by providing heatmaps for fundus images affected by severe DR and PDR (Figure 6 in [40]). However, they did not provide any evidence for mild DR or moderate DR. Therefore, it is unclear where their ADRCs focused during decision-making.

Sayres et al. [43] applied Integrated Gradients (IG) [68] to generate heatmaps for assisting DR graders. They visualized the highlights over the fundus image by converting it to grayscale and overlaying the explanation heatmap as a semi-transparent green heatmap. They investigated the accuracy, speed, and confidence of HRPs if they are provided model-predicted DR scores and heatmaps. They observed that grades plus heatmaps improved accuracy for patients with DR but reduced accuracy for patients without DR.

Torre et al. [51] developed score maps to quantify pixel contributions to the final classifications. They followed an approach similar to pixel-wise decomposition [69,70]. Their study concluded that the input space and the receptive fields of each layer were the two main factors that contributed to the output score. They generated score maps by propagating back the score part that depends on the previous input in each layer. Their investigation provided evidence that the combination of micro-information from input space maps with macro-information obtained from intermediate scores was the key to understanding the results and helping medical personnel improve their diagnosis processes.

Al-Antary et al. [53] proposed to use an attention map, one kind of heatmap, to demonstrate the importance of the input area for DR classification. Unlike in [35–37,40,43,49,51], the authors utilized attention maps during the training of their ADRC to enhance its focus on informative image regions, specifically highlighting the diseased parts in the fundus image while placing less emphasis on the normal regions. They produced attention maps from multi-level, multi-scale pyramid representations of fundus images. They demonstrated the importance of the input area for recognizing DR by providing two attention maps extracted for normal and moderate DR in Figure 11 in [53]. However, they did not provide evidence to indicate whether the highlighted area in the attention map always contains DR-grade specific lesions or biomarkers.

4. Proposed ADRC Decision Analysis

The previous works mentioned in Section 3 used XAI techniques to aid HRPs, whereas our work use them to analyze whether DNNs use the same information, i.e., the criteria specified in the ICDR grading scheme, as HRPs for DR classification. For our analysis, we followed three stages: building ADRCs, building segmentation models, and analyzing ADRCs' decisions.

Stage A. Building ADRCs

1. First, we prepared training, validation, and test sets by splitting three DR severity classification datasets: Kaggle EyePACS [71,72], IDRiD-DR [73], and Messidor-2 [74].
2. We trained DNN-based ADRCs for classifying five severity levels of DR.
3. Then, we analyzed the DR classification ability of our ADRCs by evaluating them on the test sets using different evaluation metrics and confusion matrices.

Stage B. Building Segmentation Models

1. We prepared training, validation, and test sets using three datasets: IDRiD [73], E-ophtha [75], and PALM [76].
2. We trained six binary segmentation models to segment the optic disc, macula, and four types of abnormalities, including microaneurysms, hard exudates, soft exudates, and hemorrhages. We used these segmentation models to generate segmentation masks for each retina image.

Stage C. Analyzing ADRCs' Decisions

1. To analyze the interpretability of our ADRC, we first generated heatmaps for all fundus images of the test set of the IDRiD-DR dataset using Gradient-weighted Class Activation Map (Grad-CAM) and Integrated Gradients (IG).
2. We overlayed the heatmap on the fundus images and manually identified and recorded where DNN-based ADRCs emphasize while making decisions about the DR severity levels, i.e., DR classes.
3. We automatically determined, with the aid of segmented masks and heatmaps, which parts of the fundus images played a crucial role in the final DR-level detection. For this, we first extracted the areas where XAI technique-generated heatmaps had the highest activation. Next, we performed a pixel-wise logical "AND" operation between these high-activation regions, obtained by applying percentile-based thresholding to the heatmaps and the corresponding binary masks generated by our segmentation models. After that, we determined the overlapped area between the model's predicted regions of interest and the actual pathological areas. Formally, for each image i, we define the following:
   - H_i: the XAI heatmap (Grad-CAM or IG) for image i, normalized to the range $[0, 1]$.
   - M_i: the binary segmentation mask for image i, where 1 indicates a lesion region.
   - T^{GC}, T^{IG}: the thresholds corresponding to the 90th and 97th percentiles of H_i, for Grad-CAM and IG, respectively.

   The binary activation map A_i is defined as follows:

   $$A_i(x,y) = \begin{cases} 1, & \text{if } H_i(x,y) \geq T \\ 0, & \text{otherwise} \end{cases}$$

   where $T = T^{GC}$ or T^{IG}, depending on the method.
   The intersection mask is computed as follows:

   $$I_i = A_i \wedge M_i$$

   If the intersection contains any non-zero values (i.e., $\sum_{x,y} I_i(x,y) > 0$), it implies that the heatmap has activated a pathological region. We count the number of such instances across all images and analyze the results.
4. We also applied a semi-automatic approach where we generated Grad-CAM and IG heatmaps from the Retinal-Lesions dataset and performed the same mentioned operation between the highest activation regions and their provided segmentation masks for eight types of lesions: microaneurysm, cotton wool spots, hard exudate, retinal hemorrhage, preretinal hemorrhage, vitreous hemorrhage, fibrous proliferation, and neovascularization.

5. Experimental Setup

5.1. Hardware and Software

We used a computer system equipped with an AMD Ryzen 5 3500X CPU, 32 GB of DDR4 memory, and a single NVIDIA RTX 3090 GPU with 24 GB of GDDR6 VRAM for conducting our experiments. Our system ran Ubuntu 22.04 LTS with Python 3.11. We utilized TensorFlow v2.15, along with the Keras API and other Python-based libraries, including OpenCV, Scikit-Learn, and Matplotlib.

5.2. Datasets

We used six datasets in total, as listed in Table 1. We only utilized free and publicly available datasets in our work. No additional data collected from hospitals, clinics, or other external sources were used.

Table 1. Datasets employed at each stage of our study.

Task	Datasets
Building ADRCs	Kaggle EyePACS [71,72], IDRiD-DR [73], Messidor-2 [74]
Building Segmentation Models	IDRiD [73], E-ophtha [75], PALM [76]
Analyzing ADRCs' Decision	IDRiD-DR [73], Retinal-Lesions [77]

5.2.1. Short Summary of Datasets

The Kaggle EyePACS Diabetic Retinopathy Detection (Kaggle EyePACS in short) dataset is one of the most popular datasets for detecting DR [71]. It was made publicly available on the Kaggle community platform for a competition on DR detection sponsored by the California Healthcare Foundation. Fundus images of this dataset were provided by EyePACS [72]. The images in this dataset are high-resolution color fundus photographs captured using different models and types of specialized imaging equipment with diverse resolutions, ranging from 211×320 to 3456×5184 pixels.

The Indian Diabetic Retinopathy Image Dataset (IDRiD) [73] contains images from clinical examinations conducted at an eye clinic in India. Images of this dataset are split into three subsets: (1) Disease Grading, (2) Localization, and (3) Segmentation. The subset of the IDRiD dataset intended for 'Disease Grading' contains a total of 516 high-quality fundus images for DR classification. Each image was captured with a 50° field of view and a resolution of 4288×2848 pixels, stored in JPEG format. We denote this subset as IDRiD-DR.

Messidor-2 contains a total of 1748 macula-centered fundus images collected from 874 individuals diagnosed with diabetes [74]. The images have a 45° field of view, and the dataset comes with a spreadsheet containing image pairs but does not include annotations for DR severity classification. However, third-party annotations are available.

The 'Segmentation' subset of the IDRiD dataset consists of 81 high-resolution macula-centered fundus images with a resolution of 4288×2848 pixels and a 45° Field of View (FOV). The images were captured using a Kowa VX-10 Alpha digital fundus camera. The images are divided into two distinct sets for training and testing, containing 54 and 27 images, respectively. The dataset provides pixel-wise annotations for four types of lesions: hard exudates, soft exudates, hemorrhages, and microaneurysms, as well as the optic disc. We denote this subset as IDRiD.

E-ophtha [75] is a publicly available dataset of fundus images designed for the segmentation and detection of DR lesions. It consists of two subsets: e-ophtha EX for hard exudates and e-ophtha MA for microaneurysms. The dataset contains a total of 463 images, among which 47 and 148 images contain annotations for hard exudates and microaneurysms, respectively. The images have varying resolutions and FOVs and are widely used for evaluating segmentation models.

The PAthoLogic Myopia (PALM) [76] dataset is designed for the segmentation and detection of pathological myopia (PM) and associated lesions in fundus images. It contains 400 images with a resolution of 1444 × 1444 pixels, captured with a 45° FOV. The dataset includes pixel-wise annotations for the optic disc and three types of lesions: retinal detachment, atrophy, and myopic maculopathy. The images are divided into a training set of 320 images and a test set of 80 images.

The Retinal-Lesions [77] dataset includes 12,252 fundus images sourced from the Kaggle EyePACS dataset and some local hospitals. The images were annotated for eight lesion types, including microaneurysms, intraretinal hemorrhages, hard exudates, cotton-wool spots, vitreous hemorrhages, preretinal hemorrhages, neovascularization, and fibrous proliferation. To ensure high-quality labeling, 45 ophthalmologists participated in the annotation process. The images were standardized by cropping to FOV and resizing to a 896 × 896 pixel resolution. A subset of this dataset, comprising a total of 1593 images and 4143 masks, is available upon request. Among 1593 images, 953 have the same grades as in the Kaggle EyePACS dataset, while 640 images have different grades. The images were from both the training and testing sets of the Kaggle EyePACS dataset. In addition to the lesion masks, adjusted labels are also provided for each image.

5.2.2. Datasets for Building ADRCs

For training and testing ADRCs, we used three publicly available datasets: Kaggle EyePACS [71,72], IDRiD-DR [73], and Messidor-2 [74]. Each dataset contains images categorized into five distinct retinopathy grades, according to the ICDR grading scheme [64]. Table 2 shows each dataset's number of images per DR grade. We used the first two datasets for both training and testing because they have distinct training and test sets. The third dataset (Messidor-2) lacks a separate training set and test set. Therefore, we used it solely for testing, as in previous works on DR classification [33,35,41,45]. For the Messidor-2 labels, we utilized the annotations provided by Krause et al. [38]. These annotations are accessible on the Kaggle platform.

Table 2. Distribution of images in the DR severity classification datasets used for building ADRCs.

Dataset Name	Train						Test						Total
	DR-0	DR-1	DR-2	DR-3	DR-4	Subtotal	DR-0	DR-1	DR-2	DR-3	DR-4	Subtotal	
Kaggle EyePACS	25,810	2443	5292	873	708	35,126	39,533	3762	7861	1214	1206	53,576	**88,702**
IDRiD-DR	134	20	136	74	49	413	34	5	32	19	13	103	**516**
Messidor-2	-	-	-	-	-	-	1017	270	347	75	35	1744	**1744**

5.2.3. Datasets for Building Segmentation Models

For training segmentation models, we used three datasets: IDRiD [73], E-ophtha [75], and PALM [76]. We merged and shuffled the IDRiD (both the training set and the test set), E-ophtha, and PALM datasets. Of the total images, 68% were used for training, 15% for validation, and 17% for testing. The distribution of images is shown in Table 3.

Table 3. Distribution of images used for building segmentation models. Here, Type: type of segmentation area, #Patches: number of patches per image, #Train: number of training images, #Valid: number of validation images, #Test: number of test images, MA: microaneurysms, EX: hard exudates, HE: haemorrhage, SE: soft exudates, OD: optic disc, MC: macula.

Type	Datasets	#Patches	#Train	#Valid	#Test
MA	IDRiD + E-ophtha	25	155	35	39
EX	IDRiD + E-ophtha	25	86	20	22

Table 3. *Cont.*

Type	Datasets	#Patches	#Train	#Valid	#Test
HE	IDRiD	25	53	13	14
SE	IDRiD	25	26	7	7
OD	IDRiD + PALM	1	289	64	73
MC	PALM	1	234	52	59

5.2.4. Datasets for Analyzing ADRCs' Decision

For analyzing the decisions made by our trained ADRCs, we used the IDRiD-DR test set and the Retinal-Lesions [77] dataset. We split the 1593 images of the Retinal-Lesion dataset into two subsets: Retinal-Lesion (A) and Retinal-Lesion (B), based on the Kaggle EyePACS subsets from which they were originally derived. There were 977 images in the Retinal-Lesion (A) subset and 616 images in the Retinal-Lesion (B) subset. As shown in Table 4, the Retinal-Lesion (A) set contains images of all DR grades. On the contrary, the Retinal-Lesion (B) set does not include images of DR-0 and DR-1 grades. We used the original grades provided in the Kaggle EyePACS dataset for our analysis. For manually analyzing the ADRCs' decisions, we only utilized the IDRiD-DR test set, as it is sufficiently small to allow for manual analysis. For automatic analysis, we used the IDRiD-DR test set and both of the Retinal-Lesion subsets.

Table 4. Distribution of images in our datasets used for ADRCs' decision analysis.

Dataset	DR-0	DR-1	DR-2	DR-3	DR-4	Total
IDRiD-DR	34	5	32	19	13	103
Retinal-Lesions (A)	67	184	561	120	45	977
Retinal-Lesions (B)	-	-	415	136	65	616

5.3. Training Setup for ADRCs

We used a MobileNet [78] as the backbone of our ADRCs. It was pretrained on the ImageNet-1k dataset for classifying 1000 types of objects. We discarded the fully connected layers of the pretrained MobileNet. We added two fully connected layers: a hidden layer with 2048 units having ReLU6 activation and a classification layer with softmax activation (as shown in Figure 1).

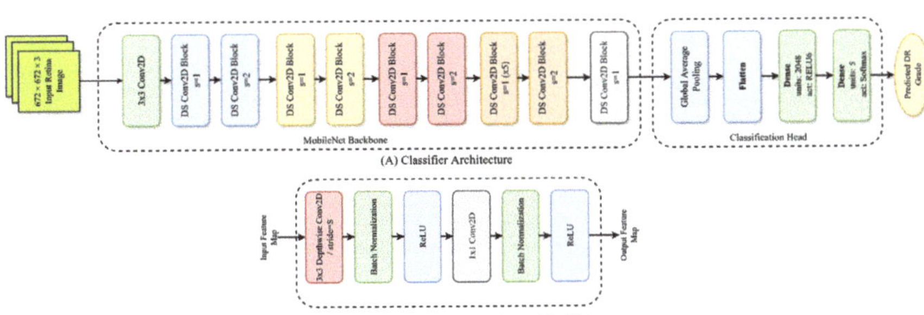

Figure 1. Architecture of our ADRCs. Here, (**A**): overall architecture; (**B**): architecture of a single Depthwise Separable (DS) block.

We first resized the fundus images to a 672 × 672 resolution and then normalized the pixel values to the range of −1 to +1. Initially, the MobileNet backbone of our model was frozen, and only the weights of the newly added layers were adjusted for 50 epochs. Then, the backbone was unfrozen, enabling the entire model's weights to be updated. Then, the model was trained for another 200 epochs. Its performance was monitored using the accuracy of the validation set. The validation set consisted of 10 and 5 images per class for the Kaggle EyePACS and IDRiD-DR datasets. We used early stopping with a patience of 50 epochs and a batch size of 16. We employed the Adam optimizer with an initial learning rate of 0.0001 to minimize the cross-entropy between the labels and targets. We also used a Plateau learning rate scheduler with a patience of 20 epochs and a reduction factor of 0.8. The weights of the best-performing epoch were saved and later used for the final evaluation of the model. We also employed data augmentation techniques, such as random flipping, rotation, and adjustments in brightness, contrast, and saturation, to improve the generalization of our models.

5.4. Training Setup for Segmentation Models

We trained a total of six segmentation models: four for segmenting lesions (MA, EX, HE, SE) and two for segmenting the optic disc and the macula. Our segmentation models were based on the U-Net [79] architecture. It had a MobileNet [78] encoder and a customized decoder (see Figure 2A). The MobileNet encoder was pretrained on the ImageNet-1k dataset. The customized decoder had five Upsampling blocks (Figure 2B). Each block contained an UpSampling2D layer, a concatenation layer, a Depthwise Separable Conv2D block (3 × 3 Depthwise Convolution, Batch Normalization, LeakyReLU), a Conv2D block (3 × 3 Convolution, Batch Normalization, LeakyReLU), and a Spatial Dropout2D layer with a rate of 10%. The output of the last Upsampling blocks was passed to another Conv2D block, and the output layer was a 1 × 1 pointwise convolution layer with softmax activation.

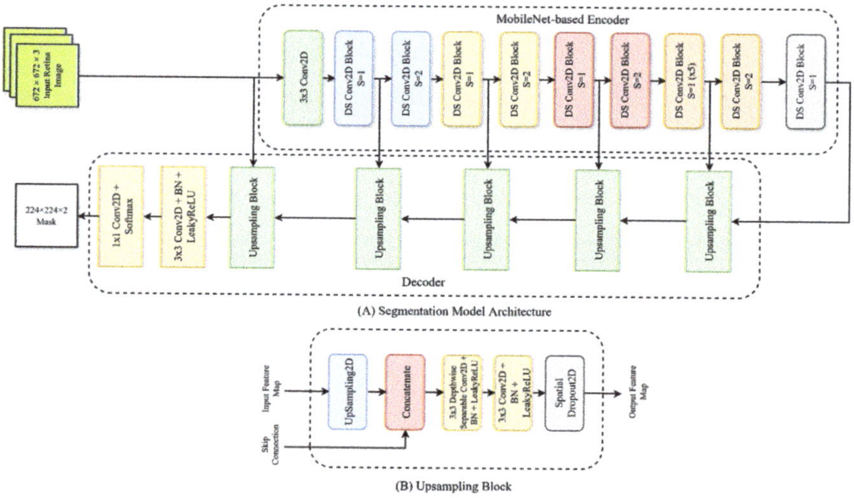

Figure 2. Architecture of our U-net-based segmentation model with the MobileNet encoder and a customized decoder. Here, (**A**): overall segmentation model architecture; (**B**): a single upsampling block.

The fundus images were preprocessed by applying Contrast Limited Adaptive Histogram Equalization (CLAHE) with a clip limit of 5 and tile grid size of 12 × 12. The images' pixel values were then normalized to a range of −1 to +1. For predicting lesion masks, the preprocessed images and lesion masks were resized to 1120 × 1120 resolution, and a total of 25 pairs (224 × 224 resolution each) of non-overlapping patches were extracted from each image-mask pair. For segmenting the optic disc and the macula, whole images resized to 224 × 224 resolution were used as a single patch. The patches were shuffled and grouped into batches, each containing 32 samples. Data augmentation techniques, random vertical and horizontal flips, rotations, translations, scaling, perspective changes, and adjustments to brightness and contrast were used to improve generalization.

We trained segmentation models in two stages. Initially, we trained by updating only the decoder weights, keeping the pretrained encoder weights frozen for 50 epochs. Then, the entire model was trained for an additional 250 epochs. In both stages, we used the cross-entropy loss function with a label smoothing value of 0.1. We used the AdamW optimizer with initial learning rates and weight decays of 1×10^{-3} and 4×10^{-3} in the first stage, and 5×10^{-4} and 1×10^{-4} in the second stage. We used a plateau learning rate scheduler with a patience of 15 epochs and a factor of 0.8.

Our goal was not to benchmark the segmentation models against previous works, but rather to develop workable models for subsequent interpretability analysis. To this end, we trained the segmentation models twice. In the first round, only the training set was used for model training, while the test set was used to evaluate the generalization ability of our models. After confirming that the models achieved reasonable performance, we retrained them on a combined set of the training and testing data. In both cases, the same validation set was used for tuning the hyperparameters.

5.5. Evaluation Metrics

For our ADRCs, we used accuracy, F1 score, specificity, and Area Under the Receiver Operating Characteristic Curve (AUC-ROC) as evaluation metrics. For our segmentation models, we used Mean Intersection Over Union (MIoU) as an evaluation metric. We also used Cohen's kappa coefficient to assess the agreement between our manual and automatic analyses. Accuracy is a measure of the overall correctness of a model. It is calculated as the proportion of correctly classified images over the total number of images. The F1 score metric is the harmonic mean of precision and recall. It provides a balance between precision and recall and is especially useful when the class distribution is imbalanced. Specificity is the proportion of actual negatives that are correctly identified as negatives. It is essential for assessing the model's performance when false positives are costly. The AUC-ROC curve plots the true positive rate versus the false positive rate for various threshold values. The AUC-ROC value ranges from 0 to 1, with higher values indicating better discriminatory performance of the model. IoU, also known as the Jaccard index in multi-label classification, compares a set of inference labels of a sample to the corresponding set of ground-truth labels. The Cohen's kappa coefficient is a statistical measure that quantifies the level of agreement between two raters or classification methods, while correcting for the agreement that could occur purely by chance. It ranges from -1 to $+1$, where values closer to 1 indicate strong agreement, values near 0 indicate agreement by chance, and negative values reflect systematic disagreement.

5.6. Interpretation Using Explainable AI

We used two XAI techniques named Gradient-weighted Class Activation Map (Grad-CAM) proposed by Selvaraju et al. in [80] and Integrated Gradients (IG) proposed by Sundararajan et al. in [68] to visualize the parts of an input fundus image that influence the decision of an ADRC. For Grad-CAM, we computed the gradient of the predicted DR class for an input fundus image with respect to the activation of the last convolutional layer (before the global average pooling layer) of our ADRC. For IG, we computed the average gradients of the model's output with respect to each input pixel along a straight-line path from a baseline image to the actual fundus image. The final attribution score for a pixel is obtained by multiplying this average gradient by the difference between the input pixel and its baseline value.

6. Results and Discussion

6.1. Results of DR Severity Classification

The performance of our ADRCs, trained on two datasets (Kaggle EyePACS and IDRiD-DR) and evaluated across three test datasets (Kaggle EyePACS, IDRiD-DR, and Messidor-2), is summarized in Table 5. In general, the results show that the ADRCs performed best when the training and test sets came from the same dataset. Conversely, when an ADRC was tested on other datasets, we observed a decline in performance. This may indicate some inter-dataset variability, i.e., there are noticeable differences in the images across different datasets. Some reasons that may be behind these differences are the use of different fundus cameras, different setups of image acquisition systems and environment (e.g., varying lighting conditions), involvement of different camera operators, different ethnicities of subjects from which images were captured, different aged subjects, and even the differences in image storing methods.

Table 5. Performance of our DNN-based ADRCs for DR severity classification.

Train Dataset	Test Dataset	Accuracy	F1 Score	Specificity	AUC
Kaggle EyePACS	Kaggle EyePACS	0.8287	0.8102	0.7246	0.8785
	IDRiD-DR	0.4951	0.4765	0.8414	0.8352
	Messidor-2	0.7466	0.7120	0.7844	0.8674
IDRiD-DR	Kaggle EyePACS	0.7042	0.6507	0.4492	0.6589
	IDRiD-DR	0.6214	0.6089	0.8655	0.8412
	Messidor-2	0.5826	0.5338	0.6575	0.7020

The results also show that the ADRC trained on the Kaggle EyePACS dataset achieved the highest overall performance across all datasets. The reason may be attributed to the large size of the training set of the Kaggle EyePACS dataset (a total of 35,126 images). The ADRC achieved its highest performance across most evaluation metrics when it was evaluated on the test set of the Kaggle EyePACS dataset. On the other hand, the performance on the IDRiD-DR dataset was significantly low, with an accuracy of only 0.4951, although the drop in AUC was not substantial. The F1 score for this dataset was also the lowest, implying poor precision and recall. We also observed a moderate performance with an accuracy of 0.7466 and an AUC of 0.8674 on the Messidor-2 dataset. The specificity for this dataset was significantly higher than on the Kaggle EyePACS dataset. Figure 3a shows the confusion matrix of the model trained on the Kaggle EyePACS dataset, evaluated over the three mentioned datasets. It highlights the overall class-wise performance for each of the DR grades. We observed that the performance for DR-0, DR-3, and DR-4 was the highest, whereas that for DR-1 and DR-2 was the lowest. The model failed to correctly identify DR-1 in most instances. Most of the time, the model predicted them as DR-0 or DR-2. The detection performance for the DR-2 grade was moderate, with nearly half (45.72%) of the overall images correctly identified. Most of the misclassified DR-2 images were classified as DR-3, with a smaller number classified as DR-1.

The ADRC trained using the IDRiD-DR dataset achieved the highest score in its test set but had lower performance on the other datasets compared to the ADRC trained using the Kaggle EyePACS dataset. This is not unexpected, as the IDRiD-DR training set is much smaller than the Kaggle EyePACS training set, with only 413 heavily class-imbalanced images. Figure 3b shows the confusion matrix of the ADRC trained on the IDRiD-DR dataset. We found that the ADRC had a bias towards the DR-0 grade. Except for DR-0, the successful detection of no other grades exceeded 45%. The ADRC predicted most of the images from the DR-0 to DR-2 grades as DR-0. Over 20% of the DR-3 and DR-4 images were misclassified as DR-0 and DR-2, respectively.

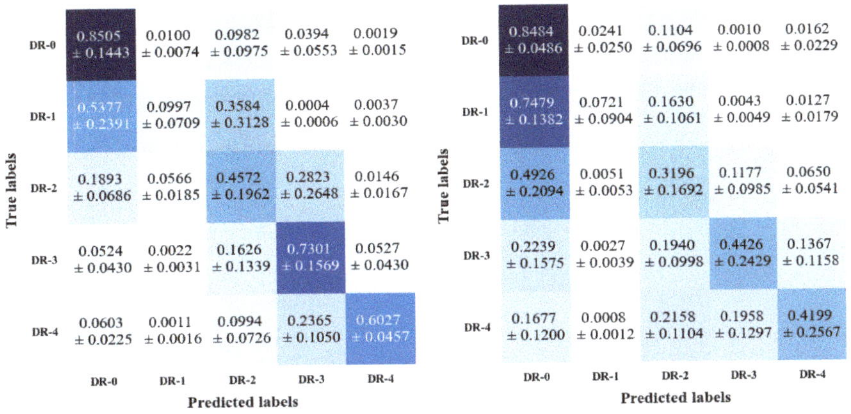

(**a**) Trained using Kaggle EyePACS (**b**) Trained using IDRiD-DR

Figure 3. Confusion matrices of our ADRCs evaluated on the Kaggle EyePACS, IDRiD, and Messidor-2 datasets. Each subfigure shows mean ± std over each class of the three datasets.

The performance differences of our ADRCs across the DR grades may be attributed to class imbalance in the Kaggle EyePACS and IDRiD-DR datasets. Both datasets contain a large proportion of DR-0 images, with substantially fewer samples in the higher grades, which biased the ADRCs towards DR-0, particularly in the IDRiD-DR dataset. However, due to the comparatively larger number of non–DR-0 images in Kaggle EyePACS, the ADRC trained on this dataset was better able to learn discriminative features for the other DR grades than the ADRC trained on IDRiD-DR. For both of our ADRCs, we observed high standard deviations in the performance for certain DR grades. This suggests substantial variability in detection across datasets. The variability may be attributed to differences in image quality, grading distributions, and other dataset-specific characteristics.

6.2. Statistical Analysis of Prediction Scores

Table 6 contains detailed statistics of our ADRCs' prediction scores. The table includes the number of predictions in each class, along with the minimum, maximum, mean, median, and standard deviation values for both the correct and incorrect predictions. Note that scores of correct predictions belong to the target class, whereas scores of incorrect predictions belong to a class other than the target class.

As shown in Table 6, for our five-class ADRCs, we observed that the model trained on the Kaggle EyePACS dataset had more correct predictions than incorrect ones for nearly all severity grades of the three datasets. For accurate predictions, the mean and median probability scores of DR-0 and DR-4 were the highest. For example, in the case of DR-0 of the Kaggle EyePACS dataset, the mean and median probability scores were 0.9422 and 0.9800 for correct predictions, and 0.6361 and 0.6107 for incorrect predictions, respectively. Similarly, for DR-4, the corresponding scores were 0.9118 and 0.9957 for correct predictions and 0.7212 and 0.7353 for incorrect predictions. This suggests that when the ADRC made correct predictions, it was more confident in its predictions than when it made incorrect ones. Conversely, DR-1 had the lowest mean and median probability scores for correct predictions. Moreover, it had the highest mean and median confidence for incorrect predictions. For example, the mean and median probability scores for correct predictions were 0.6800 and 0.6824, respectively, compared to 0.8523 and 0.9258 for incorrect predictions on the Kaggle EyePACS dataset. This indicates that the model was particularly confused when predicting DR-1. We also observed that the model was unable to correctly predict any of the DR-1 images in the IDRiD-DR dataset. The overall performance was also significantly lower for this dataset. Lastly, in the case of the Messidor-2 dataset, we observed comparable performance across all grades to that of the Kaggle EyePACS dataset. For our IDRiD-DR dataset-based ADRC, we observed decent performance on its own test set but much poorer performance on the other two datasets. The number of incorrect predictions for every DR grade was substantially higher. We again observed poor mean and median probability scores for the DR-1 grade for correct predictions and higher scores for incorrect predictions.

Table 6. Statistical analysis of the probability scores for correct and incorrect predictions of our DNN-based ADRCs for DR severity classification.

TrainDB	**TestDB**	**Grade**	**Correct Prediction**						**Incorrect Prediction**					
			Count	**Min**	**Max**	**Mean**	**Median**	**Std**	**Count**	**Min**	**Max**	**Mean**	**Median**	**Std**
Kaggle EyePACS	Kaggle EyePACS	0	38,199	0.2978	1.0000	0.9422	0.9800	0.0946	1334	0.2889	1.0000	0.6361	0.6107	0.1624
		1	596	0.3153	0.9840	0.6800	0.6824	0.1526	3166	0.2838	1.0000	0.8523	0.9258	0.1613
		2	4211	0.2988	0.9988	0.7832	0.8195	0.1606	3650	0.2977	1.0000	0.7616	0.7893	0.1841
		3	617	0.3458	0.9986	0.7663	0.7892	0.1599	597	0.3392	1.0000	0.7304	0.7376	0.1685
		4	773	0.3598	1.0000	0.9118	0.9957	0.1476	433	0.3192	0.9990	0.7212	0.7353	0.1738
	IDRiD-DR	0	22	0.5599	0.9990	0.9024	0.9313	0.1270	12	0.3918	0.6568	0.4961	0.4773	0.0895
		1	0	-	-	-	-	-	5	0.3893	0.9932	0.6285	0.5188	0.2603
		2	6	0.5877	0.8786	0.6890	0.6447	0.1233	26	0.4320	0.9823	0.7082	0.7212	0.1448
		3	16	0.4993	0.9658	0.7695	0.8338	0.1617	3	0.5036	0.9936	0.7712	0.8163	0.2481
		4	7	0.7129	0.9714	0.8496	0.8585	0.0810	6	0.5334	0.8767	0.6868	0.6334	0.1484
	Messidor-2	0	954	0.3746	0.9997	0.9283	0.9749	0.1112	63	0.3703	0.9613	0.6420	0.6235	0.1516
		1	38	0.3634	0.7954	0.5915	0.6150	0.1207	232	0.3661	0.9992	0.8076	0.8644	0.1775
		2	225	0.3180	0.9857	0.7269	0.7619	0.1581	122	0.3216	0.9979	0.7043	0.6930	0.1786
		3	63	0.5104	0.9971	0.8378	0.8931	0.1475	12	0.4974	0.9490	0.6705	0.7025	0.1558
		4	22	0.4267	0.9999	0.8174	0.9691	0.2231	13	0.4046	0.9738	0.6919	0.6238	0.2094
IDRiD-DR	Kaggle EyePACS	0	36,222	0.2848	1.0000	0.9687	0.9988	0.0853	3311	0.2825	1.0000	0.7555	0.7810	0.1939
		1	34	0.3771	0.9556	0.6723	0.6863	0.1775	3728	0.2962	1.0000	0.9564	0.9984	0.1070
		2	631	0.3211	0.9993	0.7145	0.7135	0.1764	7230	0.2881	1.0000	0.9101	0.9891	0.1496
		3	134	0.3480	0.9971	0.7244	0.7242	0.1766	1080	0.3007	1.0000	0.8238	0.8938	0.1846
		4	708	0.2761	1.0000	0.8959	0.9723	0.1474	498	0.2869	1.0000	0.8021	0.8690	0.1927
	IDRiD-DR	0	28	0.4664	1.0000	0.9627	0.9962	0.1011	6	0.5833	0.9436	0.7346	0.6735	0.1662
		1	1	0.6733	0.6733	0.6733	0.6733	0.0000	4	0.7935	0.9984	0.9250	0.9540	0.0948
		2	14	0.4390	1.0000	0.8432	0.9330	0.1813	18	0.5406	0.9997	0.8794	0.9420	0.1423
		3	13	0.5118	0.9866	0.8424	0.8855	0.1332	6	0.5114	0.9965	0.7682	0.8096	0.2134
		4	8	0.6453	0.9993	0.8231	0.8198	0.1417	5	0.4572	0.8498	0.6560	0.6431	0.1709
	Messidor-2	0	819	0.3463	1.0000	0.9308	0.9931	0.1291	198	0.3577	0.9995	0.8039	0.8493	0.1718
		1	2	0.3396	0.4064	0.3730	0.3730	0.0472	268	0.2867	1.0000	0.8680	0.9346	0.1602
		2	153	0.3798	0.9992	0.8226	0.8694	0.1586	194	0.3337	0.9999	0.8437	0.9468	0.1847
		3	40	0.4838	0.9991	0.8317	0.9486	0.1870	35	0.4035	0.9994	0.7386	0.7425	0.1873
		4	2	0.4735	0.7559	0.6147	0.6147	0.1997	33	0.4775	0.9967	0.7858	0.8327	0.1733

6.3. *Performance of Segmentation Models*

We trained six segmentation models using the datasets mentioned in Table 3. Since in many cases, multiple areas in the generated masks were detected as the optic disc and macula, which can only be one, we applied image processing techniques to retain only one optic disc and macula per image. We did not use any post-processing techniques while generating the four lesion masks. The performances of our six segmentation models are given in Table 7. We observed that segmenting the optic disc and macula was easier than segmenting four types of lesions. We also observed that the segmentation model failed to segment microaneurysms more often than other lesions.

Table 7. Mean Intersection-over-Union (mIoU) scores of our segmentation models. In Round 1, the models were trained on the training set and evaluated on the test set. In Round 2, the models were retrained on the combined training and test sets. In both rounds, the same validation set was used. Here, MA: microaneurysms, EX: hard exudates, HE: hemorrhage, SE: soft exudates, OD: optic disc, MC: macula.

Segmentation Category	Round 1		Round 2
	Val. Set	Test Set	Val. Set
MA	0.6042	0.5800	0.5996
EX	0.7904	0.7608	0.8240
HE	0.7169	0.6942	0.7344
SE	0.7513	0.6411	0.7659
OD	0.9162	0.9229	0.9382
MC	0.8159	0.8237	0.8229

7. ADRCs' Decision Analysis Using XAI Techniques

7.1. Manual Analysis

Using the IDRiD-DR dataset, we evaluated manually whether our ADRCs activated around the correct regions in fundus images when predicting DR grades. We employed Grad-CAM and IG heatmaps from our five-class ADRC, trained on the Kaggle EyePACS dataset. We then quantified the number of images in each grade where the ADRC highlighted relevant regions, such as microaneurysms, exudates, hemorrhages, neovascularization, the optic disc, or the macula. If there were more than one such region, we counted whether at least one of them was highlighted. Figure 4 shows one image from each grade with corresponding Grad-CAM and IG heatmaps and overlays, where the ADRC predicted all grades correctly except DR-1. The results of our analysis on the entire dataset, along with the number of correct and incorrect predictions, are summarized in Table 8.

Table 8. Manual analysis of the highlighted regions in Grad-CAM and IG heatmaps for our ADRC trained using the Kaggle EyePACS dataset. The analysis was performed on the IDRiD-DR test set, showing the number of images for which the ADRC paid attention to various DR lesions across the dataset. Here, # Images: total images per DR grade, OD: optic disc, MC: macula, HE: hemorrhage, EX: exudate, MA: microaneurysms, NV: neovascularization, OT: other areas of the fundus image.

Method	DR Grade	# Images	Correct Prediction								Incorrect Prediction							
			Total	MA	EX	HE	NV	MC	OD	OT	Total	MA	EX	HE	NV	MC	OD	OT
Grad-CAM	0	34	22	0	0	0	0	19	11	22	12	0	0	0	0	8	2	12
	1	5	0	-	-	-	-	-	-	-	5	4	0	0	0	0	0	1
	2	32	6	5	3	0	0	0	0	1	26	18	18	9	0	0	1	10
	3	19	16	16	14	10	0	1	6	11	3	2	1	1	0	0	0	1
	4	13	7	5	6	6	4	0	0	1	6	3	3	6	4	0	1	3
IG	0	34	22	0	0	0	0	9	13	20	12	0	0	0	0	7	1	12
	1	5	0	-	-	-	-	-	-	-	5	3	0	0	0	2	1	5
	2	32	6	5	4	0	0	5	4	6	26	25	19	9	0	13	22	24
	3	19	16	16	14	12	0	9	16	16	3	3	1	1	0	2	2	3
	4	13	7	6	5	7	2	5	7	6	6	6	4	5	1	4	6	6

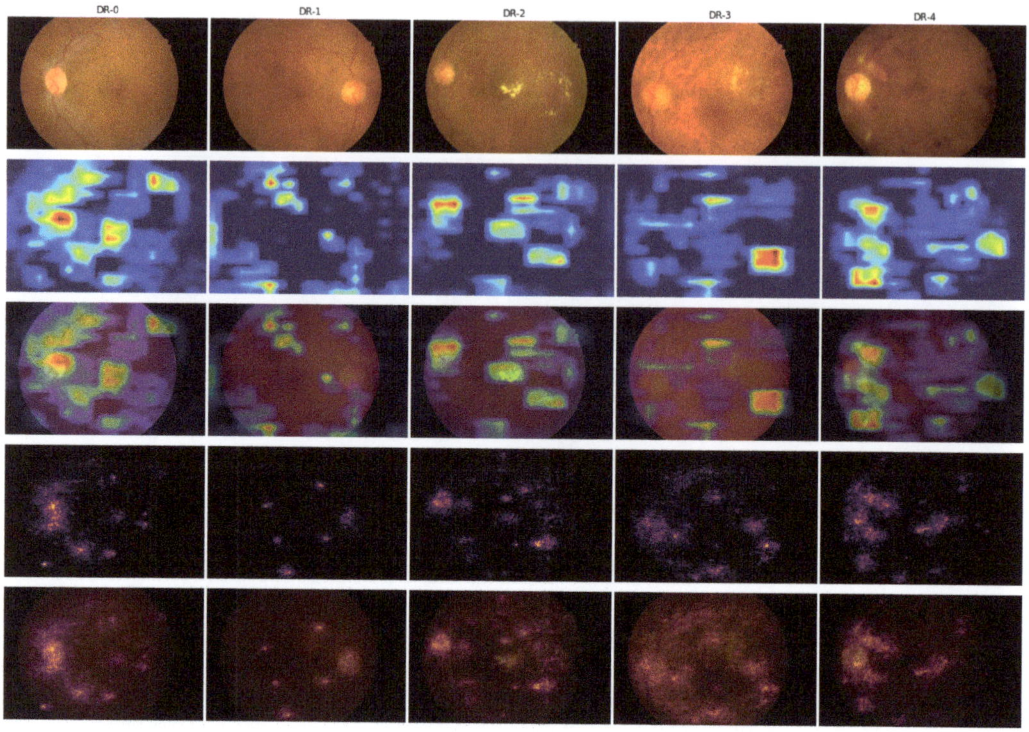

Figure 4. Visualization of Grad-CAM and IG for sample images from the IDRiD-DR dataset. Each row shows the original image, the Grad-CAM heatmap and overlay, and the IG heatmap and overlay. The five columns correspond to images from DR-0 through DR-4 grades, where the ADRC made correct predictions in all cases except DR-1.

According to the ICDR grading scheme, DR-0 cases do not contain any abnormalities. For the DR-0 example image (column 1 of Figure 4), both Grad-CAM and IG heatmaps exhibited the strongest activation around the macula and optic disc regions. Additional activation was observed along the retinal blood vessels, while the IG heatmap also highlighted some unrelated regions. Across all the analyzed images, we consistently observed the same phenomena in the majority of cases in both the Grad-CAM and IG heatmaps (rows 1 and 6 in Table 8). In the ICDR grading scheme, the presence of only microaneurysms characterizes the DR-1 grade. For the example representing DR-1 (column 2), the Grad-CAM heatmap had the strongest activations appearing in two distinct regions containing microaneurysms: one at the top and another at the bottom of the retina. Although the image contained more than two microaneurysms, only these two were consistently highlighted. In the IG heatmap, further activation appeared in other areas, some of which corresponded to microaneurysms, while others did not. Upon analyzing all the images in a similar manner, we found that the optic disc and macula regions were most frequently highlighted in both the Grad-CAM and IG heatmaps. The ADRC also highlighted microaneurysms in four out of five images in the Grad-CAM heatmap and in three out of five images in the IG heatmap. Despite these relevant activations, the ADRC still failed to classify the images correctly. As presented in Table 6, the mean and median probability scores for this grade remained very low, even in cases of misclassification. This further underscores the difficulty the ADRC faced, despite focusing on clinically important lesions. According to the ICDR grading scheme, DR-2 is characterized by the presence of one or more microaneurysms, retinal dot or blot hemorrhages, and hard or soft exudates. In the DR-2 example (column 3), the Grad-CAM heatmap strongly activated around exudate regions and the optic disc. The IG heatmap highlighted the same regions while also highlighting additional areas. Some of these areas also contained additional lesions, such as exudates and microaneurysms. Analyzing all the images, we found that the ADRC was able to highlight regions of exudates in 21 out of 32 total images in the Grad-CAM and 23 out of 32 images in the IG heatmaps, respectively. However, it still lacked confidence (see Table 6) and misclassified the majority of the images. The ICDR grading scale defines DR-3 as a stage in which each of the four quadrants contains more than 20 intraretinal hemorrhages, with either definite venous beading in at least two quadrants or prominent intraretinal microvascular abnormalities in at least one quadrant, without signs of proliferative retinopathy. For the DR-3 example image (column 4), we found the highest activation on a hemorrhage for the Grad-CAM heatmaps. The IG heatmaps also highlighted the same hemorrhage in the highest activated regions, along with some exudates. Analyzing all the images, we again observed that the ADRCs had activation on exudates and hemorrhages in most of the instances (both for Grad-CAM and IG heatmaps). These lesions are relevant for the DR-3 grade. We also found some of the images to have activation on the optic disc, macula, and microaneurysms for the Grad-CAM and IG heatmaps. According to the ICDR scale, DR-4 is characterized by the presence of neovascularization or vitreous or preretinal hemorrhages. In the DR-4 example (column 5), the ADRC primarily focused on areas containing neovascularization in both Grad-CAM and IG heatmaps. The ADRC also showed activations in other regions containing retinal hemorrhages. From the analysis of all images, we observed that the ADRC activated around hemorrhages in 12 out of 13 instances in both the Grad-CAM and IG heatmaps. The ADRC also detected and highlighted neovascularization in eight images but correctly classified only half of them. In the IG heatmap specifically, activations were observed around the neovascularization region in two correctly predicted images and one misclassified image. In both heatmap types, the ADRC additionally highlighted other lesions such as microaneurysms and exudates, while the IG heatmap further showed activations around the optic disc and macula regions.

Although our manual observation revealed that the ADRC could identify relevant lesions in most images, it still failed to classify a significant number of DR-1 and DR-2 images accurately. The boundary between these grades and the adjacent grades is subtle and often challenging, even for HRPs. We also observed that although the ADRCs' decision-making is influenced by multiple retinal regions, including areas with lesions and non-lesions, it did not consistently follow the ICDR grading criteria.

In most of the IDRiD-DR test images, we further noted spots that were in the same regions of the images. The observed spots were not related to any retinal abnormalities and appeared to be caused by imperfections in the imaging equipment (e.g., dust or smudges on the fundus camera lens). For example, there is a spot that appears consistently in the bottom portion of all the example images of row 1 in Figure 4. A close-up view of this region with the heatmap overlays is shown in Figure 5. The spot is not clearly visible in the DR-1 image because it is on a blood vessel. The ADRC also activated around these imperfections, particularly in the IG heatmaps, for all DR grades except DR-1. Similar spots like this one may have contributed to the confusion of the ADRC and may have resulted in the poor overall performance.

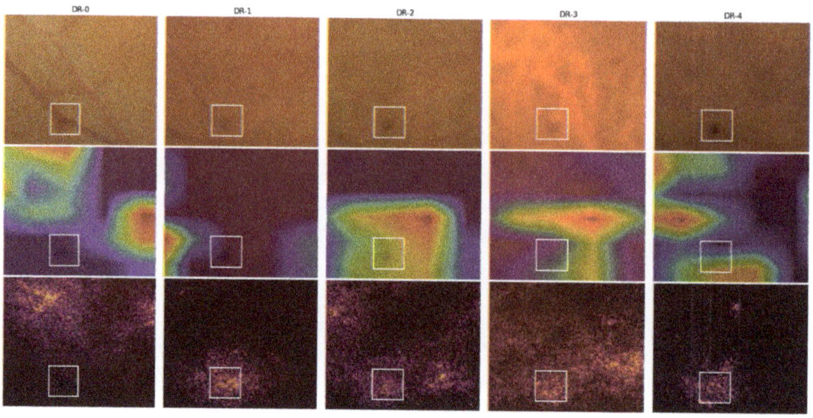

Figure 5. A close-up view of the lower region of the fundus images from Figure 4 showcasing a spot caused by an imaging equipment flaw. Here, row 1 contains the cropped position of the original images, and rows 2 and 3 contain the Grad-CAM and IG overlays, respectively. The white boxes highlight the identified spots.

7.2. Automatic Analysis

As described in Section 7.1, we conducted a manual analysis of the predictions of our ADRC model trained on the Kaggle EyePACS dataset. Our manual analysis involved careful examination of the areas highlighted by the Grad-CAM and Integrated Gradients (IGs) heatmaps for each fundus image in the IDRiD-DR test set. It required a significant amount of time and effort, and it is not feasible to analyze the performance of all our ADRCs, especially with large test sets. Therefore, we propose a new method for assessing the interpretability of ADRCs using the binary masks of different lesions (i.e., microaneurysms, soft exudates, hard exudates, and hemorrhages) and landmarks (i.e., optic disc and macula) automatically. We utilized two datasets for this analysis: the IDRiD-DR test set and the Retinal-Lesions dataset.

To automatically analyze the interpretability of our ADRCs, we checked how often the regions highlighted by Grad-CAM and IG correspond to clinically relevant lesions using the method described in Section 4. The overlapped heatmap and segmentation mask suggest that the ADRC focuses on that particular area while making predictions. It shows that the ADRC's decisions are clinically meaningful, which increases trust among medical professionals. By comparing the high-activation regions with the output of our trained segmentation models, we were able to quantify whether the ADRCs consistently focus on meaningful pathological areas. Figure 6 illustrates our process for automatically assessing the interpretability of an ADRC prediction by comparing its Grad-CAM activations with segmented lesion masks. We observe that there are only a few overlapping regions. The IoU values for the overlapping areas in Figure 6e and Figure 6g are 0.0052 and 0.0188, respectively. This result was expected, since the applied XAI methods, Grad-CAM and IG, selectively emphasized the regions most influential for predicting the DR grade rather than capturing all pathological areas. Both methods highlighted regions that contributed most to the ADRCs' decision, but those regions may not necessarily align perfectly with localized lesions (as shown in Figure 4). Grad-CAM produced coarse and low-resolution heatmaps. When the heatmap was resized to the original image's size, the activation spread across broader areas. This made fine-grained overlap with small lesions unlikely. On the other hand, IG captured pixel-level sensitivity of the prediction score. However, the highlighted pixels often extended beyond the precise lesion boundaries rather than only the annotated lesion itself. Therefore, for our analysis, we only considered whether there was any overlap between the high-activation regions and the lesion masks, instead of the amount of overlap.

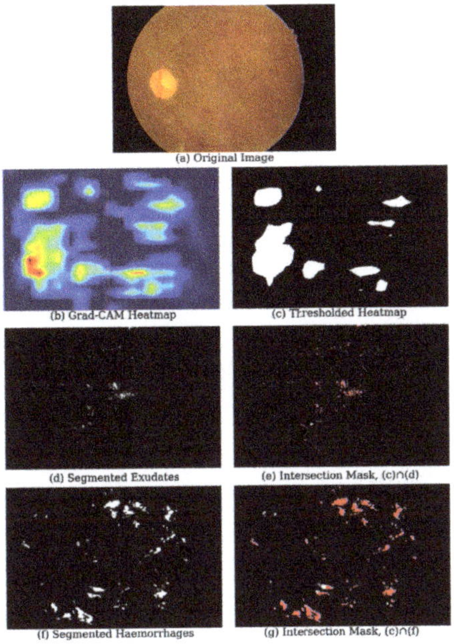

Figure 6. Our procedure of automatically evaluating ADRC interpretability. We demonstrate the process of interpreting the Grad-CAM activation around the exudates and hemorrhages. Here: (**a**) original fundus image, (**b**) Grad-CAM heatmap (with color-map) highlighting the ADRC-activated regions of the fundus image, (**c**) binary thresholded heatmap showing the highest activations (90th percentile), (**d**,**f**) segmented lesion masks from our trained segmentation models (exudates and hemorrhages), and (**e**,**g**) intersection map from the logical AND operation between the thresholded heatmap and lesion masks. The white regions represent overlapping pixels, whereas red regions correspond to non-overlapping regions. The non-overlapping regions refer to regions with lesion masks but no heatmap activation.

The results of our approach for automatic analysis are summarized in Tables 9 and 10. Our approach may have two main limitations. The first limitation is that we only trained automatic segmentation models for four types of lesions. Segmentation models of other lesions or pathologies responsible for higher levels of DR, such as retinal hemorrhages, preretinal hemorrhage, vitreous hemorrhage, fibrous proliferation, and neovascularization, were not used due to the lack of exact pixel-wise annotated training data for such segmentation tasks. The second limitation is that the intersected areas created by the false positives in the segmentation masks may lead to the incorrect conclusion that the ADRC highlighted a lesion that did not exist. To overcome the limitations, we used the ground truth masks of the eight lesion types provided by the Retinal-Lesions dataset. The results are compiled in Table 11. We found that the results are similar to those of our automatic approach.

As shown in Table 9, we observed that exudates and hemorrhages influenced the decision-making process for the DR-0 images of the IDRiD-DR dataset. Similarly, for the Retinal-Lesion (A), we observed the impact of microaneurysms, exudates, and hemorrhages for classifying images belonging to DR-0 (see Table 11). However, according to the ICDR grading scale, images of DR-0 should not have any abnormalities; i.e., the presence of exudates and hemorrhages is not possible for DR-0. One reason for this phenomenon could be that our trained segmentation models were weak and incorrectly recognized normal areas as exudates and hemorrhages. Another possible reason is the incorrect ground truth labels in the Kaggle EyePACS dataset, which may have been taken from the Retinal-Lesion dataset. Besides these, we observed that both the optic disc and macula regions had heated areas in the Grad-CAM and IG heatmaps. In Table 10, we observed uniform lesion detection across the correct and incorrect Retinal-Lesions (B) set predictions. In Table 11, we found that the semi-automatic analysis based on manually segmented ground truth masks was closely aligned with the fully automatic analysis, which was based on masks generated using the trained segmentation models for microaneurysms and hemorrhages.

Table 12 presents Cohen's kappa scores quantifying the agreement between our manual and automatic analysis on the IDRiD-DR test set. It was created using the common features from Tables 8 and 9. The results revealed significant disagreement between our manual and automatic analyses, particularly in cases where predictions by the ADRC were incorrect. In our manual analysis, we carefully inspected the overlays to identify whether high-activation regions overlapped with the presence of lesions. In contrast, our automatic analysis relied on overlap calculations using the method described in Section 4. The limited robustness of the segmentation models, which often produced false detections (including in normal fundus images) and missed true lesions, largely accounts for the low agreement observed between the two approaches.

Table 9. Automatic analysis of the ADRC highlighted areas in Grad-CAM and IG heatmaps of the IDRiD-DR test set using automatic segmentation models. It displays the number of IDRiD-DR images where the ADRC focused on various DR lesions and landmarks, as identified by masks generated using six trained segmentation models. Here, # Images: total images per DR grade, MA: microaneurysm, SE: soft exudate, EX: hard exudate, HE: hemorrhage, MC: macula, and OD: optic disc.

Method	Class ID	# Images	Correct Prediction							Incorrect Prediction						
			Total	MA	SE	EX	HE	MC	OD	Total	MA	SE	EX	HE	MC	OD
Grad-CAM	0	34	22	8	11	20	16	21	16	12	7	2	9	7	11	5
	1	5	0	-	-	-	-	-	-	5	3	0	4	4	2	5
	2	32	6	4	3	6	6	5	6	26	18	14	25	21	14	26
	3	19	16	13	8	16	16	2	15	3	2	1	2	1	2	3
	4	13	7	5	5	6	7	3	7	6	5	4	6	6	1	4
IG	0	34	22	14	12	17	18	22	22	12	11	6	8	11	11	12
	1	5	0	-	-	-	-	-	-	5	4	0	4	5	5	5
	2	32	6	5	2	6	5	6	6	26	23	17	24	25	22	26
	3	19	16	16	12	16	16	12	16	3	3	0	2	2	2	3
	4	13	7	7	5	6	7	7	7	6	5	6	6	6	3	6

Table 10. Automatic analysis of the highlighted areas in Grad-CAM and IG heatmaps of the Retinal-Lesions (B) dataset. It shows the number of Retinal-Lesions (B) images where the ADRC focused on various DR lesions and landmarks located by masks generated using six trained segmentation models. Here, # Images: total images per DR grade, MA: microaneurysm, SE: soft exudate, EX: hard exudate, HE: hemorrhage, MC: macula, and OD: optic disc.

Method	Class ID	# Images	Correct Prediction							Incorrect Prediction						
			Total	MA	SE	EX	HE	MC	OD	Total	MA	SE	EX	HE	MC	OD
Grad-CAM	2	415	238	190	157	206	189	111	229	177	132	117	147	117	83	168
	3	136	74	72	56	71	71	16	30	62	53	44	56	53	21	38
	4	65	21	17	17	18	20	10	14	44	41	36	42	41	23	20
IG	2	415	238	224	200	228	216	230	234	177	144	135	156	137	168	177
	3	136	74	74	64	72	74	69	73	62	59	50	59	58	60	62
	4	65	21	19	18	20	21	16	21	44	41	36	44	42	40	44

Table 11. Semi-automatic analysis of the highlighted areas in Grad-CAM and IG heatmaps of the Retinal-Lesions dataset. It shows the number of images of the Retinal-Lesions dataset for which our ADRC (trained on the Kaggle EyePACS dataset) paid attention to various DR lesions and landmarks located by manually prepared ground truth binary masks. Here, # Images: total images per DR grade, MA: microaneurysm, SE: soft exudates, EX: hard exudate, rHE: retinal hemorrhage, pHE: preretinal hemorrhage, vHE: vitreous hemorrhage, FP: fibrous proliferation, NV: neovascularization, Subset: subset of Retinal-Lesion dataset.

Subset	Method	DR Grade	# Images	Correct Prediction									Incorrect Prediction								
				Total	MA	SE	EX	rHE	vHE	pHE	FP	NV	Total	MA	SE	EX	rHE	vHE	pHE	FP	NV
(A)	Grad-CAM	0	67	57	9	0	4	3	0	0	1	0	10	3	0	1	1	0	0	0	0
		1	184	33	26	0	0	1	0	0	0	0	151	75	2	6	13	0	0	0	0
		2	561	320	254	80	165	235	0	0	0	2	241	140	46	65	112	0	0	2	0
		3	120	94	80	44	70	92	0	0	0	0	26	23	8	15	20	0	0	0	0
		4	45	29	18	6	15	24	8	6	10	16	16	13	5	10	15	0	0	0	2
	IG	0	67	57	24	3	9	5	0	0	1	0	10	8	2	3	3	0	0	0	0
		1	184	33	32	0	3	4	0	0	0	0	151	106	6	28	17	0	0	0	0
		2	516	320	288	91	197	251	0	0	0	2	241	166	52	79	118	0	0	2	0
		3	120	94	87	52	76	92	0	0	0	0	26	24	10	16	22	0	0	0	0
		4	45	29	23	8	20	26	8	6	12	16	16	16	6	11	15	0	0	0	2
(B)	Grad-CAM	2	415	238	194	64	127	201	0	0	0	0	177	106	40	46	97	0	0	0	1
		3	136	74	57	45	55	72	0	0	0	0	62	53	24	29	53	0	0	0	1
		4	65	21	12	5	8	16	2	4	9	13	44	31	14	32	39	2	1	2	11
	IG	2	415	238	222	81	150	215	0	0	1	0	177	131	50	68	100	0	0	1	1
		3	136	74	69	48	59	73	0	0	0	1	62	59	26	33	54	0	0	0	1
		4	65	21	15	6	9	18	2	4	8	14	44	36	21	33	41	2	1	2	13

Table 12. Kappa scores for lesion detection agreement between our manual and automatic analyses of the highlighted areas in Grad-CAM and IG heatmaps across DR grades of the IDRiD-DR test set. Here, positive values indicate better-than-chance agreement, negative values indicate disagreement, zeros represent no agreement beyond chance, dashes ("– ") denote missing cases, and "N/A" indicates that the kappa score is undefined due to no variability in the labels for that category.

Method	DR Grade	Correct Prediction					Incorrect Prediction				
		MA	EX	HE	OD	MC	MA	EX	HE	OD	MC
Grad-CAM	0	0.0000	0.0000	0.0000	0.3636	0.4634	0.0000	0.0000	0.0000	0.0625	0.3077
	1	–	–	–	–	–	−0.3636	0.0000	0.0000	0.0000	0.0000
	2	0.5714	0.0000	0.0000	0.0000	0.0000	−0.2639	0.0000	0.2239	0.0000	0.0000
	3	0.0000	0.0000	0.0000	0.0769	−0.0909	−0.5000	0.0000	1.000	0.0000	0.0000
	4	0.3000	0.0000	0.0000	0.0000	0.0000	−0.3333	0.0000	"N/A"	0.1818	0.0000
IG	0	0.0000	0.0000	0.0000	0.0000	0.0000	0.0000	0.0000	0.0000	0.0000	0.258
	1	–	–	–	–	–	0.5455	0.0000	0.0000	0.0000	0.0000
	2	1.0000	0.0000	0.0000	0.0000	0.0000	0.4694	0.1959	0.0415	0.0000	0.0000
	3	"N/A"	0.0000	0.0000	"N/A"	0.3333	"N/A"	0.4000	0.4000	0.0000	−0.5000
	4	0.0000	0.0000	"N/A"	"N/A"	0.0000	0.0000	0.0000	0.0000	"N/A"	0.6667

From Tables 9–11, we notice that our ADRCs recognized lesions, i.e., signs of diabetes, quite well. However, they did not completely rely on these lesions to determine the level of DR for fundus images. Besides different types of lesions, we found that other parts of the retina, such as the optic disc, macula, and sometimes blood vessels, also influenced ADRCs' decisions. Moreover, we observed that ADRCs did not consider all relevant lesions when predicting the DR grade. We also observed a notable disagreement between our manual and automatic analyses (see Table 12). To increase agreement, more advanced segmentation models, trained on larger and more diverse datasets with improved lesion annotations, should be employed, which would enhance the reliability of the automatic interpretability analysis. We found that the ADRC's decision-making procedure regarding the DR severity level is quite complicated. While our analysis has clarified some aspects, more details will have to be analyzed in future work. Our ADRCs also showed some transferability issues across datasets. These will also need to be analyzed in future work through domain adaptation, advanced data augmentation, or other strategies. With our current understanding of DNNs' working procedure, we could not determine the reasons behind incorrect predictions, despite observing that diabetic signs influenced the final decision in the same way as they do in correct predictions. We propose that non-lesion areas that influence the ADRCs' prediction can help HRPs.

8. Conclusions

In this paper, we investigated whether deep neural network (DNN)-based Automatic Diabetic Retinopathy Classifiers (ADRCs) follow the same criteria as human retina professionals (HRPs). In general, HRPs follow the criteria listed in the International Clinical Diabetic Retinopathy (ICDR)'s grading scheme. To this end, we trained DNN-based ADRCs for DR severity classification and six segmentation models and generated binary masks for the optic disc, macula, and four kinds of abnormalities, such as microaneurysms, hard exudates, soft exudates, and hemorrhages. Our DNN-based ADRCs were able to detect normal fundus images with high accuracy, and severe and proliferative DR with reasonable accuracy. However, the ADRCs failed to detect retinas affected by mild and moderate DR. Our segmentation models showed moderate performance.

Using Gradient-weighted Class Activation Map (Grad-CAM) and Integrated Gradients (IGs) eXplainable Artificial Intelligence (XAI) techniques, we generated heatmaps from our best-performing ADRC. Using the generated segmentation masks and heatmaps, we automatically analyzed the decision of our ADRC. Our manual observation showed that while the ADRC identified relevant lesions in most cases, it struggled to accurately classify many DR-1 and DR-2 images due to their subtle boundaries with adjacent grades. Moreover, its decision-making, though influenced by both lesion and non-lesion regions, did not consistently align with the ICDR grading criteria. Our automatic analysis also revealed that while ADRCs recognize retinopathy lesions, their decisions are also influenced by non-lesion retinal regions. In the future, we will use advanced DNN-based models for building ADRCs and segmentation models.

Author Contributions: Conceptualization, S.B. and M.A.A.K.; Methodology, S.B., M.A.A.K., M.H.A., J.R., S.P., M.I.A.K., S.K.C. and B.K.P.; Validation, S.B., M.A.A.K., M.H.A., J.R., S.P. and M.I.A.K.; Formal analysis, S.B., M.A.A.K., M.H.A. and J.R.; Investigation, S.B., M.A.A.K., M.H.A., J.R., S.P., M.I.A.K., S.K.C. and B.K.P.; Resources, S.B., M.A.A.K., M.H.A., J.R., S.P., M.I.A.K., S.K.C. and B.K.P.; Writing—original draft, S.B., M.A.A.K., M.H.A., J.R., S.P. and M.I.A.K.; Writing—review & editing, S.B., M.A.A.K., M.H.A., J.R., S.P. and M.I.A.K.; Visualization, S.B., M.A.A.K., M.H.A., J.R., S.P., M.I.A.K., S.K.C. and B.K.P.; Supervision, S.B.; Project administration, S.B.; Funding acquisition, S.B. All authors have read and agreed to the published version of the manuscript.

Funding: This research received no external funding.

Institutional Review Board Statement: Not applicable.

Informed Consent Statement: Not applicable.

Data Availability Statement: The source code used in this research is publicly available at https://github.com/ahanaf019/Interpreting-Deep-Neural-Networks-in-Diabetic-Retinopathy-Grading, (accessed on 6 September 2025).

Acknowledgments: This work relied on publicly available resources that were essential for conducting the experimental analyses. Therefore, we acknowledge the creators and maintainers of the Kaggle EyePACS, IDRiD, Messidor-2, E-ophtha, PALM, and Retinal-Lesions datasets. Furthermore, we would like to express our appreciation to the developers of the pre-trained models in the Keras repository for their significant contributions to this research.

Conflicts of Interest: The authors declare no conflicts of interest.

References

1. Santos, G.S.P.; Prazeres, P.H.D.M.; Mintz, A.; Birbrair, A. Role of pericytes in the retina. *Nat. Eye* **2018**, *32*, 483–486. [CrossRef]
2. Demir, S.; Nawroth, P.P.; Herzig, S.; Üstünel, B.E. Emerging Targets in Type 2 Diabetes and Diabetic Complications. *Adv. Sci.* **2021**, *8*, 2100275. [CrossRef]
3. Liu, R.; Li, L.; Shao, C.; Cai, H.; Wang, Z. The Impact of Diabetes on Vascular Disease: Progress from the Perspective of Epidemics and Treatments. *J. Neurol. Neurosurg. Psychiatry* **2022**, *2022*, 153128. [CrossRef] [PubMed]
4. Magliano, D.J.; Boyko, E.J.; Balkau, B.; Barengo, N.; Barr, E.; Basit, A.; Bhata, D.; Bommer, C.; Booth, G.; Cariou, B.; et al. *International Diabetes Federation (IDF) Diabetes Atlas 2021*, 10th ed.; International Diabetes Federation: Brussels, Belgium, 2021; Volume 10, pp. 34–62.
5. Teo, Z.L.; Tham, Y.C.; Yu, M.; Chee, M.L.; Rim, T.H.; Cheung, N.; Bikbov, M.M.; Wang, Y.X.; Tang, Y.; Lu, Y.; et al. Global Prevalence of Diabetic Retinopathy and Projection of Burden through 2045: Systematic Review and Meta-analysis. *Ophthalmology* **2021**, *128*, 1580–1591. [CrossRef]
6. Stitt, A.W.; Curtis, T.M.; Chen, M.; Medina, R.J.; McKay, G.J.; Jenkins, A.; Gardiner, T.A.; Lyons, T.J.; Hammes, H.P.; Simó, R.; et al. The progress in understanding and treatment of diabetic retinopathy. *Prog. Retin. Eye Res.* **2016**, *51*, 156–186. [CrossRef] [PubMed]
7. Vision Loss Expert Group of the Global Burden of Disease Study & the GBD 2019 Blindness and Vision Impairment Collaborators. Global estimates on the number of people blind or visually impaired by diabetic retinopathy: A meta-analysis from 2000 to 2020. *Eye* **2024**, *38*, 2047–2057. [CrossRef]
8. Liu, Y.; Teng, X.; Zhang, W.; Zhang, R.; Liu, W. Association between diabetic retinopathy and subclinical atherosclerosis in China: Results from a community-based study. *Diabetes Vasc. Dis. Res.* **2015**, *12*, 366–372. [CrossRef]
9. Rema, M.; Mohan, V.; Deepa, R.; Ravikumar, R. Association of Carotid Intima-Media Thickness and Arterial Stiffness with Diabetic Retinopathy: The Chennai Urban Rural Epidemiology Study (CURES-2). *Diabetes Care* **2004**, *27*, 1962–1967. [CrossRef]
10. Saif, A.; Karawya, S.; Abdelhamid, A. Retinopathy is a Strong Determinant of Atherosclerosis in Type 2 Diabetes: Correlation with Carotid Intima Media Thickness. *Endocr. Pract.* **2015**, *21*, 226–230. [CrossRef]
11. Alonso, N.; Traveset, A.; Rubinat, E.; Ortega, E.; Alcubierre, N.; Sanahuja, J.; Hernández, M.; Betriu, A.; Jurjo, C.; Fernández, E.; et al. Type 2 diabetes-associated carotid plaque burden is increased in patients with retinopathy compared to those without retinopathy. *Cardiovasc. Diabetol.* **2015**, *17*, 49. [CrossRef]
12. Targher, G.; Bertolini, L.; Zenari, L.; Lippi, G.; Pichiri, I.; Zoppini, G.; Muggeo, M.; Arcaro, G. Diabetic retinopathy is associated with an increased incidence of cardiovascular events in Type 2 diabetic patients. *Diabet. Med.* **2008**, *25*, 45–50. [CrossRef]
13. Gimeno-Orna, J.A.; Faure-Nogueras, E.; Castro-Alonso, F.J.; Boned-Juliani, B. Ability of retinopathy to predict cardiovascular disease in patients with type 2 diabetes mellitus. *Am. J. Cardiol.* **2009**, *103*, 1364–1367. [CrossRef]
14. Kawasaki, R.; Tanaka, S.; Tanaka, S.; Abe, S.; Sone, H.; Yokote, K.; Ishibashi, S.; Katayama, S.; Ohashi, Y.; Akanuma, Y.; et al. Risk of cardiovascular diseases is increased even with mild diabetic retinopathy: The Japan Diabetes Complications Study. *Ophthalmology* **2013**, *120*, 574–582. [CrossRef]
15. Guo, V.Y.; Cao, B.; Wu, X.; Lee, J.J.W.; Zee, B.C.Y. Prospective Association between Diabetic Retinopathy and Cardiovascular Disease-A Systematic Review and Meta-analysis of Cohort Studies. *J. Stroke Cerebrovasc. Dis.* **2016**, *25*, 1688–1695. [CrossRef]
16. Pearce, I.; Simó, R.; Lövestam-Adrian, M.; Wong, D.T.; Evans, M. Association between diabetic eye disease and other complications of diabetes: Implications for care. A systematic review. *Diabetes Obes. Metab.* **2019**, *21*, 467–478. [CrossRef]
17. Cheung, N.; Rogers, S.; Couper, D.J.; Klein, R.; Sharrett, A.R.; Wong, T.Y. Is diabetic retinopathy an independent risk factor for ischemic stroke? *Stroke* **2007**, *38*, 398–401. [CrossRef] [PubMed]
18. Hägg, S.; Thorn, L.M.; Putaala, J.; Liebkind, R.; Harjutsalo, V.; Forsblom, C.M.; Gordin, D.; Tatlisumak, T.; Groop, P.H. Incidence of stroke according to presence of diabetic nephropathy and severe diabetic retinopathy in patients with type 1 diabetes. *Diabetes Care* **2013**, *36*, 4140–4146. [CrossRef] [PubMed]
19. Zhu, X.R.; Zhang, Y.P.; Bai, L.; Zhang, X.L.; Zhou, J.B.; Yang, J.K. Prediction of risk of diabetic retinopathy for all-cause mortality, stroke and heart failure: Evidence from epidemiological observational studies. *Medicine* **2017**, *96*, e5894. [CrossRef] [PubMed]
20. Wong, K.H.; Hu, K.; Peterson, C.; Sheibani, N.; Tsivgoulis, G.; Majersik, J.J.; de Havenon, A.H. Diabetic Retinopathy and Risk of Stroke: A Secondary Analysis of the ACCORD Eye Study. *Stroke* **2020**, *51*, 3733–3736. [CrossRef]
21. Hu, K.; Jiang, M.; Zhou, Q.; Zeng, W.; Lan, X.; Gao, Q.; Mei, F.; Zhao, L.; Chen, F.; Wu, A.; et al. Association of Diabetic Retinopathy with Stroke: A Systematic Review and Meta-Analysis. *Front. Neurol.* **2021**, *12*, 626996. [CrossRef]

22. Hugenschmidt, C.E.; Lovato, J.F.; Ambrosius, W.T.; Bryan, R.N.; Gerstein, H.C.; Horowitz, K.R.; Launer, L.J.; Lazar, R.M.; Murray, A.M.; Chew, E.Y.; et al. The Cross-sectional and Longitudinal Associations of Diabetic Retinopathy with Cognitive Function and Brain MRI Findings: The Action to Control Cardiovascular Risk in Diabetes (ACCORD) Trial. *Diabetes Care* **2014**, *37*, 3244–3252. [CrossRef]
23. Exalto, L.G.; Biessels, G.J.; Karter, A.J.; Huang, E.S.; Jr, C.P.Q.; Whitmer, R.A. Severe diabetic retinal disease and dementia risk in type 2 diabetes. *J. Alzheimer's Dis.* **2014**, *3*, S109–S117. [CrossRef] [PubMed]
24. Frith, E.; Loprinzi, P.D. Retinopathy and Mortality. *Diabetes Spectr.* **2018**, *31*, 184–188. [CrossRef]
25. Xu, X.H.; Sun, B.; Zhong, S.; Wei, D.D.; Hong, Z.; Dong, A.Q. Diabetic retinopathy predicts cardiovascular mortality in diabetes: A meta-analysis. *BMC Cardiovasc. Disord.* **2020**, *20*, 478. [CrossRef]
26. Fadzil, M.H.A.; Izhar, L.I.; Nugroho, H.; Nugroho, H.A. Analysis of Retinal Fundus Images for Grading of Diabetic Retinopathy Severity. *Med. Biol. Eng. Comput.* **2011**, *49*, 693–700. [CrossRef]
27. Bhaskaranand, M.; Ramachandra, C.; Bhat, S.; Cuadros, J.; Nittala, M.G.; Sadda, S.; Solanki, K. Automated Diabetic Retinopathy Screening and Monitoring Using Retinal Fundus Image Analysis. *J. Diabetes Sci. Technol.* **2016**, *10*, 254–261. [CrossRef]
28. Ganesan, K.; Martis, R.J.; Acharya, U.R.; Chua, C.K.; Min, L.C.; Ng, E.Y.K.; Laude, A. Computer-aided diabetic retinopathy detection using trace transforms on digital fundus images. *Med. Biol. Eng. Comput.* **2014**, *52*, 663–672. [CrossRef]
29. Casanova, R.; Saldana, S.; Chew, E.Y.; Danis, R.P.; Greven, C.M.; Ambrosius, W.T. Application of Random Forests Methods to Diabetic Retinopathy Classification Analyses. *PLoS ONE* **2014**, *9*, e98587. [CrossRef] [PubMed]
30. Akram, M.U.; Khalid, S.; Tariq, A.; Javed, M.Y. Detection of neovascularization in retinal images using multivariate m-Mediods based classifier. *Comput. Med. Imaging Graph.* **2013**, *37*, 346–357. [CrossRef] [PubMed]
31. Abràmoff, M.D.; Niemeijer, M.; Suttorp-Schulten, M.S.; Viergever, M.A.; Russell, S.R.; van Ginneken, B. Evaluation of a System for Automatic Detection of Diabetic Retinopathy from Color Fundus Photographs in a Large Population of Patients with Diabetes. *ADA Diabetes Care* **2008**, *31*, 193–198. [CrossRef]
32. Gardner, G.G.; Keating, D.; Williamson, T.H.; Elliott, A.T. Automatic Detection of Diabetic Retinopathy Using an Artificial Neural Network: A Screening Tool. *BMJ Br. J. Ophthalmol.* **1996**, *80*, 940–944. [CrossRef]
33. Gulshan, V.; Peng, L.; Coram, M.; Stumpe, M.C.; Wu, D.; Narayanaswamy, A.; Venugopalan, S.; Widner, K.; Madams, T.; Cuadros, J.; et al. Development and Validation of a Deep Learning Algorithm for Detection of Diabetic Retinopathy in Retinal Fundus Photographs. *J. Am. Med. Assoc.* **2016**, *316*, 2402–2410. [CrossRef]
34. Pratt, H.; Coenen, F.; Broadbent, D.M.; Harding, S.P.; Zheng, Y. Convolutional Neural Networks for Diabetic Retinopathy. *Procedia Comput. Sci.* **2016**, *90*, 200–205. [CrossRef]
35. Gargeya, R.; Leng, T. Automated Identification of Diabetic Retinopathy Using Deep Learning. *Ophthalmology* **2017**, *124*, 962–969. [CrossRef]
36. Quellec, G.; Charrière, K.; Boudi, Y.; Cochener, B.; Lamard, M. Deep Image Mining for Diabetic Retinopathy Screening. *Med. Image Anal.* **2017**, *39*, 178–193. [CrossRef]
37. Wang, Z.; Yang, J. Diabetic Retinopathy Detection via Deep Convolutional Networks for Discriminative Localization and Visual Explanation. *arXiv* **2017**, arXiv:1703.10757. [CrossRef]
38. Krause, J.; Gulshan, V.; Rahimy, E.; Karth, P.; Widner, K.; Corrado, G.S.; Peng, L.; Webster, D.R. Grader Variability and the Importance of Reference Standards for Evaluating Machine Learning Models for Diabetic Retinopathy. *Ophthalmology* **2018**, *125*, 1264–1272. [CrossRef] [PubMed]
39. Lin, G.M.; Chen, M.J.; Yeh, C.H.; Lin, Y.Y.; Kuo, H.Y.; Lin, M.H.; Chen, M.C.; Lin, S.D.; Gao, Y.; Ran, A.; et al. Transforming Retinal Photographs to Entropy Images in Deep Learning to Improve Automated Detection for Diabetic Retinopathy. *J. Ophthalmol.* **2018**, *2018*, 2159702. [CrossRef]
40. Gao, Z.; Li, J.; Guo, J.; Chen, Y.; Yi, Z.; Zhong, J. Diagnosis of Diabetic Retinopathy Using Deep Neural Networks. *IEEE Access* **2019**, *7*, 3360–3370. [CrossRef]
41. Li, F.; Liu, Z.; Chen, H.; Jiang, M.; Zhang, X.; Wu, Z. Automatic Detection of Diabetic Retinopathy in Retinal Fundus Photographs Based on Deep Learning Algorithm. *ARVO Transl. Vis. Sci. Technol.* **2019**, *8*, 4. [CrossRef]
42. Sahlsten, J.; Jaskari, J.; Kivinen, J.; Turunen, L.; Jaanio, E.; Hietala, K.; Kaski, K. Deep Learning Fundus Image Analysis for Diabetic Retinopathy and Macular Edema Grading. *Nat. Sci. Rep.* **2019**, *9*, 10750. [CrossRef]
43. Sayres, R.; Taly, A.; Rahimy, E.; Blumer, K.; Coz, D.; Hammel, N.; Krause, J.; Narayanaswamy, A.; Rastegar, Z.; Wu, D.; et al. Using a Deep Learning Algorithm and Integrated Gradients Explanation to Assist Grading for Diabetic Retinopathy. *Ophthalmology* **2019**, *126*, 552–564. [CrossRef]
44. Shanthi, T.; Sabeenian, R. Modified Alexnet architecture for classification of diabetic retinopathy images. *Comput. Electr. Eng.* **2019**, *76*, 56–64. [CrossRef]
45. Voets, M.; Møllersen, K.; Bongo, L.A. Reproduction study using public data of: Development and validation of a deep learning algorithm for detection of diabetic retinopathy in retinal fundus photographs. *PLoS ONE* **2019**, *14*, e0217541. [CrossRef]

46. Zeng, X.; Chen, H.; Luo, Y.; Ye, W. Automated Diabetic Retinopathy Detection Based on Binocular Siamese-Like Convolutional Neural Network. *IEEE Access* **2019**, *4*, 30744–30753. [CrossRef]
47. Zhao, Z.; Zhang, K.; Hao, X.; Tian, J.; Chua, M.C.H.; Chen, L.; Xu, X. BiRA-Net: Bilinear Attention Net for Diabetic Retinopathy Grading. In Proceedings of the IEEE International Conference on Image Processing (ICIP), Taipei, Taiwan, 22–25 September 2019; pp. 1385–1389. [CrossRef]
48. Gayathri, S.; Gopi, V.P.; Palanisamy, P. A lightweight CNN for Diabetic Retinopathy classification from fundus images. *Biomed. Signal Process. Control* **2020**, *62*, 102115. [CrossRef]
49. Quellec, G.; Lamard, M.; Conze, P.H.; Massin, P.; Cochener, B. Automatic Detection of Rare Pathologies in Fundus Photographs Using Few-Shot Learning. *Med. Image Anal.* **2020**, *61*, 101660. [CrossRef] [PubMed]
50. Smailagic, A.; Costa, P.; Gaudio, A.; Khandelwal, K.; Mirshekari, M.; Fagert, J.; Walawalkar, D.; Xu, S.; Galdran, A.; Zhang, P.; et al. O-MedAL: Online active deep learning for medical image analysis. *Wires Data Min. Knowl. Discov.* **2020**, *10*, e1353. [CrossRef]
51. de la Torre, J.; Valls, A.; Puig, D. A deep learning interpretable classifier for diabetic retinopathy disease grading. *Neurocomputing* **2020**, *396*, 465–476. [CrossRef]
52. Kalyani, G.; Janakiramaiah, B.; Karuna, A.; Prasad, L.V.N. Diabetic retinopathy detection and classification using capsule networks. *Complex Intell. Syst.* **2021**, *9*, 2651–2664. [CrossRef]
53. Al-Antary, M.T.; Arafa, Y. Multi-Scale Attention Network for Diabetic Retinopathy Classification. *IEEE Access* **2021**, *9*, 54190–54200. [CrossRef]
54. Farag, M.M.; Fouad, M.; Abdel-Hamid, A.T. Automatic Severity Classification of Diabetic Retinopathy Based on DenseNet and Convolutional Block Attention Module. *IEEE Access* **2022**, *10*, 38299–38308. [CrossRef]
55. Prakash, J; Kumar B.V. An ensemble approach for classification of diabetic retinopathy in fundus image. *Multimed. Tools Appl.* **2024**, *84*, 10567–10586. [CrossRef]
56. Hirschberg, J. Über Diabetische Netzhautentzündung. *Dtsch Med Wochenschr* 1890. Available online: https://www.degruyterbrill.com/document/doi/10.1515/9783112339848-016/html?srsltid=AfmBOoo55W-7WBknFg_3U1EOJs6mOnEP_MKLmPWFqk25P4oVJFXEo-sf (accessed on 27 March 2024).
57. Wagener, H.P.; Dry, T.J.S.; Wilder, R.M. Retinitis in Diabetes. *N. Engl. J. Med.* **1934**, *211*, 1131–1137. [CrossRef]
58. Ballantyne, A.J.; Loewenstein, A. Exudates in Diabetic Retinopathy. *Trans. Ophthalmol. Soc. UK* **1943**, *63*, 95.
59. Scott, G.I. Ocular complications of diabetes mellitus. *Br. J. Ophthalmol.* **1953**, *37*, 705–715. [CrossRef]
60. Lee, P.F.; McMeel, J.W.; Schepens, C.L.; Field, R.A. A New SeverityScale of Diabetic Retinopathy. *Am. J. Ophthalmol.* **1966**, *62*, 207–219. [CrossRef]
61. Oakley, N.; Hill, D.W.; Joplin, G.F.; Kohner, E.M.; Fraser, T.R. Diabetic retinopathy. I. The assessment of severity and progress by comparison with a set of standard fundus photographs. *Diabetologia* **1967**, *3*, 402–405. [CrossRef]
62. Goldberg, M.F.; Jampol, L.M. Knowledge of Diabetic Retinopathy before and 18 Years after the Airlie House Symposium on Treatment of Diabetic Retinopathy. *Ophthalmology* **1987**, *94*, 741–746. [CrossRef]
63. Group, E.T.D.R.S.R. Fundus photographic risk factors for progression of diabetic retinopathy. ETDRS report number 12. *Ophthalmology* **1991**, *98*, 823–833. [CrossRef]
64. Wilkinson, C.P.; Ferris, F.L.; Klein, R.; Lee, P.P.; Agardh, C.D.; Davis, M.D.; Dills, D.G.; Kampik, A.; Pararajasegaram, R.; Verdaguer, J.T. Proposed international clinical diabetic retinopathy and diabetic macular edema disease severity scales. *Ophthalmology* **2003**, *110*, 1677–1682. [CrossRef]
65. Zachariah, S.; Wykes, W.; Yorston, D. Grading diabetic retinopathy (DR) using the Scottish grading protocol. *Community Eye Health* **2015**, *28*, 72–73.
66. Nakayama, L.F.; Ribeiro, L.Z.; Gonçalves, M.B.; Ferraz, D.A.; dos Santos, H.N.V.; Malerbi, F.K.; Morales, P.H.; Maia, M.; Regatieri, C.V.S.; Mattos, R.B., Jr. Diabetic retinopathy classification for supervised machine learning algorithms. *Int. J. Retin. Vitr.* **2022**, *8*, 1. [CrossRef]
67. Simonyan, K.; Vedaldi, A.; Zisserman, A. Deep Inside Convolutional Networks: Visualising Image Classification Models and Saliency Maps. *arXiv* **2014**, arXiv:1312.6034. [CrossRef]
68. Sundararajan, M.; Taly, A.; Yan, Q. Axiomatic Attribution for Deep Networks. In Proceedings of the International Conference on Machine Learning , Sydney, Australia, 7–9 August 2017; Volume 70, pp. 3319–3328.
69. Bach, S.; Binder, A.; Montavon, G.; Klauschen, F.; Müller, K.R.; Samek, W. On Pixel-Wise Explanations for Non-Linear Classifier Decisions by Layer-Wise Relevance Propagation. *PLoS ONE* **2015**, *10*, e0130140. [CrossRef]
70. Montavon, G.; Lapuschkin, S.; Binder, A.; Samek, W.; Müller, K.R. Explaining nonlinear classification decisions with deep Taylor decomposition. *Pattern Recognit.* **2017**, *65*, 211–222. [CrossRef]
71. Cuadros, J.; Bresnick, G. EyePACS: An Adaptable Telemedicine System for Diabetic Retinopathy Screening. *Sage J. Diabetes Sci. Technol.* **2009**, *3*, 509–516. [CrossRef] [PubMed]
72. Dugas, E.; Jared; Jorge; Cukierski, W. Diabetic Retinopathy Detection. 2015. Available online: https://kaggle.com/competitions/diabetic-retinopathy-detection (accessed on 6 September 2025).

73. Porwal, P.; Pachade, S.; Kamble, R.; Kokare, M.; Deshmukh, G.; Sahasrabuddhe, V.; Meriaudeau, F. Indian Diabetic Retinopathy Image Dataset (IDRiD). *IEEE Dataport* **2018**. [CrossRef]
74. Abràmoff, M.D.; Folk, J.C.; Han, D.P.; Walker, J.D.; Williams, D.F.; Russell, S.R.; Massin, P.; Cochener, B.; Gain, P.; Tang, L.; et al. Automated analysis of retinal images for detection of referable diabetic retinopathy. *JAMA Ophthalmol.* **2013**, *131*, 351–357. [CrossRef] [PubMed]
75. Decencière, E.; Cazuguel, G.; Zhang, X.; Thibault, G.; Klein, J.C.; Meyer, F.; Marcotegui, B.; Quellec, G.; Lamard, M.; Danno, R.; et al. TeleOphta: Machine Learning and Image Processing Methods for Teleophthalmology. *IRBM* **2013**, *34*, 196–203. [CrossRef]
76. Fu, H.; Li, F.; Orlando, J.I.; Bogunović, H.; Sun, X.; Liao, J.; Xu, Y.; Zhang, S.; Zhang, X. PALM: PAthoLogic Myopia Challenge. *IEEE Dataport* **2019**. [CrossRef]
77. Wei, Q.; Li, X.; Yu, W.; Zhang, X.; Zhang, Y.; Hu, B.; Mo, B.; Gong, D.; Chen, N.; Ding, D.; et al. Learn to Segment Retinal Lesions and Beyond. In Proceedings of the International Conference on Pattern Recognition (ICPR), Milan, Italy, 10–15 January 2021; pp. 7403–7410. [CrossRef]
78. Howard, A.G.; Zhu, M.; Chen, B.; Kalenichenko, D.; Wang, W.; Weyand, T.; Andreetto, M.; Adam, H. MobileNets: Efficient Convolutional Neural Networks for Mobile Vision Applications. *CoRR* **2017**, arXiv:1704.04861.
79. Ronneberger, O.; Fischer, P.; Brox, T. U-Net: Convolutional Networks for Biomedical Image Segmentation. In *Medical Image Computing and Computer-Assisted Intervention—MICCAI 2015, Proceedings of the 18th International Conference, Munich, Germany, 5–9 October 2015, Proceedings, Part III*; Springer: Cham, Switzerland, 2015; Volume 9351, pp. 234–241. [CrossRef]
80. Selvaraju, R.R.; Cogswell, M.; Das, A.; Vedantam, R.; Parikh, D.; Batra, D. Grad-CAM: Visual Explanations from Deep Networks via Gradient-Based Localization. In Proceedings of the IEEE International Conference on Computer Vision (ICCV), Venice, Italy, 22–29 October 2017; pp. 618–626. [CrossRef]

life MDPI

Review

Exploring Stem-Cell-Based Therapies for Retinal Regeneration

Madalina Radu [1,*], Daniel Constantin Brănișteanu [2,†], Ruxandra Angela Pirvulescu [3,4,†], Otilia Maria Dumitrescu [1], Mihai Alexandru Ionescu [1] and Mihail Zemba [1,3]

1 Department of Ophthalmology, "Dr. Carol Davila" Central Military Emergency University Hospital, 010825 Bucharest, Romania
2 Department of Ophthalmology, University of Medicine and Pharmacy "Grigore T. Popa", 700115 Iasi, Romania
3 Department of Ophthalmology, University of Medicine and Pharmacy "Carol Davila", 050474 Bucharest, Romania
4 Department of Ophthalmology, University Emergency Hospital, 050098 Bucharest, Romania
* Correspondence: madalina.rd@yahoo.com
† These authors contributed equally to this work.

Abstract: The escalating prevalence of retinal diseases—notably, age-related macular degeneration and hereditary retinal disorders—poses an intimidating challenge to ophthalmic medicine, often culminating in irreversible vision loss. Current treatments are limited and often fail to address the underlying loss of retinal cells. This paper explores the potential of stem-cell-based therapies as a promising avenue for retinal regeneration. We review the latest advancements in stem cell technology, focusing on embryonic stem cells (ESCs), pluripotent stem cells (PSCs), and mesenchymal stem cells (MSCs), and their ability to differentiate into retinal cell types. We discuss the challenges in stem cell transplantation, such as immune rejection, integration into the host retina, and functional recovery. Previous and ongoing clinical trials are examined to highlight the therapeutic efficacy and safety of these novel treatments. Additionally, we address the ethical considerations and regulatory frameworks governing stem cell research. Our analysis suggests that while stem-cell-based therapies offer a groundbreaking approach to treating retinal diseases, further research is needed to ensure long-term safety and to optimize therapeutic outcomes. This review summarizes the clinical evidence of stem cell therapy and current limitations in utilizing stem cells for retinal degeneration, such as age-related macular degeneration, retinitis pigmentosa, and Stargardt's disease.

Keywords: stem cell; stem cell therapy; retinal degenerative diseases; age-related macular degeneration; retinitis pigmentosa; Stargardt's disease; embryonic stem cells; induced pluripotent stem cell; mesenchymal stem cells

1. Introduction

Within the broad spectrum of human sensory capabilities, vision undoubtedly holds a paramount position. The clarity and acuity of vision are closely linked with an individual's quality of life, influencing everything from daily tasks to broader social interactions. Many irreversible retinal diseases are marked by the progressive degeneration of retinal neural cells. Notably, age-related macular degeneration (AMD) persists as a dominant cause of severe visual impairment, especially in industrialized nations, emphasizing the pressing need for efficacious therapeutic strategies. While AMD needs significant attention due to its prevalence, other retinal degenerative disorders, such as retinitis pigmentosa and Stargardt's disease, also inflict considerable visual deficits on affected individuals [1,2].

In the complex landscape of retinal degenerative diseases, the initiation and progression of these disorders involve a nuanced interplay between various cellular components of the retina. Notably, the retinal pigment epithelium (RPE) and photoreceptors constitute a critical partnership, essential for visual function. The RPE, a monolayered epithelium, plays a pivotal role in the survival, integrity, and functionality of photoreceptors by forming an intricate connection with their outer segments. However, the pathogenesis of retinal diseases often triggers a cascade of cellular degeneration, commencing with either the RPE or photoreceptors [3,4].

As these diseases advance, a broader spectrum of retinal components, including ganglion cells and the microvascular network comprising endothelial cells and pericytes, undergoes degeneration. This widespread cellular loss highlights the intricacies of retinal degenerative diseases and underscores the urgency for developing multifaceted therapeutic approaches. This review specifically focuses on macular degenerative diseases, including AMD, retinitis pigmentosa, and Stargardt's disease, acknowledging the unique challenges and therapeutic targets they present [5].

The pursuit of effective treatments for retinal degenerative disorders has led to the exploration of stem cell replacement therapy as a promising avenue. Stem cell therapy aims to rejuvenate the retina by replacing deteriorated cells with healthy ones and/or rescuing retinal neurons from further degeneration, potentially restoring vision [6]. The initial forays into ocular stem cell therapy were directed towards the cornea, benefiting from its accessibility and simpler structure [7]. These efforts established the foundation for future investigations focused on the retina, which is a tissue that is more intricate in structure, has a rich blood supply, and comprises an extensive variety of different types of cells [7].

The field of stem cell therapy for the retina has witnessed significant advancements since those early days. The discovery of human embryonic stem cells (ESCs) [8] in the late 1990s marked a pivotal moment, offering unprecedented possibilities for regenerative medicine. Despite ethical and practical challenges, the subsequent development of induced pluripotent stem cells (iPSCs) [9] in 2006 provided a more versatile and ethically acceptable source of stem cells. These innovations have significantly advanced the application of stem cell therapies in addressing retinal degenerative diseases.

Today, the field of stem cell treatment for retinal disorders includes a range of strategies, from transplanting ESC-derived retinal pigment epithelium cells to utilizing autologous iPSCs for personalized medicine. Recent advances also emphasize allogenic stem cell approaches, which utilize donor-derived cells to provide ready-to-use, standardized treatments. These allogenic therapies offer advantages in scalability and consistency, but they require careful management of immune rejection risks through enhanced immunomodulatory techniques [10]. Clinical trials and research efforts are increasingly focused on optimizing delivery methods, enhancing cell integration and survival, and ensuring long-term safety and efficacy. As the field continues to advance, the integration of stem cell therapy into clinical practice emerges as a significant source of optimism for individuals with retinal degenerative disorders.

2. Materials and Methods

An extensive review of the existing literature was undertaken utilizing the PubMed digital database to locate relevant publications, complemented by a systematic search of clinical trials on ClinicalTrials.gov. The investigation focused on key terms such as "stem cell", "stem cell therapy", "retinal degenerative diseases", "age-related macular degeneration", "retinitis pigmentosa, "Stargardt's disease", "embryonic stem cells", "induced pluripotent stem cell", and "mesenchymal stem cells" to encompass a broad spectrum of research in this field. All pertinent articles published in English were included in this review. Additionally, articles in other languages were considered if they were accompanied by a detailed summary and an English abstract, ensuring a comprehensive understanding of global advancements. The references of all articles were meticulously examined to uncover further significant studies.

3. Stem Cell Sources

Significant progress has been achieved in stem cell research since the isolation of embryonic stem cells from a mouse embryo in 1981 and, subsequently, from a human embryo in 1998. The advent of human embryonic stem cells was a notable milestone, notwithstanding the ethical debates and financial limitations in place in 2001. The emergence of induced pluripotent stem cells in 2006, as pioneered by Japanese scientists, presented a very promising and morally more acceptable alternative that has comparable pluripotency and self-renewal attributes [9].

In addressing the challenges posed by the microenvironment in degenerative retinal diseases, it is vital to consider the inherent properties of the retina and the pathological changes these diseases induce [11,12]. The progression disrupts the delicate architecture essential for cellular survival and integration of transplanted cells. Current stem cell therapy trials are exploring strategies such as cell preconditioning, scaffold use for structural support, and co-delivery of trophic factors to modify the adverse microenvironment [13]. These approaches are essential in overcoming the obstacles that limit the efficacy of regenerative therapies in the retina.

The initiation of the first FDA-approved clinical study with human embryonic stem cells took place in 2009, with a specific emphasis on investigating spinal cord injuries. Subsequently, in 2010, the scope of stem cell research expanded to include the exploration of therapeutic interventions for retinal diseases [14]. The retinal pigment epithelium is pivotal in the pathogenesis of several degenerative retinal disorders, marking it as a significant target for regenerative therapies. Its unique properties, such as operational independence from neuronal synapses and the ability to function within the traditionally immune-privileged environment of the eye, underscore its therapeutic promise [15]. However, the notion of immune privilege is nuanced, particularly in disease states that may compromise this status, thus heightening immune activity within the RPE and subretinal space. This complexity does not diminish the RPE's therapeutic potential but, rather, highlights the need for a sophisticated understanding of its interactions within the ocular immunological environment, especially under pathological conditions [16]. Among this, the exploration of various stem cell types, including human embryonic stem cells, induced pluripotent stem cells, and mesenchymal stem cells, continues to advance, offering hope for regenerative treatments that can address the multifaceted nature of retinal illnesses.

Within the sphere of regenerative medicine, the decision to use autologous (derived from the patient) versus allogeneic (derived from a donor) stem cells presents a nuanced set of advantages and challenges. Autologous stem cells minimize the risk of immune rejection and are considered safer for the patient. However, they might harbor inherent genetic anomalies that could potentially compromise the treatment's effectiveness, a concern particularly pertinent to retinal disorders known for their genetic foundations [17]. In contrast, allogeneic stem cells, sourced from donors, bypass the issue of the patient's genetic defects but carry the risk of immune incompatibility [18]. The advent of induced pluripotent stem cells offers a promising solution by creating patient-tailored stem cells with reduced immune reaction risk, although the reprogramming process raises concerns about possible genetic instabilities [19]. Therefore, the application of stem cells in therapies necessitates meticulous genetic assessment and profiling to ensure the cells' safety and efficacy, striving to balance the therapeutic benefits against genetic risk factors.

3.1. Human Embryonic Stem Cells

Embryonic stem cells (ESCs) are a distinct category of pluripotent cells derived from the inner cell mass of blastocysts. The cells in question have a notable capability for autonomous regeneration and possess the potential to undergo differentiation into several cell lineages seen in mature organisms. This differentiation may occur in any of the three fundamental germ layers, namely, the endoderm, ectoderm, and mesoderm layers. After the blastocyst stage, which typically occurs about 4 to 5 days post-fertilization, the embryonic cells begin the process of differentiation. This intricate process results in the development of distinct cell types that comprise diverse organs, such as the heart and nerve cells. As a result, these cells can no longer be categorized as stem cells. Since their discovery in 1998, hESCs have emerged as a key resource in regenerative medicine, showing promise in their ability to repair and transform into various cell types [8].

Human embryonic stem-cell-based therapy for retinal diseases is being looked into because ESCs have the ability to repair and change into different types of cells. They have shown increased telomerase activity, indicating their potential for prolonged lifespan. Furthermore, it has been shown that these cells exhibit certain markers often associated with undifferentiated cells [8]. According to the theory of pluripotency, these cells can change into any type of cell found in an adult organism. This includes retinal pigment epithelial (RPE) cells when they receive the right signals [20].

The survival of human embryonic stem cell-derived retinal pigment epithelium (hESC-RPE) cells in the subretinal microenvironment is crucial for the therapeutic restoration of vision in patients with retinal degenerative diseases. A significant barrier to the long-term success of these cell therapies is the immunogenicity of transplanted cells, which can lead to the rejection and failure of the implant. Traditional approaches rely heavily on systemic immunosuppression, which poses risks of increased infection and malignancy. Recent advancements in gene editing technologies, particularly CRISPR-Cas9 [21,22], offer promising strategies to mitigate these risks by reducing the immunogenicity of hESC-RPE cells. Gene editing can be employed to modify specific gene sequences responsible for eliciting immune responses. For example, disrupting the beta-2 microglobulin (B2M) gene diminishes the expression of major histocompatibility complex (MHC) class I molecules, thus reducing the visibility of these cells to cytotoxic T cells [23]. Similarly, knockdown of the Class II Major Histocompatibility Complex Transactivator (CIITA) leads to decreased expression of MHC class II genes, essential for antigen presentation to helper T cells [24]. Such modifications have shown potential in preclinical models to enhance the survival of hESC-RPE cells without necessitating prolonged immunosuppression. However, the application of gene editing raises concerns regarding off-target effects and the genetic stability of edited cells. Ongoing research is thus directed at enhancing the precision of gene-editing techniques to ensure the safety and efficacy of these hypoimmunogenic cells in clinical settings [21]. The evolving regulatory landscape will also play a critical role in the clinical translation of these gene-edited cell therapies, ensuring they meet safety standards without compromising therapeutic benefits.

3.2. Induced Pluripotent Stem Cells

Induced pluripotent stem cells (iPSCs) are obtained from adult tissue and were first documented by Takahashi and Yamanaka in 2006 [9]. Derived from adult somatic cells, such as dermal fibroblasts, iPSCs undergo retroviral transduction to express genes associated with pluripotency, known as "reprogramming factors". These factors, including transcription factors Oct 4, Sox 2, cMyc, and Klf4, enable the cells to acquire traits similar to embryonic stem cells [9].

Human-induced pluripotent stem cells (hiPSCs), derived from dermal fibroblasts, share key characteristics with hESCs, including appearance, gene expression, telomerase activity, cell division, and trilineage differentiation potential [25]. Due to the fact that induced pluripotent stem cells may be obtained from the same individual who will receive the iPSC-derived retinal pigment epithelium (iPSC-RPE) transplant, the transplanted tissue is not susceptible to immunological rejection. Consequently, this approach eliminates the need for systemic immunosuppression post-transplantation, thereby eliminating the associated risks. In vitro studies have shown that retinal pigment epithelial cells produced from induced pluripotent stem cells have the ability to partially suppress T cell proliferation and activation, perhaps mediated by the secretion of the soluble molecule transforming growth factor beta (TGFβ) [26].

Nevertheless, the use of induced pluripotent stem cells, while promising, is limited by a narrower range of experience and requires more extensive in vitro modification compared to human embryonic stem cells [27]. Additionally, the increased costs associated with collecting and manipulating autologous cells for transplantation could restrict large-scale production, thus impacting the economic viability of this method [27].

In considering cell-based therapies for retinal diseases, it is imperative to acknowledge the complexity of the retinal structure and the progressive nature of cellular degeneration, as discussed by Zhong et al. [28] and Hallam et al. [29]. These studies underline the inherent challenge in using a single type of cell for therapeutic purposes in conditions like AMD, where sequential death of retinal pigment epithelium (RPE) and photoreceptors occurs. The potential of hiPSCs and hESCs to differentiate into three-dimensional retinal structures that mimic the cellular organization of the human retina offers a promising avenue not only for studying disease mechanisms but also for developing regenerative therapies that address multiple affected cell types simultaneously [10,28]. This approach could potentially overcome the limitations of simpler, single-cell type transplants, aligning with the growing consensus that effective treatment for complex retinal diseases may necessitate the restoration of several layers of retinal cells to fully recuperate function [10].

Induced pluripotent stem cells present two notable advantages in comparison to ESCs. Firstly, iPSCs eliminate the necessity for ESCs, as they are generated through the reprogramming of mature adult somatic cells. Secondly, iPSCs offer the possibility of autologous production, allowing for the theoretical creation of individual-specific iPSC lines [25].

However, this technology is not without its drawbacks. Some of these are the relatively slow process of turning somatic cells into iPSCs, the chance of introducing genetic mutations through the transcription factors used in the process, and the chance that tumors will grow. This latter concern is particularly linked to the activation of oncogenes, both intentionally and unintentionally, by viruses utilized in the genomic modification of cells [30,31]. Recognizing these risks, recent advances have focused on developing non-viral methods for creating iPSCs and iPSC-RPE cells. Techniques such as mRNA transfection [32], which involves the introduction of synthetic mRNA sequences encoding reprogramming factors, offer a safer alternative by eliminating the risk of insertional mutagenesis associated with viral vectors. Additionally, the use of small chemical molecules that can induce reprogramming by altering cell signaling pathways, using the concept of transdifferentiation [33]—also known as direct lineage conversion—has also been explored. These methods not only reduce the potential for oncogene activation but also enhance the efficiency and safety of iPSC and iPSC-RPE generation, providing promising avenues for therapeutic applications without the drawbacks of viral integration [33]. In spite of the progress achieved with iPSCs, the sustained interest in employing embryonic stem cells for research persists, possibly attributable to the significant expense and extensive time required for the development of iPSCs.

3.3. *Mesenchymal Stem Cells*

Mesenchymal stem cells (MSCs), known for their multipotent capability, play a crucial role in regenerative medicine. Originating from stromal compartments of various tissues, including bone marrow, adipose tissue, and umbilical cord blood, MSCs exhibit remarkable plasticity, showing a unique ability to differentiate into diverse cell lineages, including osteoblasts, chondrocytes, and adipocytes. Recent advancements highlight their potential to differentiate into retinal cells, opening new avenues for treating degenerative retinal diseases [34].

The progressive loss of photoreceptors and retinal pigment epithelium cells that characterizes retinal degenerative diseases, such as AMD and RP, leads to irreversible vision impairment. Traditional therapeutic approaches have been predominantly palliative, with a limited capacity for reversing the degenerative process. The therapeutic efficacy of MSCs in retinal regeneration is primarily attributed to their secretion of a diverse array of trophic factors and cytokines, which provide neuroprotective effects and modulate the local retinal environment to support regeneration [35]. Moreover, MSCs are reported to possess anti-apoptotic properties [36] and can integrate into damaged retinal layers, potentially rejuvenating their structure and function. Notably, MSCs also exhibit immunomodulatory properties that may mitigate inflammation [37], a common pathological feature in degenerative retinal diseases. This dual functionality of MSCs—regenerative and immunomodulatory—enhances their potential as a transformative treatment for conditions that have previously been difficult to manage effectively. However, the efficacy of MSCs in treating macular degenerative diseases is significantly limited by the disease's stage at the time of treatment. For example, in cases of advanced macular geographic atrophy, extensive loss of retinal cells may prevent significant recovery, marking a primary limitation of MSC therapy. Furthermore, treatment outcomes with MSCs can vary widely; while some patients may see slight improvements in vision, others may not experience any benefit, underscoring the inconsistent results in late-stage retinal degeneration [36]. This variability underlines the urgent need for further research to refine MSC delivery techniques and determine the optimal timing for intervention, aiming to improve therapeutic outcomes [35].

Bone marrow stromal cells (BMSCs) are found in the bone marrow and represent the highest proportion of adult stem cells. Two distinct types of BMSCs have been identified, mesenchymal stem cells and hematopoietic stem cells, with the latter also known as CD34+ cells. These cells are characterized by their multipotency, which, while more restricted than that of pluripotent stem cells, enables them to differentiate into various cell types [38,39]. Additionally, they exhibit paracrine trophic effects through the secretion of neurotrophic factors or anti-inflammatory modulators. MSCs, constituting less than 0.1% of bone marrow cells, can be efficiently expanded in vitro and are also present in other tissues, including teeth and the liver [34].

BMSCs offer several advantages. They possess the innate ability to migrate towards sites of injury and have the capacity for transdifferentiation, meaning they can adapt to differentiate into cells of different organs under specific environmental conditions [40]. In the context of retinal damage, such as that affecting RPE cells, they respond to chemo-attractive cytokines/chemokines released by the injured tissue, facilitating their migration to the site of injury. Once present, they have the potential to differentiate into retinal cells, including RPE cells and photoreceptors, aiding in tissue repair. Furthermore, BMSCs can produce neurotrophic factors that support cell survival and exert anti-inflammatory effects. An additional benefit of CD34+ cells is their ease of extraction from patients, coupled with minimal manipulation requirements, making them suitable for autologous transplantation procedures [38,40].

The heterogeneity of mesenchymal stromal cells, influenced by their tissue source, presents significant challenges in their clinical application and manufacturing consistency. MSCs derived from bone marrow, adipose tissue, and umbilical cord blood, for instance, show distinct properties that affect their therapeutic potential [36]. Different sources of MSCs exhibit unique profiles impacting their therapeutic applications [36]. Given these differences, the manufacturing process, particularly under current good manufacturing practices (cGMP), must adapt to maintain the unique characteristics of MSCs from each source [41]. This involves stringent control over the entire manufacturing process, from cell isolation through to expansion and storage, to ensure product safety, potency, and efficacy.

The integration of automated and closed-system bioreactors into MSC manufacturing is transforming production by standardizing procedures and significantly enhancing consistency across different production runs. Advanced automation technologies enable precise control of culture conditions—such as pH, temperature, and nutrient supply—enhancing cell quality and yield [41,42]. These systems also facilitate scalable production platforms that can accommodate the high throughput demands of clinical application, ensuring that consistent MSC products are manufactured across different batches. Such developments not only streamline the manufacturing process but also minimize variations introduced by manual handling, thereby enhancing the consistency of MSC batches irrespective of their source.

4. Clinical Studies

Clinical trials of stem cell therapy in degenerative macular diseases represent a cutting-edge frontier in ophthalmological research, focusing on retinal regeneration. Stem cells have the unique ability to grow new cells. This means that these trials could lead to new treatments for conditions like age-related macular degeneration and hereditary retinal dystrophies [43,44].

The core principle of these trials is the transplantation of stem cells into the retinal space, aiming to replace or repair the damaged retinal pigment epithelium and photoreceptors [6]. Various stem cell types—including embryonic stem cells, induced pluripotent stem cells, and adult stem cells—have been explored, each with their advantages and challenges [6].

Currently, the majority of conducted human clinical studies (Table 1) fall into the initial phase, focusing on safety assessment and lacking the statistical power necessary to discern functional outcomes. These studies have consistently employed the eye with poorer vision as the subject, using the other eye as a comparative but less-than-ideal control [45]. In the context of retinal pigment epithelium cells, their distinct pigmentation facilitates identification through ophthalmoscopy. Nonetheless, similar pigmentation is also observed in macrophages that are involved in the engulfment of deteriorating RPE cells [44].

Table 1. Summary table of the current human clinical trials.

Number	Disease	Cell Type	Phase	No. of Patients	Administration Method	Status
NCT03305029	AMD (GA)	hESC-RPE (SCNT-hES-RPE)	I	3	N/A	Unknown
NCT02903576	AMD; SMD	hESC-RPE	I/II	15	suspension vs. scaffold	Completed
NCT03046407	AMD (GA)	hESC-RPE	I	10	N/A	Unknown
NCT02755428	AMD	hESC-RPE (MA09-hRPE)	I	10	N/A	Unknown
NCT02286089	AMD (GA)	hESC-RPE	I/IIa	24	suspension	Ongoing
NCT01344993	AMD	hESC-RPE (MA09-hRPE)	I/II	13	suspension	Completed
NCT02463344	AMD	hESC-RPE	I/II	11	subretinal injection	Completed
NCT01674829	AMD	hESC-RPE	I/IIa	10	N/A	Completed
NCT01691261	AMD	hESC-RPE	I	10	N/A	Recruiting
NCT03102138	AMD	hESC-RPE	Obs.	10	N/A	Ongoing
NCT02590692	AMD (GA)	hESC-RPE	I/IIa	16	scaffold (parylene membrane)	Unknown
NCT02749734	AMD; SMD	hESC-RPE	I/II	15	N/A	Unknown
NCT01345006	SMD	hESC-RPE	I/II	13	subretinal injection	Completed
NCT01469832	SMD	hESC-RPE	I/II	12	N/A	Completed
NCT01625559	SMD	hESC-RPE	I	3	N/A	Completed
NCT02941991	SMD	hESC-RPE	Obs.	12	subretinal injection	Completed
NCT02445612	SMD	hESC-RPE	Obs.	13	subretinal injection	Completed
NCT03944239	RP	hESC-RPE	I	10	N/A	Unknown
NCT03963154	RP	hESC-RPE	I/II	7	N/A	Ongoing
NCT05991986	AMD	iPSC	Obs.	10	N/A	Ongoing
NCT04339764	AMD (GA)	iPSC-RPE	I/II	20	scaffold	Recruiting

Table 1. *Cont.*

Number	Disease	Cell Type	Phase	No. of Patients	Administration Method	Status
NCT02464956	AMD	iPSC-RPE	Obs.	3	N/A	Completed
NCT05445063	AMD (GA)	iPSC-RPE	I	10	N/A	Recruiting
NCT03372746	AMD	iPSC	Obs.	187	N/A	Completed
NCT02016508	AMD	hBM-MSC	I/II	1	intravitreal injection	Unknown
NCT05712148	RP	MSC	I/II	15	suprachoroidal implantation	Completed
NCT05786287	RP	UC-MSC	Obs.	18	N/A	Ongoing
NCT04315025	RP	UC-MSC	I/II	18	suspension (peribulbar injection)	Completed
NCT01531348	RP	hBM-MSC	I	14	subretinal injection	Completed
NCT04763369	RP	UC-MSC	II	50	sub-tenon space injection	Unknown
NCT01736059	RP; AMD	BM-CD34+	I	15	intravitreal injection	Ongoing

Abbreviation: AMD, age-related macular degeneration; GA, geographic atrophy; SMD, Stargardt's macular dystrophy; RP, retinitis pigmentosa; hESC, human embryonic stem cell; iPSC, induced pluripotent stem cell; MSC, mesenchymal stem cell; RPE, retinal pigment epithelium; hBM, human bone marrow; UC, umbilical cord; Obs, Observational; N/A, Not Applicable.

It has been established through preclinical research that the detection of viable transplanted RPE cells is most reliably accomplished using immunohistochemistry [46]. As a result, in addition to documenting new pigmentation zones, there has been an increased reliance on diverse imaging modalities. Optical coherence tomography (OCT) has proven effective in revealing structural alterations within the subretinal space and the photoreceptor outer segments [47]. Fundus autofluorescence depends on lipofuscin, which is a byproduct of normal retinal processes and is missing in atrophied areas. This makes it hard to figure out what is going on after a transplant [47]. Various forms of electroretinogram (ERG), including multifocal and full-field, have been deployed to assess retinal function. Techniques such as visual field testing and microperimetry are instrumental in mapping scotomas resulting from atrophy. While tests for visual acuity and reading speed are feasible, their utility is reduced in studies focusing on patients with severely compromised baseline vision. There is a growing need to refine visual function testing methods, particularly for patients with low vision, to evaluate the efficacy of these treatments more effectively in cases of advanced-stage diseases [48].

4.1. Clinical Trials Using hESCs

Subsequent to the encouraging outcomes from preclinical investigations, the United States Food and Drug Administration sanctioned the commencement of Phase I/II clinical trials in 2010, focusing on stem cell therapies for retinal pathologies in human subjects. These trials employed retinal pigment epithelium cells derived from human embryonic stem cells. In 2012, Schwartz and colleagues (NCT01345006; NCT01344993) disseminated the inaugural findings of this pivotal study [49,50]. The procedure entailed the administration of hESC-RPE cells into the subretinal space, specifically targeting a pericentral region, accomplished via pars plana vitrectomy. This intervention was applied to individuals diagnosed with age-related macular degeneration and Stargardt's disease (SD). The initial report, covering a follow-up period of four months, highlighted one patient with AMD and another with SD. Notably, in this duration, neither patient exhibited adverse effects such as tumorigenic proliferation, the development of ectopic tissues, nor signs of graft rejection, thereby indicating a preliminary safety profile of the procedure [49]. Subsequently, Schwartz provided an elucidation of the outcomes following a 22-month follow-up period involving 18 patients, comprising nine individuals with AMD and an equal number with SD. During this period, an improvement in the Best Corrected Visual Acuity was observed in ten cases, whereas it remained stable in seven instances.

However, there was a notable deterioration exceeding 10 letters in visual acuity in one case. In approximately 72% of the instances, there was an observable increase in pigmentation at the periphery of macular atrophy, aligning with the locations of the retinal pigment epithelium transplants. Adverse effects recorded in the course of the study included the development of endophthalmitis in one patient, cataract formation in four cases, and complications associated with the immunosuppressive treatment [50].

Participants enrolled in the study necessitated immunosuppressive therapy, entailing a regimen of low-dose tacrolimus, with targeted blood concentrations between 3 and 7 ng/mL, and mycophenolate mofetil, administered orally in doses varying from 0.25 to 2.00 g daily. This immunosuppressive protocol was initiated a week preceding the surgical intervention and sustained for a duration of 12 weeks post-therapy [50]. The rationale for this approach stemmed from the heterologous origin of the transplanted cells. Notwithstanding the implementation of immunosuppression, the patients demonstrated a notable enhancement in visual acuity.

Concurrently, Song et al. [51] (NCT01674829) conducted a one-year post-transplantation follow-up of four patients, comprising two with dry age-related macular degeneration and two with Stargardt's disease, who underwent human embryonic stem cell-derived retinal pigment epithelium transplantation. Their observations revealed an absence of adverse proliferation or tumorigenicity. Echoing the findings of Schwartz et al. [50], this study noted an enhancement in visual acuity in three of the patients, while one patient exhibited stable vision.

Mehat et al. [52] (NCT01469832) conducted a comprehensive investigation to assess the safety of human embryonic stem cell-derived retinal pigment epithelium in the context of Stargardt's Macular Dystrophy. This study encompassed 12 patients, each of whom received subretinal injections of hESC-RPE cells, with the cell count ranging from 50,000 to 200,000. Accompanying the procedure, all patients were administered a combination of Tacrolimus and Mycophenolate Mofetil (MMF) as immunosuppressants, with no reported complications stemming from their use [52]. Surgical challenges were noted in four patients, including instances of retinal dialysis, subretinal hemorrhage, and vitreous hemorrhage. Notably, there were no indications of immune rejection or undesirable proliferation of the transplanted RPE cells. Observation of subretinal pigmentation was uniform across all study participants, and optical coherence tomography imaging revealed a hyper-reflective layer congruent with the presence of RPE cells. The study also elucidated a dose-dependent relationship between the quantity of injected cells and the extent of pigmentation observed.

In terms of visual acuity, electroretinography testing, and microperimetry sensitivity, there was no significant change detected across the patient cohort. This lack of notable improvement or decline in visual functions was hypothesized to be associated with the advanced stage of the disease present in all study participants.

In the field of regenerative medicine, significant research has been directed toward evaluating the safety and tolerability of various scaffolds for human embryonic stem cell-derived retinal pigment epithelium cells. One pioneering approach involved the transplantation of hESC-RPE cells cultured on a synthetic parylene substrate, known as CPCB-RPE1, designed to emulate the characteristics of Bruch's membrane (NCT02590692) [53]. This novel intervention did not raise any safety concerns in early assessments.

In the majority of subjects (four out of five) receiving this transplant, OCT images revealed alterations indicative of integration between the hESC-RPE cells and the host's photoreceptors. Notably, none of the eyes that underwent this procedure exhibited a progression in vision loss. Furthermore, one eye recorded a significant improvement, manifesting as a 17-letter increase in visual acuity, while two other eyes showed enhanced fixation capabilities. The researchers posited that the observed structural and functional enhancements might indicate that CPCB-RPE1 has the potential to ameliorate visual function, at least in the short term. This potential benefit was particularly noted in patients suffering from severe vision impairment due to advanced dry AMD [53]. After one year of follow-up, the implant was proven to be safe and well-tolerated by participants with advanced dry AMD [54].

Furthermore, in a more recent longitudinal study of five years, Li et al. [55] thoroughly investigated the long-term safety and tolerability of subretinal transplantation using human embryonic stem cell-derived retinal pigment epithelium in patients diagnosed with Stargardt's macular dystrophy. Their findings provided substantial evidence supporting the sustained safety and tolerability of this innovative therapeutic approach in treating SMD over an extended period of time [55].

Human embryonic stem cell-derived retinal pigment epithelium cells have been demonstrated to be viable for plating onto a vitronectin-coated polyester membrane, a technique facilitating their transplantation into the subretinal space. In an initial assessment of this method [56], researchers observed its efficacy and safety when applied to two patients with severe wet age-related macular degeneration, monitored over a 12-month period (NCT01691261). The immunosuppressive regime in this case involved the use of fluocinolone. The study noted improvements in both patients in terms of best corrected visual acuity, microperimetry, and reading speed. However, alongside these positive outcomes, the procedure was not without significant complications. These included exposure to the fluocinolone suture, instances of retinal detachment, and an exacerbation of diabetes, attributed to the use of systemic corticoids [56]. This juxtaposition of favorable visual outcomes and serious adverse events highlights the complex balance between therapeutic benefits and potential risks in advanced ocular interventions.

4.2. Clinical Trials Using hiPSCs

Following promising outcomes in preclinical studies, human clinical trials were strategically designed and initiated. In 2014, the RIKEN research institute in Japan embarked on a pioneering human clinical trial employing autologous induced pluripotent stem cells to treat a patient with neovascular age-related macular degeneration [57]. The unique aspect of this trial was the autologous nature of the cell transplantation, which eliminated the need for a scaffold and systemic immunosuppression. However, the trial faced an interruption due to the implementation of a new regulatory framework for regenerative medicine in Japan in 2014, despite the patient not experiencing any serious adverse effects [57]. Over a year of follow-up, the patient's visual acuity remained stable, with no noted improvement. Concerns regarding tumor formation were not realized during the trial. The transplantation of a second patient was not pursued due to the detection of genetic discrepancies, specifically single-nucleotide variations and copy number variants in the hiPSCs, which were absent in the patient's original somatic cells [30,57].

Subsequently, Mandai et al. [58] (UMIN 000011929) demonstrated the feasibility of transplanting a sheet of autologous RPE cells derived from hiPSCs, sourced from skin fibroblasts, into a patient with wet AMD. Over a 25-month observation period, no adverse events were reported, nor was there any improvement in visual acuity [58].

In a distinct clinical trial, conducted by Sugita et al. [59] (UMIN 000026003), five patients diagnosed with neovascular age-related macular degeneration were enrolled. Induced pluripotent stem cells utilized in this study were derived from a donor with a homozygous human leukocyte antigen (HLA) match. Following pars plana vitrectomy, a suspension of iPSC-derived retinal pigment epithelium cells was administered subretinally. The immunosuppressive regimen was limited to the administration of sub-tenon's triamcinolone. Over a 52-week monitoring period, no adverse events were reported among the patients. However, all participants developed epiretinal membranes; upon examination, these membranes were found to contain pigmented cells positive for RPE markers. Increased subretinal pigmentation was observed in all subjects, yet in most cases, the pigment deposition was not predominantly located in the macula, likely due to a less than ideal injection location or technique. Sugita and colleagues further acknowledged the complication of graft cells backflowing into the vitreous, highlighting this as an area requiring additional investigation [59].

4.3. Clinical Trials Using MSCs

A distinct strategy in stem cell therapy involves the facilitation of functional recovery in the retina's compromised cells via the introduction of stem cells that exert a paracrine trophic effect. This method, achievable through the employment of mesenchymal stem cells, is not confined to a particular disease, thereby offering a wide range of clinical applications [6].

The encouraging outcomes from experimental studies have paved the way for the initiation of clinical trials. In a prospective phase I study [60], a singular dose of intravitreally administered autologous bone-marrow-derived mesenchymal stem cells was given to three patients with retinitis pigmentosa and two with cone–rod dystrophy. Over a follow-up period of ten months, no significant structural or functional toxic effects were observed in the retina. In this research, conducted by Siqueira et al. [60], four patients with an advanced stage of the disease demonstrated an improvement of one row in best corrected visual acuity one-week post-injection, a benefit that persisted throughout the follow-up period [60].

In a subsequent extension of this study, intravitreal MSCs were administered to 20 patients, who were then monitored for a year. The researchers noted a statistically significant elevation in the patient's vision-related quality of life scores at the three-month mark. However, by the 12-month evaluation, these scores had reverted to their initial levels, suggesting that the observed improvements were transient [61].

In a separate investigation conducted by Park et al. (NCT01736059) [62], a total of 3.4 million bone-marrow-derived mesenchymal stem cells were intravitreally administered into six eyes suffering from irreversible vision loss due to various conditions, including retinal vascular diseases, hereditary or non-exudative age-related macular degeneration, and retinitis pigmentosa. This therapeutic approach was found to be well-tolerated, with no incidents of intraocular inflammation or proliferation observed. Additionally, there were no declines in electroretinography and best corrected visual acuity results over a six-month follow-up period [62].

With the increasing utilization of mesenchymal stem cells in treatments, there has been a concurrent rise in reported ocular complications associated with this therapy, such as elevated intraocular pressure, hemorrhagic retinopathy, and vitreous hemorrhage [63]. In one particular study [64], the application of autologous bone-marrow-derived MSCs resulted in enhanced visual acuity in two out of three patients with advanced retinitis pigmentosa. However, complications arose in the third patient from the second week post-treatment, including the development of preretinal and vitreal fibrous tissue, shallowing of the anterior chamber, and the formation of a cyclitic membrane, leading to ocular hypotonia. This patient experienced total tractional retinal detachment and a consequent complete loss of vision within three months [64].

In contrast, the suprachoroidal approach proposed by Limoli et al. [65] might mitigate the vitreoretinal complications observed in intravitreal and subretinal MSC applications. In their study, no complications were reported, and visual function was improved in 36 eyes of 25 patients with dry age-related macular degeneration. This was achieved six months after MSCs were administered under a deep scleral flap into the suprachoroidal space, highlighting the potential benefits and reduced risks of this technique [65].

Beyond the direct application of MSCs, recent advancements have highlighted the potential of MSC-derived exosomes and vesicles in retinal therapies [66,67]. MSC-exosomes, which are extracellular vesicles released by MSCs, encapsulate a variety of bioactive molecules including proteins, mRNAs, and microRNAs that can modulate inflammation, angiogenesis, and cellular repair processes [68]. These vesicles harness the paracrine effects of MSCs, potentially offering a cell-free option for treating retinal diseases. For instance, studies have shown that MSC-derived exosomes can promote neuroprotection and angiogenic responses in degenerative conditions of the retina, enhancing retinal cell survival under stress conditions [69] (Table 2). This approach could mitigate some risks associated with direct stem cell transplantation, such as cell proliferation and immune rejection, while still delivering therapeutic benefits.

Table 2. Summary of published clinical study results on stem cell therapies.

Disease	Stem Cell Type	Clinical Trials and Phases	Key Findings and Outcomes	Challenges and Considerations
Age-Related Macular Degeneration (AMD)	ESCs, iPSCs, MSCs	NCT01345006 (Phase I/II), NCT01344993 (Phase I/II), [49,50] NCT01691261 (Phase I) [56]	Trials indicate safety and efficacy of ESCs and iPSCs in replacing damaged RPE cells. Some improvements in visual acuity noted.	Managing immune rejection, ensuring integration and long-term safety, ethical concerns with ESCs.
Retinitis Pigmentosa	ESCs, iPSCs, MSCs	NCT01531348 (Phase I) [70], NCT01736059 (Phase I) [62]	Stem cell therapies shown to slow disease progression with potential restoration of some visual functions. MSCs highlighted for their neuroprotective effects.	Genetic stability of iPSCs, ethical considerations, technical delivery challenges.
Stargardt's Disease	ESCs, iPSCs, MSCs	NCT01345006 (Phase I/II) [49,50], NCT01469832 (Phase I/II) [52]	iPSC trials show potential in restoring visual function. Positive safety profiles and functional improvements in early results	Addressing immune rejection, long-term viability of transplanted cells, ethical and technical challenges.

5. Stem Cell Administration Method

Contemporary approaches for stem cell delivery to the ocular region encompass intravitreal, subretinal, and suprachoroidal injections, each with distinct advantages and challenges. Intravitreal injection, a prevalent and relatively straightforward procedure, is extensively utilized for treating retinal diseases, such as exudative AMD. Nevertheless, the integrity of the blood–retinal barrier poses limitations on the effective transport of transplanted stem cells and their secreted neurotrophic factors [71]. Additionally, there is a risk of the therapeutic agents diffusing to non-target areas, potentially inducing fibrous tissue proliferation, retinal detachment, and epiretinal membrane formation [72]. Despite some clinical studies affirming the general safety of stem cell therapy for retinitis pigmentosa patients via this method, it necessitates meticulous consideration prior to application [44].

Conversely, subretinal injection targets the potential space between the retinal pigment epithelium and photoreceptors, providing a more direct approach to the retina. This method, however, entails a pars plana vitrectomy, introducing the risks of RD and associated complications. The successful application of human embryonic stem cell-derived RPE in the subretinal space attests to its relative safety when executed with precision [73]. The delivery of cell suspensions and cells adhered to scaffolds represent two distinct techniques within subretinal injections. While the former is less invasive, the latter, despite necessitating a larger retinotomy for cell delivery, may be secured in place using intraoperative devices, mitigating the risk of postoperative complications such as cell migration, trans-differentiation, and uncontrolled proliferation [74].

The suprachoroidal space (SCS) introduces a novel and less-invasive administration route, accurately targeting the choroid, RPE, and neuroretina, and ensuring high bioavailability [75]. Limoli et al. pioneered the suprachoroidal implantation of stem cells, highlighting its safety profile with no reported ocular adverse events in contrast to the other methods. Furthermore, the SCS facilitates the sustained release of stem cell-derived growth factors, promoting constant secretion to the choroid and retina, which is advantageous for patients requiring multiple cell suspension injections [76,77].

Overall, while each method presents its own set of advantages and potential risks, careful consideration and precision in execution are paramount to optimizing therapeutic outcomes and minimizing complications in stem-cell-based interventions for retinal diseases.

6. Cell Suspension or Reconstructed Tissue

The design of a treatment plan remains a critical factor, particularly when dealing with pathologies that include alterations to the Bruch's membrane. Initially, early methodologies relied on the use of cell suspension [10]. This approach involves delivering stem cells directly into the affected retinal area in a fluid medium, allowing for a diffuse distribution across the retina. The primary mechanism of action for stem cells administered in this manner hinges on their ability to secrete paracrine factors that can modulate the local environment, fostering repair and regeneration. These factors include a range of cytokines and growth factors that promote cell survival, reduce inflammation, and stimulate the resident retinal cells towards repair processes [10,78]. However, science in this field is currently moving towards the use of more sophisticated tissue formation techniques. In order to facilitate the transfer of a preformed epithelium, it is necessary to use a supporting matrix that enables the safe removal of the sheet from the culture plate and subsequent loading into the transplantation device, and to maintain the integrity of the polarized RPE monolayer [13]. This technique, known as the scaffold-based approach, involves attaching stem-cell-derived retinal pigment epithelium cells to a biocompatible scaffold, which not only serves as a structural support but also promotes cell adhesion, proliferation, and differentiation in a controlled manner. Therefore, several characteristics are necessary to facilitate the functioning and viability of human pluripotent stem cell-derived retinal pigment epithelium: thickness, mechanical features such as flexibility and ease of manipulation, permeability, and potential for biodegradation [79]. Various types of scaffolds have been used for different purposes, including the utilization of synthetic polymers, biological materials like Descemet's membrane or human amniotic membranes, or even the absence of support altogether. In the latter case, retinal pigment epithelial (RPE) cells are permitted to form an epithelium on a collagen covering, which is then broken down to release the sheet [80].

Furthermore, scaffolds have the potential to not only serve as a mechanical support system for transplanted cells but also provide trophic support via the inclusion of substances that enhance cell survival and development. Notwithstanding these benefits, other factors to be taken into account for the effective utilization of scaffolds in the subretinal region include the need for a slender layer measuring between 5 and 90 microns, as well as the possibility of an inflammatory reaction to the implanted substance [81].

Table 3 encapsulates a summary of the materials utilized as scaffolds for retinal pigment epithelium cells. Several research groups are exploring a variety of materials for this purpose, including parylene and various forms of polyester, such as polyethylene terephthalate, lactic-co-glycolic acid, polycaprolactone, poly-L-lactic acid, and vitronectin-coated polyester membranes, to fabricate scaffolds for RPE sheets [56,82]. Concurrently, there is an ongoing evaluation of temporary, biodegradable scaffolds. These are intended to aid in the attachment of RPE cells to the native Bruch's membrane while concurrently minimizing sustained inflammation [83,84]. In addition, other research factions investigating the use of different substrates, such as amniotic membrane [85] and lenticules derived from femtosecond laser intrastromal lenticule extraction [86]. There is also interest in the creation of scaffold-free sheets using innovative materials like peptide-modified alginate hydrogels [87,88].

Table 3. Materials used as scaffolds for RPE sheets.

Types of Materials
Parylene [82]
Polyethylene terephthalate [79]
Lactic-co-glycolic acid [79]
Polycaprolactone [79]
Poly-L-lactic acid [79]
Vitronectin-coated polyester membrane [56]
Amniotic membrane [85]
Femtosecond-derived lenticule [86]

Increasingly, cell-derived vesicles, particularly mesenchymal stem cell-derived exosomes [66], are being recognized for their potential in retinal therapies due to their ability to encapsulate and deliver a range of therapeutic molecules. These vesicles are nanosized extracellular vesicles that transport proteins, lipids, and nucleic acids, capable of influencing cell behavior and tissue repair without the complexities and risks associated with whole-cell therapies [66,89]. MSC-derived exosomes have been shown to play a significant role in modulating inflammation, preventing apoptosis, and enhancing angiogenesis, which are crucial for the repair of damaged retinal tissues. Scientific investigations have demonstrated that when these exosomes are introduced into the retinal environment, either independently or in conjunction with scaffold-based systems, they can significantly enhance the therapeutic efficacy by providing localized delivery of growth factors and cytokines directly to the damaged cells [89]. This cell-free approach not only mitigates the risks of cell transplantation, such as immune rejection and tumor formation, but also offers a controlled and sustained release of bioactive factors, potentially improving the integration and functionality of transplanted cells or supporting intrinsic repair mechanisms. The incorporation of MSC-derived vesicles into scaffold systems could offer a dual mechanism of action: structural support from the scaffold and bioactive molecular delivery through the vesicles, thereby enhancing the regenerative capacity of the treatment strategy [89,90].

7. Ethical and Safety Issues of Stem-Cell-Based Therapy

The exploration of stem-cell-based therapies for retinal regeneration, particularly involving human embryonic stem cells and induced pluripotent stem cells, is surrounded by complex ethical and safety considerations. The use of hESCs has historically sparked considerable ethical debate due to their derivation methods, which require the destruction of human embryos. This issue has led to significant ethical concerns and diverse political and policy responses across different countries [91–93]. In some regions, this has led to strict regulations or outright bans on hESC research, while in others, it has been politically charged, used in broader debates over human rights and medical ethics. Furthermore, the advent of iPSC technology, which involves reprogramming adult cells to an embryonic-like state, was initially perceived as a less ethically fraught alternative. However, iPSCs also harbor the potential for unlimited differentiation and could theoretically be used for human cloning, raising new ethical dilemmas concerning identity and the potential creation of human embryos for research purposes [92].

A significant safety issue with iPSC transplantation and the use of iPSC-derived cells is the possibility of unintended differentiation and malignant transformation. To address this, there is a crucial need to refine and optimize protocols for iPSC differentiation. Ensuring the purity of iPSC-derived differentiated cell populations is essential before they can be deemed safe for clinical application. This is vital to prevent any oncogenic potential these cells might harbor, which could have detrimental effects on patients [93].

Mesenchymal stem cells have also emerged as a popular choice in stem cell therapy, often touted as a universal remedy in various medical treatments across the globe. However, their safety profile is not fully understood, particularly concerning their potential to promote tumor growth and metastasis. Therefore, research employing MSCs must prioritize continuous monitoring and extensive long-term follow-up, especially in animal models. This vigilance is necessary to uncover any pro-tumorigenic or other ad-verse effects of MSC-based therapy. Such comprehensive evaluations are essential to establish a robust understanding of the implications of stem cell therapies and to ensure that their therapeutic benefits do not come at the cost of patient safety [60,93].

While the field of stem cell therapy for retinal regeneration is burgeoning with potential, it navigates a complex landscape of ethical dilemmas and safety challenges. Another ethical and safety concern is raised by the use of stem cells in treating degenerative retinal diseases in clinics not officially approved for such therapies. While the potential of stem cell therapy offers hope for conditions with limited treatment options, administering these treatments outside of approved clinical trials or recognized hospitals can jeopardize patient safety. Without rigorous regulatory oversight, patients may be exposed to unproven interventions lacking evidence of efficacy or safety, risking possible adverse effects without the guarantee of therapeutic benefit [94]. It underscores the necessity of adhering to established clinical guidelines and regulatory approvals, ensuring that stem cell therapies are both safe and effective before applying them in a broader clinical setting. Addressing these concerns requires a balanced approach, ensuring that these innovative therapies are both ethically sound and safe for clinical use [94].

8. Conclusions

Over the last two to three decades, a substantial body of research has convincingly shown that transplantation of the retinal pigment epithelium (RPE) can at least partially restore retinal structure, function, and subjective visual perception in individuals with retinal degenerative diseases. Breakthroughs in foundational sciences and translational research domains, including stem cell biology, retinal surgery, non-invasive retinal imaging, retinal physiology, and vision science, have brought the field to the brink of human clinical trials capable of offering therapies to restore vision. A number of early-phase human clinical trials are currently underway globally, and their outcomes are anticipated to be groundbreaking. However, ongoing research and collaboration among funding bodies, academic institutions, and industry partners are imperative to ensure successful outcomes.

Primary concerns in human clinical trials related to RPE cell layer transplantation involve several key aspects: ensuring the longevity of the donor RPE in the host to justify the risks associated with implantation and cell-based therapy; maintaining the polarity and functional integrity of the donor RPE akin to normal RPE cells; preventing further degeneration of the donor RPE cells that may be associated with the disease process; and determining the most effective technique for delivering the RPE cells into the subretinal space. It is anticipated that the numerous concerns presently raised will find their resolutions in the forthcoming period.

Moreover, as the field advances, the importance of rigorous patient selection, ethical considerations, and long-term follow-up cannot be overstated. Technological innovations and improvements in surgical techniques promise to refine the delivery and integration of RPE cells, potentially enhancing treatment outcomes. Concurrently, ethical practices and thorough patient education will remain paramount to navigate the complex landscape of stem cell therapy with transparency and integrity. Ultimately, the success of these pioneering therapies will depend on sustained collaborative efforts, ensuring that the vision of restoring sight through stem cell therapy becomes a safe and accessible reality for those in need.

Author Contributions: Conceptualization, M.R., M.Z., D.C.B. and R.A.P.; methodology, M.Z. and M.R.; software, M.R.; validation, M.Z., D.C.B. and R.A.P.; formal analysis, M.Z.; investigation, M.R., O.M.D. and M.A.I.; writing—original draft preparation M.R., O.M.D. and M.A.I.; writing—review and editing, M.Z., D.C.B., R.A.P. and M.R.; visualization, M.R. and D.C.B.; supervision, M.Z. and R.A.P.; project administration, M.Z., M.R., O.M.D. and M.A.I. All authors have read and agreed to the published version of the manuscript.

Funding: This research received no external funding.

Institutional Review Board Statement: Not applicable.

Informed Consent Statement: Not applicable.

Data Availability Statement: Not applicable.

Conflicts of Interest: The authors declare no conflicts of interest.

References

1. Verbakel, S.K.; van Huet, R.A.C.; Boon, C.J.F.; den Hollander, A.I.; Collin, R.W.J.; Klaver, C.C.W.; Hoyng, C.B.; Roepman, R.; Klevering, B.J. Non-Syndromic Retinitis Pigmentosa. *Prog. Retin. Eye Res.* **2018**, *66*, 157–186. [CrossRef] [PubMed]
2. Voisin, A.; Gaillard, A.; Balbous, A.; Leveziel, N. Proteins Associated with Phagocytosis Alteration in Retinal Pigment Epithelial Cells Derived from Age-Related Macular Degeneration Patients. *Antioxidants* **2022**, *11*, 713. [CrossRef]
3. Yang, S.; Zhou, J.; Li, D. Functions and Diseases of the Retinal Pigment Epithelium. *Front. Pharmacol.* **2021**, *12*, 727870. [CrossRef] [PubMed]
4. Lyle, W.M. The Retinal Pigment Epithelium: Function and Disease. *Optom. Vis. Sci.* **1999**, *76*, 193. [CrossRef]
5. Somasundaran, S.; Constable, I.J.; Mellough, C.B.; Carvalho, L.S. Retinal Pigment Epithelium and Age-Related Macular Degeneration: A Review of Major Disease Mechanisms. *Clin. Exp. Ophthalmol.* **2020**, *48*, 1043–1056. [CrossRef] [PubMed]
6. Schwartz, S.D.; Pan, C.K.; Klimanskaya, I.; Lanza, R. Retinal Degeneration. In *Principles of Tissue Engineering*, 4th ed.; Elsevier Inc.: Amsterdam, The Netherlands, 2013; pp. 1427–1440, ISBN 9780123983589.
7. Pellegrini, G.; De Luca, M.; Arsenijevic, Y. Towards Therapeutic Application of Ocular Stem Cells. *Semin. Cell Dev. Biol.* **2007**, *18*, 805–818. [CrossRef]
8. Thomson, J.A. Embryonic Stem Cell Lines Derived from Human Blastocysts. *Science* **1998**, *282*, 1145–1147. [CrossRef] [PubMed]
9. Takahashi, K.; Yamanaka, S. Induction of Pluripotent Stem Cells from Mouse Embryonic and Adult Fibroblast Cultures by Defined Factors. *Cell* **2006**, *126*, 663–676. [CrossRef] [PubMed]
10. Voisin, A.; Pénaguin, A.; Gaillard, A.; Leveziel, N. Stem Cell Therapy in Retinal Diseases. *Neural Regen. Res.* **2023**, *18*, 1478–1485. [CrossRef]
11. Ikelle, L.; Al-Ubaidi, M.R.; Naash, M.I. Pluripotent Stem Cells for the Treatment of Retinal Degeneration: Current Strategies and Future Directions. *Front. Cell Dev. Biol.* **2020**, *8*, 743. [CrossRef]
12. Huo, D.M.; Dong, F.T.; Gao, F. Differentiation of Mesenchymal Stem Cell in the Microenviroment of Retinitis Pigmentosa. *Int. J. Ophthalmol.* **2010**, *3*, 216–219. [CrossRef]
13. Liu, H.; Jing, L.; Sun, J.; Huang, D. An Overview of Scaffolds for Retinal Pigment Epithelium Research. *Procedia Manuf.* **2021**, *53*, 492–499. [CrossRef]
14. Shintani, K.; Shechtman, D.L.; Gurwood, A.S. Review and Update: Current Treatment Trends for Patients with Retinitis Pigmentosa. *Optometry* **2009**, *80*, 384–401. [CrossRef]
15. Wang, X.; Wang, T.; Lam, E.; Alvarez, D.; Sun, Y. Ocular Vascular Diseases: From Retinal Immune Privilege to Inflammation. *Int. J. Mol. Sci.* **2023**, *24*, 12090. [CrossRef]
16. Du, Y.; Yan, B. Ocular Immune Privilege and Retinal Pigment Epithelial Cells. *J. Leukoc. Biol.* **2023**, *113*, 288–304. [CrossRef] [PubMed]
17. Sharma, A.; Jaganathan, B.G. Stem Cell Therapy for Retinal Degeneration: The Evidence to Date. *Biologics* **2021**, *15*, 299–306. [CrossRef]
18. Nair, D.S.R.; Thomas, B.B. Stem Cell-Based Treatment Strategies for Degenerative Diseases of the Retina. *Curr. Stem Cell Res. Ther.* **2021**, *17*, 214–225. [CrossRef]
19. Jin, Z.B.; Okamoto, S.; Osakada, F.; Homma, K.; Assawachananont, J.; Hirami, Y.; Iwata, T.; Takahashi, M. Modeling Retinal Degeneration Using Patient-Specific Induced Pluripotent Stem Cells. *PLoS ONE* **2011**, *6*, e17084. [CrossRef]
20. Tibbetts, M.D.; Samuel, M.A.; Chang, T.S.; Ho, A.C. Stem Cell Therapy for Retinal Disease. *Curr. Opin. Ophthalmol.* **2012**, *23*, 226–234. [CrossRef] [PubMed]
21. Lotfi, M.; Morshedi Rad, D.; Mashhadi, S.S.; Ashouri, A.; Mojarrad, M.; Mozaffari-Jovin, S.; Farrokhi, S.; Hashemi, M.; Lotfi, M.; Ebrahimi Warkiani, M.; et al. Recent Advances in CRISPR/Cas9 Delivery Approaches for Therapeutic Gene Editing of Stem Cells. *Stem Cell Rev. Rep.* **2023**, *19*, 2576–2596. [CrossRef]
22. Carlson-Stevermer, J.; Goedland, M.; Steyer, B.; Movaghar, A.; Lou, M.; Kohlenberg, L.; Prestil, R.; Saha, K. High-Content Analysis of CRISPR-Cas9 Gene-Edited Human Embryonic Stem Cells. *Stem Cell Rep.* **2016**, *6*, 109–120. [CrossRef] [PubMed]

23. Wang, D.; Quan, Y.; Yan, Q.; Morales, J.E.; Wetsel, R.A. Targeted Disruption of the β 2-Microglobulin Gene Minimizes the Immunogenicity of Human Embryonic Stem Cells. *Stem Cells Transl. Med.* **2015**, *4*, 1234–1245. [CrossRef] [PubMed]
24. Petrus-Reurer, S.; Winblad, N.; Kumar, P.; Gorchs, L.; Chrobok, M.; Wagner, A.K.; Bartuma, H.; Lardner, E.; Aronsson, M.; Plaza Reyes, Á.; et al. Generation of Retinal Pigment Epithelial Cells Derived from Human Embryonic Stem Cells Lacking Human Leukocyte Antigen Class I and II. *Stem Cell Rep.* **2020**, *14*, 648–662. [CrossRef]
25. Takahashi, K.; Tanabe, K.; Ohnuki, M.; Narita, M.; Ichisaka, T.; Tomoda, K.; Yamanaka, S. Induction of Pluripotent Stem Cells from Adult Human Fibroblasts by Defined Factors. *Cell* **2007**, *131*, 861–872. [CrossRef] [PubMed]
26. Sugita, S.; Kamao, H.; Iwasaki, Y.; Okamoto, S.; Hashiguchi, T.; Iseki, K.; Hayashi, N.; Mandai, M.; Takahashi, M. Inhibition of T-Cell Activation by Retinal Pigment Epithelial Cells Derived from Induced Pluripotent Stem Cells. *Investig. Ophthalmol. Vis. Sci.* **2015**, *56*, 1051–1062. [CrossRef]
27. Bharti, K.; Rao, M.; Hull, S.C.; Stroncek, D.; Brooks, B.P.; Feigal, E.; van Meurs, J.C.; Huang, C.A.; Miller, S.S. Developing Cellular Therapies for Retinal Degenerative Diseases. *Investig. Ophthalmol. Vis. Sci.* **2014**, *55*, 1191–1201. [CrossRef]
28. Zhong, X.; Gutierrez, C.; Xue, T.; Hampton, C.; Vergara, M.N.; Cao, L.H.; Peters, A.; Park, T.S.; Zambidis, E.T.; Meyer, J.S.; et al. Generation of Three-Dimensional Retinal Tissue with Functional Photoreceptors from Human IPSCs. *Nat. Commun.* **2014**, *5*, 4047. [CrossRef] [PubMed]
29. Hallam, D.; Hilgen, G.; Dorgau, B.; Zhu, L.; Yu, M.; Bojic, S.; Hewitt, P.; Schmitt, M.; Uteng, M.; Kustermann, S.; et al. Human-Induced Pluripotent Stem Cells Generate Light Responsive Retinal Organoids with Variable and Nutrient-Dependent Efficiency. *Stem Cells* **2018**, *36*, 1535–1551. [CrossRef] [PubMed]
30. Pera, M.F. Stem Cells: The Dark Side of Induced Pluripotency. *Nature* **2011**, *471*, 46–47. [CrossRef]
31. Rohowetz, L.J.; Koulen, P. Stem Cell-Derived Retinal Pigment Epithelium Cell Therapy: Past and Future Directions. *Front. Cell Dev. Biol.* **2023**, *11*, 1098406. [CrossRef]
32. Warren, L.; Lin, C. MRNA-Based Genetic Reprogramming. *Mol. Ther.* **2019**, *27*, 729–734. [CrossRef] [PubMed]
33. Zhang, Y.; Li, X.; Xing, J.; Zhou, J.; Li, H. Chemical Transdifferentiation of Somatic Cells: Unleashing the Power of Small Molecules. *Biomedicines* **2023**, *11*, 2913. [CrossRef]
34. Caplan, A.I. Mesenchymal Stem Cells. *J. Orthop. Res.* **1991**, *9*, 641–650. [CrossRef] [PubMed]
35. Han, Y.; Li, X.; Zhang, Y.; Han, Y.; Chang, F.; Ding, J. Mesenchymal Stem Cells for Regenerative Medicine. *Cells* **2019**, *8*, 886. [CrossRef]
36. Ezquerra, S.; Zuleta, A.; Arancibia, R.; Estay, J.; Aulestia, F.; Carrion, F. Functional Properties of Human-Derived Mesenchymal Stem Cell Spheroids: A Meta-Analysis and Systematic Review. *Stem Cells Int.* **2021**, *2021*, 8825332. [CrossRef]
37. Alvites, R.; Branquinho, M.; Sousa, A.C.; Lopes, B.; Sousa, P.; Maurício, A.C. Mesenchymal Stem/Stromal Cells and Their Paracrine Activity—Immunomodulation Mechanisms and How to Influence the Therapeutic Potential. *Pharmaceutics* **2022**, *14*, 381. [CrossRef]
38. Hong, Y.; Xu, G.X. Proteome Changes during Bone Mesenchymal Stem Cell Differentiation into Photoreceptor-like Cells in Vitro. *Int. J. Ophthalmol.* **2011**, *4*, 466–473. [CrossRef]
39. Aboutaleb Kadkhodaeian, H.; Tiraihi, T.; Ahmadieh, H.; Ziaei, H.; Daftarian, N.; Taheri, T. Generation of Retinal Pigmented Epithelium-Like Cells from Pigmented Spheres Differentiated from Bone Marrow Stromal Cell-Derived Neurospheres. *Tissue Eng. Regen. Med.* **2019**, *16*, 253–263. [CrossRef] [PubMed]
40. Park, U.C.; Park, S.S.; Kim, B.H.; Park, S.W.; Kim, Y.J.; Cary, W.; Anderson, J.D.; Nolta, J.A.; Yu, H.G. Subretinal versus Intravitreal Administration of Human CD34+ Bone Marrow-Derived Stem Cells in a Rat Model of Inherited Retinal Degeneration. *Ann. Transl. Med.* **2021**, *9*, 1275. [CrossRef]
41. Sanz-Nogués, C.; O'Brien, T. Current Good Manufacturing Practice Considerations for Mesenchymal Stromal Cells as Therapeutic Agents. *Biomater. Biosyst.* **2021**, *2*, 100018. [CrossRef]
42. Ochs, J.; Barry, F.; Schmitt, R.; Murphy, J.M. Advances in Automation for the Production of Clinical-Grade Mesenchymal Stromal Cells: The AUTOSTEM Robotic Platform. *Cell Gene Ther. Insights* **2017**, *3*, 739–748. [CrossRef]
43. Cotrim, C.C.; Jorge, R.; de Oliveira, M.C.; Pieroni, F.; Messias, A.M.V.; Siqueira, R.C. Clinical Studies Using Stem Cells for Treatment of Retinal Diseases: State of the Art. *Arq. Bras. Oftalmol.* **2020**, *83*, 160–167. [CrossRef] [PubMed]
44. Chen, X.; Xu, N.; Li, J.; Zhao, M.; Huang, L. Stem Cell Therapy for Inherited Retinal Diseases: A Systematic Review and Meta-Analysis. *Stem Cell Res. Ther.* **2023**, *14*, 286. [CrossRef] [PubMed]
45. Oswald, J.; Baranov, P. Regenerative Medicine in the Retina: From Stem Cells to Cell Replacement Therapy. *Ther. Adv. Ophthalmol.* **2018**, *10*, 2515841418774433. [CrossRef]
46. Lu, B.; Malcuit, C.; Wang, S.; Girman, S.; Francis, P.; Lemieux, L.; Lanza, R.; Lund, R. Long-Term Safety and Function of RPE from Human Embryonic Stem Cells in Preclinical Models of Macular Degeneration. *Stem Cells* **2009**, *27*, 2126–2135. [CrossRef] [PubMed]
47. Takagi, S.; Mandai, M.; Gocho, K.; Hirami, Y.; Yamamoto, M.; Fujihara, M.; Sugita, S.; Kurimoto, Y.; Takahashi, M. Evaluation of Transplanted Autologous Induced Pluripotent Stem Cell-Derived Retinal Pigment Epithelium in Exudative Age-Related Macular Degeneration. *Ophthalmol. Retin.* **2019**, *3*, 850–859. [CrossRef]
48. Hinkle, J.W.; Mahmoudzadeh, R.; Kuriyan, A.E. Cell-Based Therapies for Retinal Diseases: A Review of Clinical Trials and Direct to Consumer "Cell Therapy" Clinics. *Stem Cell Res. Ther.* **2021**, *12*, 538. [CrossRef] [PubMed]

49. Schwartz, S.D.; Hubschman, J.-P.; Heilwell, G.; Franco-Cardenas, V.; Pan, C.K.; Ostrick, R.M.; Mickunas, E.; Gay, R.; Klimanskaya, I.; Lanza, R. Embryonic Stem Cell Trials for Macular Degeneration: A Preliminary Report. *Lancet* **2012**, *379*, 713–720. [CrossRef]
50. Schwartz, S.D.; Regillo, C.D.; Lam, B.L.; Eliott, D.; Rosenfeld, P.J.; Gregori, N.Z.; Hubschman, J.P.; Davis, J.L.; Heilwell, G.; Spirn, M.; et al. Human Embryonic Stem Cell-Derived Retinal Pigment Epithelium in Patients with Age-Related Macular Degeneration and Stargardt's Macular Dystrophy: Follow-up of Two Open-Label Phase 1/2 Studies. *Lancet* **2015**, *385*, 509–516. [CrossRef]
51. Song, W.K.; Park, K.M.; Kim, H.J.; Lee, J.H.; Choi, J.; Chong, S.Y.; Shim, S.H.; Del Priore, L.V.; Lanza, R. Treatment of Macular Degeneration Using Embryonic Stem Cell-Derived Retinal Pigment Epithelium: Preliminary Results in Asian Patients. *Stem Cell Rep.* **2015**, *4*, 860–872. [CrossRef]
52. Mehat, M.S.; Sundaram, V.; Ripamonti, C.; Robson, A.G.; Smith, A.J.; Borooah, S.; Robinson, M.; Rosenthal, A.N.; Innes, W.; Weleber, R.G.; et al. Transplantation of Human Embryonic Stem Cell-Derived Retinal Pigment Epithelial Cells in Macular Degeneration. *Ophthalmology* **2018**, *125*, 1765–1775. [CrossRef] [PubMed]
53. Kashani, A.H.; Lebkowski, J.S.; Rahhal, F.M.; Avery, R.L.; Salehi-Had, H.; Dang, W.; Lin, C.-M.; Mitra, D.; Zhu, D.; Thomas, B.B.; et al. A bioengineered retinal pigment epithelial monolayer for advanced, dry age-related macular degeneration. *Sci. Transl. Med.* **2018**, *10*, eaao4097. [CrossRef] [PubMed]
54. Kashani, A.H.; Lebkowski, J.S.; Rahhal, F.M.; Avery, R.L.; Salehi-Had, H.; Chen, S.; Chan, C.; Palejwala, N.; Ingram, A.; Dang, W.; et al. One-Year Follow-up in a Phase 1/2a Clinical Trial of an Allogeneic Rpe Cell Bioengineered Implant for Advanced Dry Age-Related Macular Degeneration. *Transl. Vis. Sci. Technol.* **2021**, *10*, 13. [CrossRef] [PubMed]
55. Li, S.Y.; Liu, Y.; Wang, L.; Wang, F.; Zhao, T.T.; Li, Q.Y.; Xu, H.W.; Meng, X.H.; Hao, J.; Zhou, Q.; et al. A Phase I Clinical Trial of Human Embryonic Stem Cell-Derived Retinal Pigment Epithelial Cells for Early-Stage Stargardt Macular Degeneration: 5-Years' Follow-Up. *Cell Prolif.* **2021**, *54*, e13100. [CrossRef] [PubMed]
56. Da Cruz, L.; Fynes, K.; Georgiadis, O.; Kerby, J.; Luo, Y.H.; Ahmado, A.; Vernon, A.; Daniels, J.T.; Nommiste, B.; Hasan, S.M.; et al. Phase 1 Clinical Study of an Embryonic Stem Cell-Derived Retinal Pigment Epithelium Patch in Age-Related Macular Degeneration. *Nat. Biotechnol.* **2018**, *36*, 328–337. [CrossRef]
57. Garber, K. RIKEN Suspends First Clinical Trial Involving Induced Pluripotent Stem Cells. *Nat. Biotechnol.* **2015**, *33*, 890–891. [CrossRef] [PubMed]
58. Mandai, M.; Watanabe, A.; Kurimoto, Y.; Hirami, Y.; Morinaga, C.; Daimon, T.; Fujihara, M.; Akimaru, H.; Sakai, N.; Shibata, Y.; et al. Autologous Induced Stem-Cell–Derived Retinal Cells for Macular Degeneration. *N. Engl. J. Med.* **2017**, *376*, 1038–1046. [CrossRef] [PubMed]
59. Sugita, S.; Mandai, M.; Hirami, Y.; Takagi, S.; Maeda, T.; Fujihara, M.; Matsuzaki, M.; Yamamoto, M.; Iseki, K.; Hayashi, N.; et al. HLA-Matched Allogeneic IPS Cells-Derived Rpe Transplantation for Macular Degeneration. *J. Clin. Med.* **2020**, *9*, 2217. [CrossRef] [PubMed]
60. Siqueira, R.C.; Messias, A.; Voltarelli, J.C.; Scott, I.U.; Jorge, R. Intravitreal Injection of Autologous Bone Marrow-Derived Mononuclear Cells for Hereditary Retinal Dystrophy: A Phase i Trial. *Retina* **2011**, *31*, 1207–1214. [CrossRef]
61. Siqueira, R.C.; Messias, A.; Messias, K.; Arcieri, R.S.; Ruiz, M.A.; Souza, N.F.; Martins, L.C.; Jorge, R. Quality of Life in Patients with Retinitis Pigmentosa Submitted to Intravitreal Use of Bone Marrow-Derived Stem Cells (Reticell-Clinical Trial). *Stem Cell Res. Ther.* **2015**, *6*, 29. [CrossRef]
62. Park, S.S.; Bauer, G.; Abedi, M.; Pontow, S.; Panorgias, A.; Jonnal, R.; Zawadzki, R.J.; Werner, J.S.; Nolta, J. Intravitreal Autologous Bone Marrow Cd34+ Cell Therapy for Ischemic and Degenerative Retinal Disorders: Preliminary Phase 1 Clinical Trial Findings. *Investig. Ophthalmol. Vis. Sci.* **2015**, *56*, 81–89. [CrossRef] [PubMed]
63. Öner, A. Stem Cell Treatment in Retinal Diseases: Recent Developments. *Turk. J. Ophthalmol.* **2018**, *48*, 33–38. [CrossRef] [PubMed]
64. Satarian, L.; Nourinia, R.; Safi, S.; Kanavi, M.R.; Jarughi, N.; Daftarian, N.; Arab, L.; Aghdami, N.; Ahmadieh, H.; Baharvand, H. Intravitreal Injection of Bone Marrow Mesenchymal Stem Cells in Patients with Advanced Retinitis Pigmentosa; A Safety Study. *J. Ophthalmic Vis. Res.* **2017**, *12*, 58–64. [CrossRef] [PubMed]
65. Limoli, P.G.; Limoli, C.; Vingolo, E.M.; Scalinci, S.Z.; Nebbioso, M. Cell Surgery and Growth Factors in Dry Age-Related Macular Degeneration: Visual Prognosis and Morphological Study. *Oncotarget* **2016**, *7*, 46913–46923. [CrossRef] [PubMed]
66. Yu, B.; Li, X.R.; Zhang, X.M. Mesenchymal Stem Cell-Derived Extracellular Vesicles as a New Therapeutic Strategy for Ocular Diseases. *World J. Stem Cells* **2020**, *12*, 178–187. [CrossRef] [PubMed]
67. Lotfy, A.; AboQuella, N.M.; Wang, H. Mesenchymal Stromal/Stem Cell (MSC)-Derived Exosomes in Clinical Trials. *Stem Cell Res. Ther.* **2023**, *14*, 66. [CrossRef] [PubMed]
68. Tan, F.; Li, X.; Wang, Z.; Li, J.; Shahzad, K.; Zheng, J. Clinical Applications of Stem Cell-Derived Exosomes. *Signal Transduct. Target. Ther.* **2024**, *9*, 17. [CrossRef] [PubMed]
69. Wu, K.Y.; Ahmad, H.; Lin, G.; Carbonneau, M.; Tran, S.D. Mesenchymal Stem Cell-Derived Exosomes in Ophthalmology: A Comprehensive Review. *Pharmaceutics* **2023**, *15*, 1167. [CrossRef]
70. Tuekprakhon, A.; Sangkitporn, S.; Trinavarat, A.; Pawestri, A.R.; Vamvanij, V.; Ruangchainikom, M.; Luksanapruksa, P.; Pongpaksupasin, P.; Khorchai, A.; Dambua, A.; et al. Intravitreal Autologous Mesenchymal Stem Cell Transplantation: A Non-Randomized Phase I Clinical Trial in Patients with Retinitis Pigmentosa. *Stem Cell Res. Ther.* **2021**, *12*, 52. [CrossRef] [PubMed]
71. Kim, Y.C.; Chiang, B.; Wu, X.; Prausnitz, M.R. Ocular Delivery of Macromolecules. *J. Control. Release* **2014**, *190*, 172–181. [CrossRef]

72. Kim, J.Y.; You, Y.S.; Kim, S.H.; Kwon, W. Epiretinal membrane formation after intravitreal autologous stem cell implantation in a retinitis pigmentosa patient. *Retin. Cases Brief Rep.* **2017**, *11*, 227–231. [CrossRef] [PubMed]
73. Peng, Y.; Tang, L.; Zhou, Y. Subretinal Injection: A Review on the Novel Route of Therapeutic Delivery for Vitreoretinal Diseases. *Ophthalmic Res.* **2017**, *58*, 217–226. [CrossRef] [PubMed]
74. Zarbin, M. Cell-Based Therapy for Retinal Disease: The New Frontier. In *Methods in Molecular Biology*; Humana Press Inc.: Totowa, NJ, USA, 2019; Volume 1834, pp. 367–381.
75. Chiang, B.; Jung, J.H.; Prausnitz, M.R. The Suprachoroidal Space as a Route of Administration to the Posterior Segment of the Eye. *Adv. Drug Deliv. Rev.* **2018**, *126*, 58–66. [CrossRef] [PubMed]
76. Limoli, P.G.; Vingolo, E.M.; Limoli, C.; Scalinci, S.Z.; Nebbioso, M. Regenerative Therapy by Suprachoroidal Cell Autograft in Dry Age-Related Macular Degeneration: Preliminary in Vivo Report. *J. Vis. Exp.* **2018**, *2018*, e56469. [CrossRef] [PubMed]
77. Limoli, P.G.; Vingolo, E.M.; Morales, M.U.; Nebbioso, M.; Limoli, C. Preliminary Study on Electrophysiological Changes after Cellular Autograft in Age-Related Macular Degeneration. *Medicine* **2014**, *93*, e355. [CrossRef] [PubMed]
78. Puertas-Neyra, K.; Usategui-Martín, R.; Coco, R.M.; Fernandez-Bueno, I. Intravitreal Stem Cell Paracrine Properties as a Potential Neuroprotective Therapy for Retinal Photoreceptor Neurodegenerative Diseases. *Neural Regen. Res.* **2020**, *15*, 1631–1638.
79. Kador, K.E.; Goldberg, J.L. Scaffolds and Stem Cells: Delivery of Cell Transplants for Retinal Degenerations. *Expert. Rev. Ophthalmol.* **2012**, *7*, 459–470. [CrossRef]
80. Nazari, H.; Zhang, L.; Zhu, D.; Chader, G.J.; Falabella, P.; Stefanini, F.; Rowland, T.; Clegg, D.O.; Kashani, A.H.; Hinton, D.R.; et al. Stem Cell Based Therapies for Age-Related Macular Degeneration: The Promises and the Challenges. *Prog. Retin. Eye Res.* **2015**, *48*, 1–39. [CrossRef]
81. Rajendran Nair, D.S.; Seiler, M.J.; Patel, K.H.; Thomas, V.; Camarillo, J.C.M.; Humayun, M.S.; Thomas, B.B. Tissue Engineering Strategies for Retina Regeneration. *Appl. Sci.* **2021**, *11*, 2154. [CrossRef]
82. Lu, B.; Zhu, D.; Hinton, D.; Humayun, M.S.; Tai, Y.C. Mesh-Supported Submicron Parylene-C Membranes for Culturing Retinal Pigment Epithelial Cells. *Biomed. Microdevices* **2012**, *14*, 659–667. [CrossRef]
83. Liu, Z.; Yu, N.; Holz, F.G.; Yang, F.; Stanzel, B.V. Enhancement of Retinal Pigment Epithelial Culture Characteristics and Subretinal Space Tolerance of Scaffolds with 200 Nm Fiber Topography. *Biomaterials* **2014**, *35*, 2837–2850. [CrossRef] [PubMed]
84. Hotaling, N.A.; Khristov, V.; Wan, Q.; Sharma, R.; Jha, B.S.; Lotfi, M.; Maminishkis, A.; Simon, C.G.; Bharti, K. Nanofiber Scaffold-Based Tissue-Engineered Retinal Pigment Epithelium to Treat Degenerative Eye Diseases. *J. Ocul. Pharmacol. Ther.* **2016**, *32*, 272–285. [CrossRef]
85. Zhang, S.; Ye, K.; Gao, G.; Song, X.; Xu, P.; Zeng, J.; Xie, B.; Zheng, D.; He, L.; Ji, J.; et al. Amniotic Membrane Enhances the Characteristics and Function of Stem Cell-Derived Retinal Pigment Epithelium Sheets by Inhibiting the Epithelial–Mesenchymal Transition. *Acta Biomater.* **2022**, *151*, 183–196. [CrossRef]
86. Gu, J.; Wang, Y.; Cui, Z.; Li, H.; Li, S.; Yang, X.; Yan, X.; Ding, C.; Tang, S.; Chen, J. The Construction of Retinal Pigment Epithelium Sheets with Enhanced Characteristics and Cilium Assembly Using IPS Conditioned Medium and Small Incision Lenticule Extraction Derived Lenticules. *Acta Biomater.* **2019**, *92*, 115–131. [CrossRef]
87. Soroushzadeh, S.; Karamali, F.; Masaeli, E.; Atefi, A.; Nasr Esfahani, M.H. Scaffold Free Retinal Pigment Epithelium Sheet Engineering Using Modified Alginate-RGD Hydrogel. *J. Biosci. Bioeng.* **2022**, *133*, 579–586. [CrossRef]
88. Fernandes, R.A.B.; Lojudice, F.H.; Zago Ribeiro, L.; Santos Da Cruz, N.F.; Polizelli, M.U.; Cristovam, P.C.; Innocenti, F.; Morimoto, L.; Magalhães, O.; Ferraz Sallum, J.M.; et al. Transplantation of subretinal stem cell-derived retinal pigment epithelium for stargardt disease: A phase I clinical trial. *Retina* **2023**, *43*, 263–274. [CrossRef] [PubMed]
89. Zhang, X.; Liu, J.; Yu, B.; Ma, F.; Ren, X.; Li, X. Effects of Mesenchymal Stem Cells and Their Exosomes on the Healing of Large and Refractory Macular Holes. *Graefe's Arch. Clin. Exp. Ophthalmol.* **2018**, *256*, 2041–2052. [CrossRef] [PubMed]
90. Re, F.; Gabusi, E.; Manferdini, C.; Russo, D.; Lisignoli, G. Bone Regeneration Improves with Mesenchymal Stem Cell Derived Extracellular Vesicles (Evs) Combined with Scaffolds: A Systematic Review. *Biology* **2021**, *10*, 579. [CrossRef]
91. Alahmad, G.; Aljohani, S.; Najjar, M.F. Ethical Challenges Regarding the Use of Stem Cells: Interviews with Researchers from Saudi Arabia. *BMC Med. Ethics* **2020**, *21*, 35. [CrossRef] [PubMed]
92. Weiss, A.M.; Breitenbach, M.; Rinnerthaler, M.; Virt, G. Ethical Considerations on Stem Cell Research. In *Pluripotent Stem Cells*; IntechOpen: Rijeka, Croatia, 2013.
93. Volarevic, V.; Markovic, B.S.; Gazdic, M.; Volarevic, A.; Jovicic, N.; Arsenijevic, N.; Armstrong, L.; Djonov, V.; Lako, M.; Stojkovic, M. Ethical and Safety Issues of Stem Cell-Based Therapy. *Int. J. Med. Sci.* **2018**, *15*, 36–45. [CrossRef]
94. Master, Z.; Matthews, K.R.W.; Abou-el-Enein, M. Unproven Stem Cell Interventions: A Global Public Health Problem Requiring Global Deliberation. *Stem Cell Rep.* **2021**, *16*, 1435–1445. [CrossRef] [PubMed]

life

MDPI

Review

Infantile Nystagmus Syndrome—Associated Inherited Retinal Diseases: Perspectives from Gene Therapy Clinical Trials

Xiaoming Gong [1,2,*] and Richard W. Hertle [1,2]

1 Department of Ophthalmology, Akron Children's Hospital, Akron, OH 44308, USA; rhertle@akronchildrens.org
2 Vision Center of Excellence, Rebecca D. Considine Research Institute, Akron Children's Hospital, Akron, OH 44308, USA
* Correspondence: xgong@akronchildrens.org

Abstract: Inherited retinal diseases (IRDs) are a clinically and genetically diverse group of progressive degenerative disorders that can result in severe visual impairment or complete blindness. Despite their predominantly monogenic inheritance patterns, the genetic complexity of over 300 identified disease-causing genes presents a significant challenge in correlating clinical phenotypes with genotypes. Achieving a molecular diagnosis is crucial for providing patients with definitive diagnostic clarity and facilitating access to emerging gene-based therapies and ongoing clinical trials. Recent advances in next-generation sequencing technologies have markedly enhanced our ability to identify genes and genetic defects leading to IRDs, thereby propelling the development of gene-based therapies. The clinical success of voretigene neparvovec (Luxturna), the first approved retinal gene therapy for *RPE65*-associated Leber congenital amaurosis (LCA), has spurred considerable research and development in gene-based therapies, highlighting the importance of reviewing the current status of gene therapy for IRDs, particularly those utilizing adeno-associated virus (AAV)-based therapies. As novel disease-causing mutations continue to be discovered and more targeted gene therapies are developed, integrating these treatment opportunities into the standard care for IRD patients becomes increasingly critical. This review provides an update on the diverse phenotypic–genotypic landscape of IRDs, with a specific focus on recent advances in the understanding of IRDs in children with infantile nystagmus syndrome (INS). We highlight the complexities of the genotypic–phenotypic landscape of INS-associated IRDs, including conditions such as achromatopsia, LCA, congenital stationary night blindness, and subtypes of retinitis pigmentosa. Additionally, we provide an updated overview of AAV-based gene therapies for these diseases and discuss the potential of gene-based therapies for underlying IRDs that lead to INS, offering a valuable resource for pediatric patients potentially eligible for ongoing clinical trials.

Keywords: genotype–phenotype; inherited retinal disease; adeno-associated virus; gene therapy; clinical trials; infantile nystagmus syndrome; pediatric ophthalmology

1. Introduction

Infantile nystagmus syndrome (INS) is a unique ocular motor disorder characterized by involuntary spontaneous oscillation of the eyes, typically associated with congenital or early onset defects in the visual sensory system within the first six months of life [1]. It is the most prevalent form of nystagmus in infancy and childhood, with an incidence of approximately 1.7 per 1000 live births [2,3]. INS is a common clinical sign of many ocular conditions in children and is frequently indicative of underlying ocular, neurological, and systemic diseases [4]. Evaluating INS in infants or children is challenging due to its association with both idiopathic oculomotor disorder and early-onset defects in the visual sensory system and systemic disorders [5]. Approximately 90% of INS cases are associated with visual sensory disorders that are either anatomical or functional [6], including various inherited retinal disorders (IRDs) [7].

INS can manifest as an isolated, inherited, or idiopathic oculomotor disorder without other associated ocular or neurological conditions. However, it is frequently observed alongside ocular disorders such as albinism, aniridia, congenital retinal dystrophies, early-onset retinal degeneration, congenital cataracts, optic nerve hypoplasia, and corneal dystrophies (Figure 1) [8,9]. INS may also be associated with neurological syndromes or conditions, including brain tumors, gliomas, spinocerebellar ataxia, and structural brain malformations [10]. Clinically, INS associated with IRDs is more common than idiopathic infantile nystagmus, although both share similar oculomotor characteristics. While the underlying "cause" of INS, such as a poorly calibrated smooth pursuit system, is consistent across cases, the diagnosis of associated visual and systemic conditions can be challenging due to the broad array of underlying genetic disorders. Given that approximately 70% of INS cases are associated with IRDs [11], identifying the genetic spectrum of IRDs in children with INS—referred to as INS-associated IRDs (INS-IRDs)—is crucial. This identification provides pediatric patients and their families with definitive diagnostic clarity, enables the early detection of INS-IRDs and tailored management strategies, and facilitates access to emerging gene-based therapies and the ever-increasing number of ongoing clinical trials.

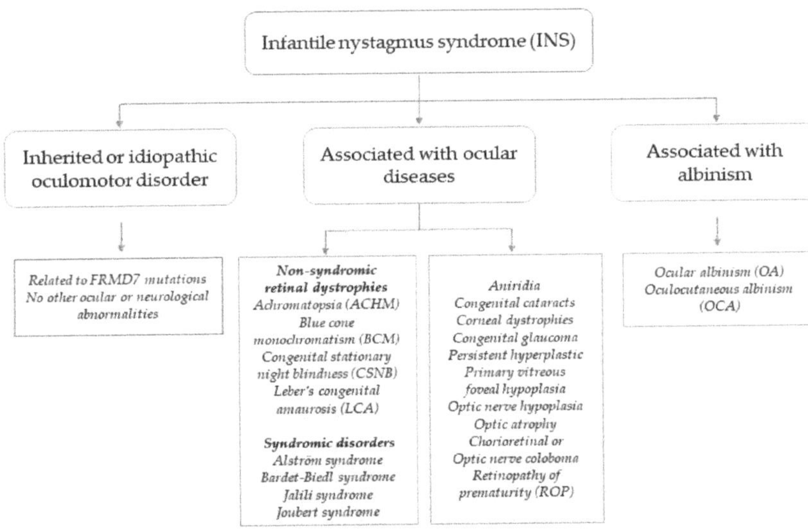

Figure 1. Clinical compositions of infantile nystagmus syndrome and its associated conditions.

IRDs are a clinically and genetically diverse group of conditions that lead to vision loss due to progressive retinal degeneration, often manifesting in childhood or early life and causing significant vision impairment and blindness in the pediatric population [12]. IRDs exhibit considerable genetic heterogeneity, with both monogenic and more complex inheritance patterns such as digenic, biallelic, and triallelic inheritance, particularly seen in conditions like achromatopsia, albinism, and Bardet–Biedl syndrome [13]. Over the past two decades, more than 100 causative genes have been identified in INS-associated IRDs [14], leading to significant phenotypic overlap (Tables S1 and S2). Mutations in the same gene can result in diverse phenotypes inherited through autosomal recessive, autosomal dominant, X-linked, or mitochondrial patterns, further complicating clinical diagnoses [15,16]. Despite this complexity, IRDs are excellent candidates for gene therapy, offering potential long-term solutions through localized delivery methods such as intravitreal, subretinal, or suprachoroidal injections [17]. The eye's immune-privileged environment and the compartmentalized nature of ocular anatomy facilitate efficient gene delivery with minimal immune response [18,19]. Advances in adeno-associated virus (AAV)-based gene therapy have demonstrated safety and efficacy in treating various IRDs, as highlighted by the FDA approval of voretigene neparvovec-rzyl (Luxturna) for RPE65-associated retinal dystrophy [20]. With over 300 clinical trials underway, nearly 100 of which focus on ocular disorders, AAV-based gene therapies offer promising therapeutic approaches for monogenic and multigenic conditions. The following review will explore the genotype–phenotype complexities of INS-associated IRDs and provide an updated analysis of AAV-based gene therapies in pediatric ophthalmology, focusing on ongoing clinical trials targeting these conditions.

2. Phenotypic–Genotypic Complexity of IRDs in Children with INS

IRDs are primarily caused by mutations in more than 300 genes (RetNet; http://sph.uth.edu/retnet, accessed on 1 May 2024), each of which may harbor many disease-causing variants with distinct clinical phenotypes, underscoring the significant genetic diversity of these conditions. In children with INS, over 100 causal genes linked to IRDs have been identified, along with over 2500 mutations within these genes (RetinoGenetics: http://www.retinogenetics.org/, accessed on 1 June 2024). This extensive genetic variability makes INS-associated IRDs some of the most genetically heterogeneous Mendelian disorders, leading to a broad spectrum of clinical manifestations and symptoms. Examples of such IRDs include achromatopsia (ACHM), Leber congenital amaurosis (LCA), Bardet–Biedl syndrome (BBS), cone–rod dystrophies (CRD), and retinitis pigmentosa (RP). Clinically, INS-associated IRDs are classified into two phenotypic categories: non-syndromic and syndromic. Most cases are non-syndromic, primarily caused by mutations in retina-specific genes, and include conditions such as ACHM, LCA, RP, and congenital stationary night blindness (CSNB). Syndromic INS-associated IRDs, which involve multiple organ systems beyond the eyes, include disorders such as Alström syndrome, BBS, Jalili syndrome, and Joubert syndrome (Figure 1, Tables S1 and S2) [21].

The phenotypic presentation of INS-associated IRDs evolves over time, influenced by genetic and environmental factors, leading to significant variability even among family members with identical genetic mutations. The degree of rod and cone involvement drives clinical presentation, with predominant rod dysfunction causing impaired night vision and cone dysfunction leading to reduced visual acuity, color vision impairment, photophobia, and INS, especially in early-onset cases. In this review, we provide an in-depth analysis of the phenotypic–genotypic spectrum of common IRDs associated with INS, focusing on conditions such as LCA, ACHM, CSNB, various subtypes of RP, and X-linked retinoschisis.

2.1. The Phenotypic–Genotypic Spectrum of Leber Congenital Amaurosis (LCA)

LCA is one of the most severe forms of early-onset IRDs and the most prevalent type of INS-associated IRDs. It encompasses a group of severe, early-onset retinal disorders affecting the pediatric population [22]. Children with LCA frequently exhibit significant visual impairment from infancy or early childhood. This condition is characterized by almost non-detectable or severely abnormal full-field electroretinogram (ERG), infantile nystagmus (reflecting poor fixation ability), and amaurotic pupils (indicating poor pupillary light responses). While the retina may initially appear normal, pigmentary retinopathy reminiscent of retinitis pigmentosa is frequently observed later in childhood [23]. Other common phenotypic features include high refractive errors (hyperopia), sensitivity to light (photophobia), night blindness (nyctalopia), oculodigital reflex, peripheral chorioretinal atrophy, intraretinal pigment migration, drusen-like deposits, keratoconus, and cataracts [24]. From a clinical standpoint, LCA exhibits an extremely heterogeneous phenotype, ranging from an essentially normal retina to a variable degree of vessel attenuation, bone spicule pigmentation, pseudopapilledema, macular coloboma, salt and pepper pigmentation, yellow confluent peripheral spots, white retinal spots, preserved para-arteriolar retinal pigment epithelium (RPE), and Coats reaction, with some gene-specific features.

A milder form within this disease spectrum has been described using various terms, including early-onset severe retinal dystrophy (EOSRD), severe early-childhood-onset retinal dystrophy (SECORD), and early-onset retinitis pigmentosa (EORP) [25]. Prominent clinical features of severe-onset cone–rod dystrophy (CORD), which is related to early cone photoreceptor involvement, include reduced visual acuity and INS. The ocular fundus may appear normal or show signs like bull's-eye or atrophic maculopathy, pigmented retinal stippling, or optic nerve pallor. EOSRD/SECORD is characterized by the onset of visual impairment typically emerging after infancy but before the age of five years, with variably preserved visual acuity and a better preserved full-field ERG response [25]. Collectively, LCA and EOSRD/SECORD are the most severe of the early onset forms of IRDs, affecting 20% of blind children and accounting for approximately 5% of all retinal dystrophies [24].

The molecular diagnosis of LCA/EOSRD is complicated by substantial genetic diversity and clinical variability. More than 28 genes have been implicated in these conditions, each associated with different phenotypes and pathogenic mechanisms that lead to various forms of retinal dysfunction [26]. These mechanisms impact several pathways, such as phototransduction, the visual cycle, and photoreceptor development and maintenance. The genes associated with LCA/EOSRD can be categorized based on the functions of their encoded proteins (Table 1). For example, *CCT2*, *CEP290*, *IQCB1*, *LCA5*, *RPGRIP1*, *SPATA7*, and *TULP1* are linked to intra-photoreceptor ciliary transport, whereas *CRB1*, *CRX*, *GDF6*, *CLUAP1*, and *PRPH2* are related to photoreceptor morphogenesis. *AIPL1*, *GUCY2D*, and *RD3* are involved in phototransduction, while *LRAT*, *RDH12*, and *RPE65* play roles in the visual cycle. *CABP4* and *KCNJ13* are important for signal transduction, *OTX2* for retinal development, *IMPDH1* for guanine synthesis, *MERTK* for outer segment phagocytosis, and *NMNAT1* for coenzyme NAD biosynthesis, whereas *DTHD1* has an unknown function. Most LCA/EOSRD genes follow an autosomal recessive inheritance pattern, though autosomal dominant mutations have been identified in *IMPDH1*, *OTX2*, and *CRX*. Certain genes, such as *GUCY2D*, *CEP290*, *NMNAT1*, and *AIPL1*, are predominantly associated with LCA, while others, including *RPE65* and *RDH12*, are more frequently linked to EOSRD [26].

The most prevalent causative genes associated with LCA/EOSRD include *GUCY2D*, *CEP290*, *CRB1*, *RDH12*, *NMNAT1*, and *RPE65*, which together account for 60% of cases. The variants in these genes lead to a broad spectrum of clinical phenotypes, which will be discussed in the following sections.

In LCA/EOSRD, *GUCY2D* mutations (associated with the LCA1 locus) are responsible for 6–21% of autosomal recessive LCA and up to 40% of autosomal dominant CORD and cone dystrophy [27]. The majority of *GUCY2D* mutations lead to a truncated protein, impairing guanylate cyclase 1 function. There are over 248 identified variants in the *GUCY2D* gene that cause the majority of autosomal recessive LCA/EOSRD, and 30 variants are linked to autosomal dominant CORD [28]. The majority of mutations in *GUCY2D* are missense, whereas nonsense, frameshift, and splice site mutations are also common. Patients with *GUCY2D*-related LCA often exhibit significantly reduced vision, INS, hyperopia, photophobia, and diminished or absent ERG responses. Despite severe visual impairment, these patients generally show normal fundi and a preserved central macular outer retinal structure on optical coherence tomography (OCT), unlike most other LCA/EOSRD genotypes [29].

Mutations in *CEP290* (associated with LCA10 locus) are responsible for >20% of LCA cases. These mutations are also implicated in a spectrum of more severe systemic conditions, including Joubert syndrome, Meckel–Gruber syndrome, Senior–Løken syndrome, and Bardet–Biedl syndrome [30]. Despite some variability among and within families, individuals with *CEP290* mutations typically exhibit a consistent and distinctive set of symptoms. Beyond the common LCA manifestations, these patients often experience significant early-onset vision loss, with most showing only light perception or complete blindness from birth. Specific LCA10 phenotypes include small atrophic spots in the RPE layer, a unique tapetal-like reflex, a prominent yellow scleral rim, pseudopapillary edema, and cyst-like macular lesions. Among the various *CEP290* variants, the pathogenic intronic nonsense mutation c.2991+1655A>G is the most prevalent, found in 60–80% of LCA patients with at least one affected allele [31]. This variant introduces a premature stop codon and a cryptic exon into the *CEP290* mRNA. However, as only about 50% of the mRNA includes the cryptic exon, this variant is considered a hypomorphic mutation, contributing to phenotypic variability [32].

Table 1. Causative genes implicated in Leber congenital amaurosis (LCA).

Diseases	Gene	Protein	Function
LCA1	*GUCY2D*	Retinal guanylate cyclase-1	Phototransduction
LCA2	*RPE65*	Retinoid isomerohydrolase RPE65	Vision (retinoid) cycle
LCA3	*SPATA7*	Spermatogenesis associated protein 7	Photoreceptor ciliary transport
LCA4	*AIPL1*	Aryl-hydrocarbon interacting protein-like 1	Phototransduction
LCA5	*LCA5*	Lebercilin	Photoreceptor ciliary transport
LCA6	*RPGRIP1*	Retinitis pigmentosa GTPase regulator-interacting protein 1	Photoreceptor ciliary transport
LCA7	*CRX*	Cone–rod homeobox	Photoreceptor morphogenesis
LCA8	*CRB1*	Crumbs homologus 1	Photoreceptor morphogenesis
LCA9	*NMNAT1*	Nicotinamide nucleotide adenyltransferase 1	Coenzyme NAD biosynthesis
LCA10	*CEP290*	Centrosomal protein 290	Photoreceptor ciliary transport
LCA11	*IMPDH1*	Inosine 5′-monophosphate dehydrogenase 1	Guanine synthesis
LCA12	*RD3*	Retinal Degeneration 3	Photoreceptor ciliary transport
LCA13	*RDH12*	Retinol dehydrogenase 12	Visual (retinoid) cycle
LCA14	*LRAT*	Lecithin:retinol acyltransferase	Visual (retinoid) cycle
LCA15	*TULP1*	Tubby-like protein	Photoreceptor ciliary transport
LCA16	*KCNJ13*	Potassium inwardly rectifying channel subfamily J member 13	Signal transduction
LCA17	*CABP4*	Calcium-binding protein 4	Signal transduction
	GDF6	Growth differentiation factor 6	Photoreceptor morphogenesis
LCA18	*PRPH2*	Peripherin 2	Photoreceptor morphogenesis
LCA19	*USP45*	ubiquitin specific peptidase 45	deubiquitylation
Unclassified	*CCT2*	Chaperonin Containing TCP1 Subunit 2	Photoreceptor ciliary transport
	CLUAP1	Clusterin associated protein 1	Photoreceptor morphogenesis
	DTHD1	Death-domain containing protein 1	unknown
	IQCB1	IQ motif containing B1 protein	Photoreceptor ciliary transport
	OTX2	Orthodenticle homeobox 2 protein	Photoreceptor morphogenesis

Biallelic mutations in *CRB1* (associated with the LCA8 locus) have been associated with a wide range of retinal phenotypes, including early-onset severe macular atrophy, RP both with and without Coats-like exudative vasculopathy, CORD, and foveal retinoscisis. However, the most commonly reported is LCA or EOSRD [33]. Approximately 9–17% of LCA cases are related to *CRB1* mutations. Typical clinical and ophthalmological manifestations are progressive macular atrophy, with nummular pigmentation, and relative preservation of the para-arteriolar RPE. Unlike the retinal thinning typically seen in LCA, patients with *CRB1* mutations often exhibit increased retinal thickness with loss of the outer limiting membrane. Some patients may also develop retinal telangiectasis, exudative retinal detachment, and neovascular glaucoma [34].

Mutations in the *RDH12* gene (associated with the LCA13 locus) contribute to approximately 4–5% of all LCA cases and often lead to an EOSRD phenotype characterized by early-

dense intraretinal pigmentation and maculopathy [35]. Patients with *RDH12* mutations typically have poor visual function early in life, show no ERG responses beyond 20 years of age, and do not experience photophobia. Clinical features include chorioretinopathy with dense pigmentation, bone spicules, minimal to no autofluorescence in the macula, and common night blindness. Spectral-domain OCT reveals significant macular thinning and a loss of foveal laminar structure. Hypomorphic alleles of *RDH12* are associated with milder retinopathy, primarily affecting the macula. The phenotypic range of *RDH12* mutations has recently been broadened to include later onset and milder presentations [36]. There are currently more than 130 disease-causing mutations in *RDH12* that have been identified in association with LCA, RP, and other retinal dystrophy phenotypes. The mutation spectrum includes missense mutations, truncating mutations (nonsense, frameshift, and consensus splice sites), and gross deletions.

Mutations in *NMNAT1* (associated with the LCA9 Locus) account for 4–14% of LCA cases. There are 75 identified mutations in *NMNAT1*, including 54 missense mutations, seven frameshift indel, six nonsense mutation, three intronic mutations, and five other mutations, such as exon deletion, exon duplication, and large duplication [37]. Extreme early onset and rapid progression are characteristics of *NMNAT1*-associated LCA. Patients with *NMNAT1* mutations show severe visual impairment at birth and coloboma-like macular dystrophy with peripheral retinal degeneration and optic atrophy since early infancy. Attenuated vessels, pallor optic disc, and peripheral pigmentation are observed during fundus examination [38]. No correlation has been found between genotype and clinical phenotype.

LCA associated with *RPE65* (the LCA2 locus) is particularly notable due to available treatments for patients with biallelic variants in this gene. Pathogenic mutations in the *RPE65* gene account for 8–16% of LCA cases and around 2% of recessive RP cases [39]. Most children with *RPE65*-associated retinopathy present with severe night blindness from birth. Patients who carried at least one nonsense variant had more progressive deterioration in retinal sensitivity. The clinical presentation aligns with that typical of LCA features, as mentioned above. Additionally, patients with *RPE65*-related retinopathy may also present with SECORD/EOSRD.

2.2. *Achromatopsia (ACHM)*

Achromatopsia (ACHM) is a congenital recessive IRD that primarily affects the cone cells in the retina [40]. This disorder is defined by dysfunctional cone photoreceptors and can manifest in either complete or incomplete forms. In complete ACHM, patients experience deficits in all three color perception axes (red, green, and blue) and typically exhibit significantly reduced visual acuity, infantile nystagmus, sensitivity to light (photophobia), and a congenital inability to differentiate colors. In the incomplete form, some cone function persists, resulting in better visual acuity and partial color discernment. Both forms of ACHM are marked by difficulties in color differentiation across all three color vision axes, INS with a dual jerk waveform, photophobia, central visual field defects (scotomata), and frequently hypermetropic refractive errors [41].

The majority of individuals with ACHM has the complete form. Full-field ERG usually shows no detectable cone-mediated responses, but rod responses remain normal or near-normal [42]. OCT can present a broad spectrum of findings, from a normal ellipsoid zone (EZ) to complete outer retinal atrophy, including the retinal pigment epithelium (RPE). OCT may also reveal hypo-reflective cavities in place of the EZ [43]. Fundus autofluorescence (FAF) imaging can show hyper-autofluorescence in zones with preserved EZ, indicating early disease stages before photoreceptor loss, or hypo-autofluorescence in areas with photoreceptor loss or RPE atrophy [44].

ACHM is a genetically well characterized condition. Six genes are currently recognized as causes of ACHM. Five of these genes, including cyclic nucleotide-gated cation channel alpha-3 (*CNGA3*), cyclic nucleotide-gated cation channel beta-3 (*CNGB3*), guanine nucleotide-binding protein G (t) subunit alpha-2 (*GNAT2*), the catalytic alpha-subunit of cone cyclic nucleotide phosphodiesterase (*PDE6C*), and the inhibitory gamma-subunit of cone phosphodiesterase (*PDE6H*), encode critical elements of the cone phototransduction pathway, which converts light signals into electrical signals. The sixth gene, activating transcription factor 6 (*ATF6*), is a more recently identified gene that codes for a transcription factor crucial for cone photoreceptor development [45]. No clear genotype–phenotype correlation has been observed in ACHM.

Genetic studies of ACHM patients have shown that many cases have pathogenic mutations in the *CNGA3* and *CNGB3* genes. Loss-of-function mutations in *CNGA3* and *CNGB3* account for over 70% of ACHM cases [46]. More than 150 mutations in *CNGA3* and 140 mutations in *CNGB3* have been linked to ACHM [47]. The majority of CNGB3 mutations are nonsense, frameshift, or splicing variants resulting in truncated or dysfunctional CNG channel proteins, while most *CNGA3* mutations are missense variants that impair or abolish CNG channel function. Patients with homozygous or compound heterozygous mutations exhibit typical ACHM symptoms, whereas carriers maintain normal vision. Interestingly, some ACHM patients exhibit a digenic or triallelic inheritance pattern involving mutations in both genes [48].

Mutations in *GNAT2* are responsible for a rare form (<2%) of ACHM [49]. Similarly rare are forms caused by mutations in *PDE6C* and *PDE6H* [50]. *ATF6* encodes a transmembrane transcription factor that activates target genes involved in the unfolded protein response during endoplasmic reticulum (ER) stress. This function potentially increases susceptibility to ER-stress-induced damage and death during cone photoreceptor development. The majority of complete ACHM patients with pathogenic mutations in *ATF6*, *CNGA3*, *CNGB3*, *GNAT2*, and *PDE6C* exhibit similar clinical phenotypes. Although a significant genotype–phenotype correlation cannot be established, the frequent observation of foveal hypoplasia with a poorly formed or absent foveal pit in *ATF6*-associated ACHM suggests the critical role of this gene in foveal development [45]. Pathogenic mutations in the *ATF6* gene predominantly include truncating mutations such as splice site, nonsense, and frameshift mutations, whereas missense mutations have also been reported.

2.3. Congenital Stationary Night Blindness (CSNB)

Congenital stationary night blindness (CSNB) is a group of IRDs that primarily manifest as night blindness in childhood. CSNB displays diverse genetic, electrophysical, and clinical characteristics. This retinal dystrophy is defined by the dysfunction of rod photoreceptors and impaired signal transduction between photoreceptor cells and bipolar cells. Clinically, patients present with night blindness, myopia, strabismus, and/or nystagmus. The nystagmus associated with CSNB is typically described as pendular, dysconjugate, and oblique, characterized by high frequency and low amplitude movements [51]. Several of these symptoms overlap with other IRDs, such as cone–rod dystrophies. CSNB can be classified into four types: Riggs, Schubert–Bornschein, fundus albipunctatus, and Oguchi disease [52]. On fundus examination, fundus albipunctatus is identified by the presence of small white dots scattered across the posterior pole, sparing the fovea. Oguchi disease, in contrast, features a distinctive gray–white metallic sheen on the retina, which disappears after dark adaptation—a phenomenon known as the Mizuo–Nakamura phenomenon. Unlike these two diseases, the Riggs and Schubert–Borstein types have normal fundi and can be distinguished using full-field ERG. In Riggs-type CSNB, a non-recordable rod ERG and a reduced a-wave in the dark-adapted combined rod–cone response indicates impaired rod phototransduction. The Schubert–Bornschein type of CSNB, on the other hand, is characterized by a normal dark-adapted a-wave but a severely reduced b-wave, resulting in an "electronegative" ERG. The Schubert–Bornschein type can be further subdivided into complete and incomplete forms. Complete CSNB presents with a normal a-wave and a reduced or absent b-wave under scotopic conditions, while photopic conditions elicit a near-normal b-wave in response to bright flashes or 30 Hz flicker stimuli. In contrast, incomplete CSNB shows a reduced b-wave under scotopic conditions but also has a near-normal response to bright flashes or flickers at 30 Hz under photopic conditions.

CSNB follows an autosomal dominant inheritance pattern in 2% of cases, an autosomal recessive pattern in 40%, and an X-linked recessive pattern in 58% [53]. To date, more than 300 mutations in 18 genes have been associated with CSNB. Specifically, GNAT1, PDE6B, and RHO are implicated in the autosomal dominant form, while CACNA1F and NYX are linked to the X-linked form. The autosomal recessive form involves 13 genes, including CABP4, GNAT1, GNB3, GPR179, GRK1, GRM6, LRIT3, RDH5, RIMS2, RPE65, SAG, SLC24A1, and TRPM1 (Table 2). The complete form of CSNB is associated with mutations in NYX, GRM6, TRPM1, SLCC24A1, GPR179, LRIT3, GNAT1, GNB3, RHO, and RPE65, which all lead to a loss of function of rod and cone ON bipolar cells. This results in a characteristic squared-off appearance of the b-wave, with a normal OFF response driven by cone OFF bipolar cells. Incomplete CSNB, attributed to mutations in CACNA1F, CABP4, and PDE6B, presents with reduced a-wave and b-wave amplitudes in light-adapted ERG and markedly decreased 30 Hz flicker amplitude, indicating the severe dysfunction of both rods and cones.

Table 2. Types and targeted gene inheritance patterns of congenital stationary night blindness (CSNB).

Inheritance Pattern	Types	Genes	Gene ID (OMIM#)
Autosomal Dominant	Riggs	*RHO*	180380
		GNAT1	610444
		PDE6B	180072
Autosomal Recessive	Abnormal fundus	*SAG*	181031
		GRK1	613411
		RDH5	610617
		RLBP1	180090
		RPE65	180069
	Complete	*GRM6*	604096
		TRPM1	603576
		RIMS2	606630
		GPR179	414515
		GNB3	617024
		LRIT3	615004
	Incomplete	*CABP4*	608965
		CACNA2D4	608171
	Riggs	*SLC24A1*	613830
		GNAT1	610444
X-linked	Complete	*NYX*	300278
	Incomplete	*CACNA1F*	300110

2.4. *Retinitis Pigmentosa (RP)*

RP is a heterogeneous group of IRDs caused by the progressive degeneration of rod and cone photoreceptors. It is the most frequent form of IRD, affecting between one in 3000 to 4000 individuals. Clinically, RP is characterized by night blindness, peripheral vision loss, and ultimately, total blindness [54]. The disease often begins with the degeneration of rod photoreceptors, which predominantly affects night vision and adaptation to low light. Early symptoms include night blindness (nyctalopia) and difficulties adjusting to changes in light sensitivity. As rod degeneration advances, it causes a progressive constriction of the visual field, initially affecting mid-peripheral vision and eventually leading to "tunnel vision" as the central vision becomes compromised [55]. In later stages, RP may present as cone–rod degeneration and cone dystrophy, where cone photoreceptors are primarily affected, resulting in significant loss of visual acuity. Some RP subtypes exhibit concurrent degeneration of both rod and cone photoreceptors.

The underlying molecular causes of RP are highly intricate, involving mutations in over 200 genes, which account for its genetic diversity and clinical variability. RP can be categorized into three main inheritance patterns [56]: autosomal dominant RP (adRP), accounting for 15–25% of cases; autosomal recessive RP (arRP), comprising 5–20% of cases; and X-linked RP (XL-RP), representing 10–15% of RP cases. Approximately 40–50% of RP cases lack a clear inheritance pattern and are thus classified as sporadic.

Several key genes are frequently mutated in RP, including *RHO* (rhodopsin), responsible for about 25% of adRP cases; *USH2A*, associated with 20% of arRP cases; and *RPGR* (retinitis pigmentosa GTPase regulator), which accounts for over 80% of XL-RP cases. XL-RP is one of the most severe forms of rod–cone dystrophy, accounting for 10–20% of all RP cases [57]. So far, mutations in three genes have been identified as causes of XL-RP: *RPGR*, retinitis pigmentosa 2(*RP2*), and oral–facial–digital syndrome type 1 (*OFD1*).

Mutations in *RPGR*, located in the RP3 region of Xp21.1, is the most common genetic cause of XL-RP. In addition to typical RP symptoms, patients with *RPGR* mutations often present with severe myopia and can display sectoral RP [58]. The severity in female carriers varies widely, ranging from no symptoms to severe RP, with some displaying a tapetal-like reflex (TLR). Microperimetry has shown that most RPGR patients experience a rapid decline in retinal sensitivity during their second and third decades of life [59].

Mutations in *RP2* are the second most common genetic cause of XL-RP, accounting for 5–20% of cases. The majority of affected males typically show early-onset, severe retinal degeneration, with significant macular involvement and a complete loss of the foveal photoreceptor layer by the third decade of life [60]. Female carriers of *RP2* mutations exhibit a range of symptoms, from a normal fundus to TLR to peripheral pigmentary changes, generally with a favorable prognosis.

The *RPGR* gene comprises nineteen exons and undergoes extensive alternative splicing, resulting in two main isoforms: the full-length $RPGR^{1-19}$ isoform and the $RPGR^{ORF15}$ isoform. The full-length $RPGR^{1-19}$ isoform, which contains all 19 exons, is widely expressed in various tissues. Conversely, the $RPGR^{ORF15}$ isoform consists of exons 1 to 14 from $RPGR^{1-19}$ plus a unique 3′ terminal exon (ORF15, open reading frame 15) that encodes two critical domains: a regulator of the chromosome condensation 1 (RCC1)-like domain (rich in glycine and glutamic acid) and a C-terminal basic domain with homology to tubulins [61]. The $RPGR^{ORF15}$ isoform is predominantly expressed in rod photoreceptor outer segments. Truncations in this terminal exon disrupt the vital C-terminal domain, with about 80% of *RPGR* frameshift mutations occurring in the ORF15 region [62]. The most frequent pathogenic mutations in $RPGR^{ORF15}$ are small deletions, whereas splice site mutations are rare across the *RPGR* gene [63]. Pathogenic mutations in exons 1–14 of *RPGR* can affect the RCC1-like domain, which is crucial for protein stability and interactions, leading to a loss of function and progressive retinal degeneration [64]. Mutations in the exons specific to the constitutive variant are primarily associated with XL-RP, while mutations in the ORF15 exon, a known mutational hotspot, are also associated with cone dystrophy (COD) and cone–rod dystrophy (CRD) [65].

Mutations in the *RPGR* gene lead to various phenotypes, such as rod–cone dystrophy (RCD) (70%), CRD (6–23%), and COD (7%) [66]. Most mutations in exons 1–14 and the 5′end of ORF15 are associated with RCD, whereas those causing COD or CRD tend to be located at the 3′end of the ORF15 [67]. However, there is no clear consensus on genotype–phenotype correlation. Some studies indicate that mutations in exons 1–14 are associated with more severe disease phenotypes than those in ORF15 [68], while others suggest the opposite [69].

2.5. X-Linked Retinoschisis (XL-RS)

X-linked retinoschisis (XL-RS) is an IRD caused by mutations in the *retinoschisin 1* (RS1) gene, which encodes Retinoschisin-1, a protein essential for the retinal structure and cell adhesion [70]. It is the most prevalent form of juvenile macular degeneration in males [71]. Symptoms generally manifest in early childhood and include reduced visual acuity, strabismus, anisometropia, and progressive vision deterioration. While most patients are diagnosed within the first decade of life, severe visual loss in infancy may present with INS and poor fixation [72,73]. The characteristic clinical sign of XL-RS is foveal schisis, which appears as small cystoid spaces in a spoke-wheel configuration on fundoscopy. Approximately half of the affected males also exhibit peripheral retinoschisis, pigmentary changes, white spiculations, and a metallic sheen. This disorder disrupts the communication between photoreceptors and bipolar cells, causing a diminished b-wave and often an electro-negative ERG [74].

XL-RS typically leads to declining visual acuity in the first or second decade of life, with progressive macular atrophy continuing until the fifth or sixth decade, potentially resulting in legal blindness [75]. ERG is a critical diagnostic tool, showing a reduced b-wave with a relatively preserved a-wave, characteristic of an "electronegative" ERG. Multimodal imaging is often necessary to identify macular schisis, with structural OCT revealing schisis cavities and OCT angiography showing foveal vascular impairment.

3. AAV-Based Gene Therapy for INS-Associated IRDs

Advances in next-generation sequencing (NGS) have significantly enhanced our understanding of the genetic underpinnings of IRDs in children with INS. These developments facilitate the early diagnosis of INS-associated IRDs and allow for the timely recruitment of eligible patients for gene-based therapy clinical trials during the early stages of the disease. Gene therapy, an innovative treatment modality, involves modifying or replacing defective genes or delivering therapeutic molecules such as small interfering RNAs (siRNAs) and proteins [76]. Various strategies are under investigation for delivering therapeutic genes to target cells in vivo, including both viral and non-viral vectors. Among these, viral vectors, particularly adeno-associated viruses (AAVs), are the most commonly employed in both preclinical studies and clinical applications [77].

Discovered in the 1960s, AAVs have emerged as ideal vectors for therapeutic gene delivery due to their minimal integration into host genomes, ability to transduce non-dividing cells, prolonged expression, low immunogenicity, versatile tissue tropism, and relatively ease of production [78,79]. AAVs exhibit a favorable benefit–risk profile, providing long-term gene expression tailored to specific serotypes and tissue tropism [80]. Recombinant AAV vectors (rAAVs), including serotypes AAV2, 5, 6, 8, and 9, along with their engineered capsid variants such as AAV2tYF, AAV2.7m8, and 4D-R100, particularly show promise for in vivo retinal gene therapy, especially for treating various IRDs [20,81,82]. The safety and efficacy of AAV-based retinal gene therapies have been well documented in numerous clinical studies, including the prominent example of Luxturna [20]. Despite their many advantages, AAV vectors have a limited packaging capacity of approximately 4.7 kilobases, posing a challenge for large gene delivery [83]. To overcome this constraint, the co-delivery of dual or triple AAVs, such as those explored for delivering the *ABCA4* gene in Stargardt disease, has been developed [84,85]. However, these delivery systems face significant challenges, particularly the requirement for the co-transduction of target cells.

Depending on the genotypic and phenotypic spectrum of INS-IRDs, as well as the pathogenic mechanisms involved, various DNA- or RNA-based therapeutic strategies are employed in AAV-mediated gene therapy. The therapeutic approach is generally tailored to the specific mutation. For loss-of-function mutations—commonly found in autosomal recessive or X-linked recessive IRDs—AAV-based gene augmentation or replacement is used to deliver a functional copy of the affected gene to the retina. This strategy's efficacy has been validated, notably with the FDA's approval of Luxturna for RPE65-related retinal dystrophy, and is supported by multiple clinical trials. The developmental pathway established by Luxturna has been adapted to explore additional gene therapy techniques for IRDs, such as mitochondrial gene delivery, gene editing tools, RNA interference (RNAi), and microRNA therapies.

In contrast, some autosomal dominant IRDs involve gain-of-function mutations or dominant-negative mutations, which lead to toxic effects. The therapeutic goal for these cases is to inhibit the expression of the mutated genes. Ribozyme-based, interfering RNA-based, and CRISPR-based methods have been developed for this purpose. These approaches can target the disease-causing gene through two strategies: allele-specific inhibition, which selectively silences the mutant allele while allowing the normal allele to be expressed; and allele non-specific inhibition combined with gene augmentation, where both alleles are silenced, and a normal gene is introduced using dual AAV vectors. While this second method is broadly applicable, it faces challenges with vector packaging constraints [86,87]. Additionally, given the complexity and cost of developing therapies for each disease-causing gene, gene-agnostic approaches like optogenetics and gene modifiers offer a promising alternative by targeting common pathways across multiple IRDs, potentially reducing development costs and broadening treatment access [88,89].

3.1. AAV-Based Gene Augmentation Therapy for INS-Associated IRDs in Clinical Trials

As of the second quarter of 2024, more than 60 clinical trials investigating gene therapy for IRDs are listed on Clinicaltrials.gov, accessed on 1 July 2024. This includes 29 trials for RP, 10 for LCA, 4 for achromatopsia, and 6 for X-linked retinoschisis (Table 3). Many of these trials utilize AAV-based gene augmentation or replacement approaches. Additionally, RNA-based strategies, such as RNA interference (RNAi) or microRNA, antisense oligonucleotides (AON), and gene editing techniques like CRISPR-based therapy, are explored for the treatment of IRDs. This review focuses on recent and ongoing AAV-based gene therapy clinical trials targeting INS-associated IRDs. Additionally, we briefly discuss the current status of RNA-based and CRISPR-based therapeutic strategies for IRDs.

Table 3. Clinical trials of targeted gene therapy for INS-associated IRDs.

Gene	Mutation	Treatment (Sponsor)	Agent (Constructs)	Function	Delivery Route	Clinical Trial ID	Trial Phase
Leber Congenital Amaurosis (LCA)							
GUCY2D	Loss of Function	Atsena Therapeutics (ATSN-101)	AAV8-GRK1-GUCY2D	Restores function	Subretinal	NCT03920007	I/II
CEP290	c.2991+1655A>G	Editas (EDIT-101)	AAV5-GRK1-Cas9	Gene editing	Subretinal	NCT03872479 BRILLANCE	I/II
CEP290	c.2991+1655A>G	ProQR Therapeutics	QR-110 (Sepofarsen)	Normal mRNA (AON)	Intravitreal	NCT03913143NCT04855045 NCT03140969	II/III II/III I/II
LCA5	Loss of Function	Opus Genetics	AAV8.hLCA5 (OPGx-001)	Restores function	Subretinal	NCT05616793	I/II
RPE65	Loss of Function	Janssen/MeiraGTx UK II Ltd.	AAV5-PRE65	Restores function	Subretinal	NTC02781480	I/II
			AAV6-RPE65			NCT02946879	I/II
RPE65	Loss of Function	HuidaGene Therapeutics	rAAV2-RPE65 rAAV9-RPE65	Restores function	Subretinal	NCT05906953 NCT06088992	I/II I
RPE65	Loss of Function	Spark Therapeutics	AAV2-hRPE65v2	Restores functional protein	Subretinal	NCT00999609	III
			AAV2-hRPE65v2			NCT01208389	I/II
			AAV2-hRPE65v2			NCT03602820	III
RPE65	Loss of Function	AGTC	rAAV2-CB-hRPE65	Restores function	Subretinal	NCT00749957	I/II
RPE65	Loss of Function	Nantes University Hospital	rAAV2/4.hRPE65	Restores function	Subretinal	NCT01496040	I/II
RPE65	Loss of Function	University of Pennsylvania	rAAV-CBSB-hRPE65	Restores function	Subretinal	NCT00481546	I/II
Achromatopsia (ACHM)							
GNCA3	Loss of Function	Janssen/STZ eyetrial	rAAV2/8.hCNGA3	Restores function	Subretinal	NCT02610582	I/II
GNCA3	Loss of Function	AGTC (Beacon Therapeutics)	rAAV2tYF-PR1.7-hCNGA3	Restores function	Subretinal	NCT02935517	I/II
GNCA3	Loss of Function	Janssen/MeiraGTx UK II Ltd.	AAV2/8-hG1.7p.coCNGA3	Restores function	Subretinal	NCT03758404	I/II
GNCB3	Loss of Function	Janssen/MeiraGTx UK II Ltd.	rAAV2/8.hCNGB3	Restores function	Subretinal	NCT03758404	I/II
GNCB3	Loss of Function	AGTC (Beacon Therapeutics)	rAAV2tYF-PR1.7-hCNGB3	Restores function	Subretinal	NCT02599922	I/II
GNCB3	Loss of Function	Janssen/MeiraGTx UK II Ltd.	AAV2/8-hG1.7p.coCNGB3	Restores function	Subretinal	NCT03001310	I/II

Table 3. *Cont.*

Gene	Mutation	Treatment (Sponsor)	Agent (Constructs)	Function	Delivery Route	Clinical Trial ID	Trial Phase
X-linked Retinoschisis (XL-RS)							
RS1	Loss of Function	AGTC	rAAV2YF-CB-hRS1	Restores function	Intravitreal	NCT02416622	I/II
RS1	Loss of Function	Atsena Therapeutics	ATSN-201 (AAV.SPR.hRS1)	Restores function	Intravitreal	NCT05878860	I/II
RS1	Loss of Function	InnoVec Biotherapeutics	IVB102	Restores function	Intravitreal	NCT06289452	I/II
RS1	Loss of Function	Shanghai General Hospital	LX103	Restores function	Intravitreal	NCT05814952	I/II
RS1	Loss of Function	West China Hospital	JWK002	Restores function	Intravitreal	NCT06345898	I/II
RS1	Loss of Function	National Eye Institute	AAV8-scRS/ IRBPhRS	Restores function	Intravitreal	NCT02317887	I/II
X-linked Retinitis Pigmentosa (XL-RP)							
RPGR	Loss of Function	Janssen	AAV5-hRKp.RPGR	Restores function	Subretinal	NCT04671443 (LUMEOS)	III
RPGR	Loss of Function	AGTC (Beacon Therapeutics)	rAAV2tYF-GRK1-RPGRco	Restores function	Subretinal	NCT06333249 (SKYLINE)	II
RPGR	Loss of Function	AGTC (Beacon Therapeutics)	rAAV2tYF-GRK1-RPGRco	Restores function	Subretinal	NCT04850118 (VISTA)	II/III
RPGR	Loss of Function	4D Molecular Therapeutics	AAV.R100-hcoRPGR (4D-125)	Restores function	Intravitreal	NCT04517149 (EXCEL)	I/II
RPGR	Loss of Function	Janssen	AAV5-hRKp.RPGR	Restores function	Subretinal	NCT05926583 NCT04794101	III III
RPGR	Loss of Function	Frontera Therapeutics	FT-002	Restores function	Intravitreal	NCT06492850	I/II
RPGR	Loss of Function	Biogen	AAV8-RPGR (BIIB112)	Restores function	Subretinal	NCT03584165 (SOLSTICE)	III
Autosomal dominant Retinitis Pigmentosa (adRP)							
RHO	P23H	ProQR Therapeutics	QR-1123	Gene silencing	Intravitreal	NCT04123626 (AURORA)	I/II
RHO	Optogenetic	AbbVie	AAV2-ChR2 (RST-001)	Gene agnostic	Intravitreal	NCT02556736	I/II
RHO	Agnostic	SparingVision	AAV-RdCVF-RdCVFL (SPVN06)	Gene agnostic	Subretinal	NCT05748873	I/II
RHO	Genetic modifier	Ocugen	AAV5-NR2E3 (OCU400-301)	Gene agnostic	Subretinal	NCT06388200	III
Autosomal recessive Retinitis Pigmentosa (arRP)							
RLBP1	Loss of Function	Novartis Therapeutics	AAV8-RLBP1 (CPK850)	Restores function	Subretinal	NCT03374657	I/II
USH2A	USH2A (*exon 3*)	Laboratoires Thea	Ultevursen (QR-421a)	Gene silencing	Intravitreal	NCT05176717 NCT05158296	II/III
PDE6A	Loss of Function	STZ eyetrial	rAAV-hPDE6A	Restores function	Subretinal	NCT04611503	I/II
PDE6B	Loss of Function	Coave Therapeutics	AAV5-hPDE6B	Restore function	Subretinal	NCT03328130	I/II
CNGA1	Loss of Function	ViGeneron GmbH	AAV2.NN-CNGA1 (VG-901)	Restore function	Intravitreal	NCT06291935	I
MERTK	Loss of Function	Fowzan Alkuraya	AAV2-MERTK	Restores function	Subretinal	NCT01482195	I

3.1.1. AAV-Based Gene Augmentation Therapy for Achromatopsia

Achromatopsia (ACHM) is a recessive disorder resulting from loss-of-function mutations in any of the six genes responsible for cone photoreceptor function. Currently, AAV-based gene therapies are being developed focusing on two of the most frequently affected genes, *CNGA3* and *CNGB3* [90]. The initial clinical trial assessed the safety and efficacy of the subretinal injection of AAV8.hCNGA3 in nine individuals with CNGA3-associated ACHM (NCT02610582) [91]. Over a one-year period, the treatment was well tolerated, with no serious adverse events reported. Despite the congenital loss of cone-driven light signaling in patients with CNGA3-ACHM, AAV8.CNGA3 treatment led to improvements in secondary endpoints related to cone function. These include increases in visual acuity and contrast sensitivity compared to the baseline in all nine treated patients, persisting for at least three years post-treatment [91]. An ongoing phase IIb clinical trial is targeting the treatment of the second eye in the initial patients and the treatment of children aged 6–12 years.

At present, four additional phase I/II AAV-based gene therapy trials are actively recruiting adults and children with CNGA3- and CNGB3-related ACHM. Two phase I/II open-label, dose-escalation trials for CNGA3 (NCT02935517) and CNGB3 (NCT02599922) are utilizing subretinal injections of AAV2 variant vectors. These vectors employ an engineered cone opsin promoter to drive the expression of CNGA3 (rAAV2tYF-PR1.7-hCNGA3) in patients with CNGA3-associated ACHM and CNGB3 (rAAV2tYF-PR1.7-hCNGB3) in patients with CNGB3-associated ACHM. Participants in both studies were sequentially assigned to one of four dose groups. The current data suggest that rAAV2tYF-PR1.7-hCNGB3 treatment has improved photosensitivity in some patients, while the effect of rAAV2tYF-PR1.7-hCNGA3 appears less encouraging.

Additionally, two similar phase I/II, open-label, dose-escalation clinical trials are underway for CNGA3 (NCT03758404) and CNGB3 (NCT03001310). One trial is evaluating AAV2/8-hG1.7p.coCNGA3 (AAV-CNGA3) in adults and children with CNGA3-associated ACHM, while the other is assessing AAV2/8-hG1.7p.coCNGB3 (AAV-CNGB3) in adults and children with CNGB3-associated ACHM. The primary outcome measure for each of these trials is the incidence of treatment-related adverse events at six months. Secondary outcomes include improvements in visual function, retinal function, and quality of life.

3.1.2. AAV-Based Gene Augmentation Therapy for X-Linked Retinoschisis

There are currently two ongoing clinical trials using AAV-based gene augmentation therapy for X-linked RS. Due to the retinal fragility and tendency for retinal detachments associated with XL-RS, the AAV vectors are administered via intravitreal (IVT) injection rather than subretinal injection, which is the method typically used in other retinal gene therapy. Preclinical research has demonstrated that the internal limiting membrane (ILM), which poses a key barrier to IVT AAV-mediated gene transfer, is compromised in XL-RS, thereby enhancing the effectiveness of IVT injections [92].

The National Eye Institute (NEI) conducted a phase I/IIa trial (NCT02317887) to assess the safety and efficacy of AAV8-RS1 gene therapy in nine male patients with XLRS. At the 18-month follow-up, the treatment and vector were generally well tolerated. However, functional outcome measures such as BCVA, microperimetry retinal sensitivity, and ERG response showed no significant visual improvements from baseline. One patient who receiving a high dose (1×10^{11} vg compared to 1×10^{10} vg) exhibited temporary closure of macular cavities two weeks post-injection, likely due to RS1 protein activity [92].

Another phase I/II trial (NCT02416622) utilized an AAV2 vector variant encoding the RS-1 gene (rAAV2tYF-CB-hRS1) for IVT delivery. Preclinical data suggested AAV2 more effectively transduces ganglion cells than AAV8 following IVT injection [93]. The six-month results indicated a favorable safety profile but no functional improvements, and no measurable benefits were observed at the 12-month mark [94]. Given the slow progression of XL-RS, longer-term follow-up is essential to determine potential functional gains. Recently, a new phase I/II open-label trial (NCT05878860) has commenced. This trial aims to evaluate the safety and tolerability of ATSN-201 (AAV.SPR-hRS1) in male patients aged 6 to 64 with XL-RS.

3.1.3. AAV-Based Gene Augmentation Therapy for Leber Congenital Amaurosis

LCA2 resulting from RPE65-mutation was the first IRD to undergo gene therapy trials. Biallelic disease-causing variants in RPE65 are responsible for 5–10% of LCA cases. To date, more than 10 clinical trials have evaluated the efficacy and safety of a single subretinal injection of AAV2-hRPE65 in patients with confirmed biallelic RPE65 mutations. These studies primarily assess functional vision improvements rather than structural changes.

A phase I/II trial (NCT00481546) involving 15 young patients receiving subretinal rAAV2-CBSB-hRPE65 injection reported no serious adverse events, with all patients experiencing varying degrees of visual function improvement, which persisted for up to 36 months post-injection [95]. In a phase III randomized controlled trial (RCT) (NCT00999609), 31 participants (21 in the intervention group and 10 in the control group) with confirmed RPE65-associated retinal dystrophy received bilateral sequential subretinal injections of AAV2-hRPE65v2 (voretigene neparvovec, Luxturna) [96,97]. This phase III trial demonstrated significant improvements in multi-luminance mobility test (MLMT) scores, the full-field light sensitivity threshold (FST), and visual fields (VFs) in the treatment group compared to the controls. These clinically meaningful effects were maintained for at least one year, with no serious adverse events or harmful immune responses reported [96].

A recent review of six studies (five prospective and one RCT) on RPE65-LCA gene therapy indicated that visual function improvements were generally limited to two years post-treatment [98]. However, BCVA and FST were the only visual function outcomes consistently analyzed across these studies. Other visual function measures, such as MLMT and VF testing, might show sustained improvement beyond two years. While previous safety analyses showed a favorable adverse event profile, they noted a thinning of central retinal thickness in treated eyes compared to untreated eyes 2–3 years after the intervention [98,99].

Beyond *RPE65*, several AAV-based gene therapy trials are in progress for LCA, targeting different gene mutations including *GUCY2D*, *CRB1*, and *CEP290*. *GUCY2D* is associated with LCA1, with biallelic mutations accounting for 10% to 20% of cases. A phase I/II trial (NCT03920007) was launched to evaluate the safety and tolerability of ascending doses of AAV8-hGRK1-GUCY2D, administered via subretinal injection in patients with GUCY2D-associated LCA1 [100]. This ongoing trial enrolled 15 patients aged six years or older, with nine in dose-escalation groups and six in dose-expansion groups. Early results from the first three patients treated with AAV8-hGRK1-GUCY2D indicated rod photoreceptor vision improvement by FST testing in the treated eye, with one patient experiencing a 0.3 LogMAR improvement in vision [100]. Two patients showed improvements in visual function and functional vision, reaching more than 3 log units and nearing healthy rod vision [101].

3.1.4. AAV-Based Gene Augmentation Therapy for X-Linked Retinitis Pigmentosa

XL-RP is primarily driven by mutations in several genes, with RPGR accounting for 80% of all XL-RP cases and approximately 11% of all RP cases [62]. The majority of XL-RP patients have a mutated RPGR gene, making it a key target for AAV-based gene therapy. The initial phase I/II clinical trial for XL-RP aimed to evaluate the safety and efficacy of a codon-optimized AAV8-RPGR, known as cotoretigene tiparvovec (BIIB112) (NCT03116113, XIRIUS). This trial involved subretinal injections of low, intermediate, and high doses of BIIB112 in 18 XL-RP patients. The six-month follow-up indicated no significant dose-limiting safety issues, except for steroid-responsive retinal inflammation at higher doses [102]. Some patients showed improved visual acuity and retinal structure stability. A post hoc analysis of the XIRIUS and natural history study (XOLARIS) for XL-RP revealed that the four patients who received the highest doses of BIIB112 exhibited early improvements in retinal sensitivity and low-luminance visual acuity at one year [102]. However, in the following phase II/III of the XIRIUS study, the primary endpoint of statistically significant improvement in microperimetry was not achieved. Positive trends were observed in the visual acuity of treated eyes under low luminance vision measurements. Patients exiting the trial have been offered enrollment in a separate phase III trial (NCT03584165) to monitor long-term outcomes over five years.

Additionally, two other phase I/II clinical trials have been conducted to evaluate AAV-based gene augmentation therapy for XL-RP [103]. One phase I/II, open-label, dose escalation study (NCT03252847) assessed the efficacy and safety of AAV5-hRKp.RPGR over 18 months in adults and children over 5 years old with XL-RP. The one-year data from this trial demonstrated that AAV5-hRKp.RPGR was generally well tolerated and led to significant vision improvements. Currently, a phase III randomized, controlled study (NCT04671433) of AAV5-hRKp.RPGR and a phase III trial assessing patient safety for up to 60 months (NCT04312672) are underway.

The second trial, which is also an open-label phase I/II dose escalation study (NCT03316560), involves 29 patients receiving subretinal injections of rAAV2tYF-GRK1-RPGR. This gene therapy, driven by the GRK1 promoter for targeted photoreceptor expression, is being assessed for safety and efficacy in both adults and children with XL-RP. Participants were divided into five cohorts: 21 received injections in the central macula, while 8 were injected in the peripheral retinal regions. Surgery-related adverse events were mild to moderate. The twelve-month follow-up data in male patients with XL-RP indicated significant improvements in visual function, particularly retinal sensitivity. An ongoing phase II/III trial (NCT04850118) aims to further evaluate the therapy's efficacy, safety, and tolerability by comparing two doses of rAAV2tYF-GRK1-RPGR with an untreated control group. Additionally, a phase III trial (NCT04794101) involving the bilateral administration of two vector genome doses (2×10^{11} and 4×10^{11}) is currently in progress, along with a follow-up study.

While other RPGR clinical trials utilize subretinal injections for AAV-mediated RPGR delivery, 4DMT employs intravitreal injections. The EXCEL trial (NCT04517149), a phase I/II dose-escalation study, used a capsid-engineered AAV2 vector (4D-R100) developed by 4DMT to deliver a functional copy of the RPGR gene to the retina. The study aimed to evaluate the efficacy, safety, and maximum tolerated dose of 4D-125 in XL-RP patients. In the phase I portion, participants received one of two doses (3×10^{11} vg/eye or 1×10^{12} vg/eye), and in the phase II expansion, the higher dose (1×10^{12} vg/eye) was administered. Results from the phase I/II trial in advanced RPGR patients, with limited or no measurable photoreceptor regions and low or no retinal sensitivity, showed no dose-limiting toxicity or serious adverse events. Two patients exhibited increased retinal sensitivity in treated eyes at 6 and 9 months follow-up.

3.1.5. AAV-Based Gene Therapy for MERTK-, RLBP1-, and PDE6B-Associated RP

MERTK (MER Proto-Oncogene, Tyrosine Kinase) is associated with RP and encodes a transmembrane protein found in RPE cells. This protein plays a role in the uptake of photoreceptor outer segments by RPE cells. Mutations in the *MERTK* gene hinder this phagocytic process, leading to photoreceptor cell degeneration and ultimately RP [104]. A phase I clinical trial (NCT01482195) evaluated the effects of subretinal injections of rAAV2-VMD2-hMERTK in six patients with MERTK-associated RP [105]. Over a two-year follow-up, no significant ocular or systemic adverse events were observed. Three patients exhibited measurable visual acuity improvements, but only one patient maintained these gains over two years. Two patients who had temporary visual improvements developed bilateral cataracts, which may have influenced the assessment of the treatment's efficacy.

Retinaldehyde binding protein 1 (RLBP1) is a gene associated with autosomal recessive RP, encoding a protein expressed by Müller glial and RPE cells [106]. Mutations in *RLBP1* impair the visual cycle between the RPE and photoreceptors, similar to the dysfunction seen in RPE65-associated LCA. An ongoing phase I/II clinical trial (NCT03374657) is examining the safety, tolerability, and efficacy of subretinal administration of AAV8-RLBP1 (CPK850) in patients with RLBP1-RP, though no data are yet available.

Phosphodiesterase 6 (PDE6) is a complex enzyme that hydrolyzes cGMP in rod photoreceptors, reducing its concentration in response to light activation of the G-protein-coupled receptor during phototransduction. The PDE6 complex consists of alpha, beta, and two gamma subunits, with mutations in the alpha and beta subunits each accounting for approximately 4% of all RP cases [107]. A phase I/II clinical trial (NCT03328130) evaluated the safety and efficacy of AAV5-hPDE6B via subretinal injections at three different doses. The trial showed significant visual function improvement at the highest dose with a favorable safety profile [108]. Recently, STZ eyetrial has been conducting a phase I/II study with their vector, rAAV-hPDE6A, to evaluate its safety and efficacy in patients with *PDE6A* mutations (NCT04611503, PIGMENT). No data are currently available for this trial.

3.2. AAV-Based Gene Silencing for INS-Associated IRDs

As previously mentioned, gene augmentation using AAV vectors is a well-established method for treating AR-associated IRDs. However, this technique is not effective for AD-associated retinal diseases, which constitute approximately 15–20% of all IRDs. In such cases, delivering a normal copy of the mutated gene does not suffice; instead, it is necessary to silence the dominant allele. Various methods can achieve this silencing. One option is RNA-targeting therapy, including RNA interference (RNAi), microRNAs, and antisense oligonucleotides (AONs), which work to silence gene expression. For IRDs caused by splicing defects, AONs have been effective in restoring proper pre-mRNA splicing, such as in the case of the common *CEP290* gene mutation (c.2991+1655A>G) associated with LCA10 [109,110]. In a phase I/II clinical trial (NCT03140969), an AON-based drug (QR-110) was administered via intravitreal injections to 10 patients with CEP290-LCA every three months for up to 12 months [109]. Results from the phase I/II trial demonstrated significant improvements across all outcome measures, including BCVA, OCT structure, mobility, and nystagmus measurements [111]. Following a post-surgical follow-up of at least three months, the study is now enrolling patients with the compound heterozygous or homozygous intron 26 variant c.2991+1655A>G in *CEP290* for a phase III trial (NCT03913143). Additionally, AON therapy has shown potential in correcting splicing signals to induce in-frame skipping of exon 13 in the *USH2A* gene, resulting in the production of a shorter yet functional protein in Usher syndrome [112].

Gene editing techniques, such as CRISPR/Cas9 and transcription activator-like effector nucleases (TALENs), can also be utilized to silence mutant alleles that cause toxic gain-of-function effects. The CRISPR/Cas9 method has been successfully applied in studies on autosomal dominant RP associated with rhodopsin (*RHO*) mutations (RHO-adRP) [113,114]. Moreover, correcting splice defects using this strategy has been effective in preclinical models with a deep intronic mutation in *CEP290*, showing both safety and efficacy in the retina [115]. The development of a self-limiting CRISPR/Cas9 system has further minimized the immune response by reducing the duration of Cas9 expression [115].

3.3. AAV-Based Gene Editing for INS-Associated IRDs

Precise genome editing technologies, particularly the CRISPR/Cas9 system, are becoming promising alternatives to traditional gene therapy approaches. The CRISPR/Cas9 system is especially valuable for therapeutic editing due to its adaptability in targeting virtually any gene. This system includes CRISPR RNA, which contains spacer sequences that pair with the target DNA sequence, and a tracer-RNA that binds to Cas endonucleases. Accurate target recognition relies on the complementary base pairing between the guide RNA (gRNA) and the target genomic region, necessitating the presence of a protospacer adjacent motif (PAM). PAM sequences are essential for Cas proteins to locate their target DNA, and different Cas variants have been designed to recognize a variety of PAM sequences, increasing the system's flexibility [116,117].

Cas endonuclease, such as *Streptococcus pyogenes* (SpCas9) or *Staphylococcus aureus* (SaCas9), are guided to specific DNA target sites in the genome through a gRNA, inducing DNA double-strand breaks (DSBs). These DSBs subsequently activate one of several DNA repair mechanisms: non-homologous end joining (NHEJ), homology-directed repair (HDR), or microhomology-mediated end joining (MMEJ) [118]. Among these, the NHEJ pathway is the most readily used by cells to repair DSBs. During the repair process, NHEJ often introduces random elements into the genome, resulting in substitutions, insertions, and/or deletions (indels) which can generate disruptive mutations. This system works well for disrupting a gene or eliminating a segment of DNA, both of which require DSBs in the target genome.

Unlike NHEJ, the HDR pathway can repair DSBs with high fidelity during the S and G2 phases of the cell cycle, enabling precise gene corrections using DNA donor templates. HDR employs a homologous single- or double-stranded DNA template to guide the repair process, which can be sourced from the host genome or provided externally for targeted genome editing. Research has shown that targeted in vivo gene integration can rectify mutations in genes such as *PDE6B*, *NR2E3*, *RPGR*, and *RPE65* [119–121]. However, HDR has limitations that make it less ideal for precise gene corrections in the retina. Cas9-induced DSBs often lead to significant indels, undermining the therapeutic potential of HDR. Additionally, the HDR pathway is mainly active in dividing cells, resulting in low efficiency in post-mitotic cells like photoreceptors and RPE cells. Consequently, HDR is generally unsuitable for precise genome editing in the eye.

Addressing IRDs caused by mutations in large genes exemplifies an advanced Cas9-based therapeutic approach in the retina. Mutations in the 7.4 kb *CEP290* gene are among the most frequent causes of LCA identified. The most common mutation in *CEP290* (c.2991+1655A>G) is a point mutation within an intron that creates an abnormal splice donor site, leading to the inclusion of 128 bp and the formation of a premature stop codon. Recent studies have shown that a single AAV-mediated delivery of Staphylococcus aureus Cas9 (SaCas9) with two sgRNAs can excise the aberrant splice donor created by the *CEP290* mutation in mouse and non-human primate models [122].

Several gene editing-based clinical trials for IRDs are currently in progress. One such trial involves using AAV vectors to deliver SaCas9 and CEP290-specific gRNAs to photoreceptor cells via subretinal injection. Editas Medicine sponsored a phase I/II clinical trial (NCT03872479) for LCA10, focusing on the safety, tolerability, and efficacy of their CRISPR/Cas9-based treatment, EDIT-101 [123]. This gene editing therapy employs AAV5 to deliver SaCas9 and CEP290-specific gRNAs, aiming to correct intron mutations in the *CEP290* gene [124].

The trial enrolled 14 patients with subretinal injections of EDIT-101 at varying doses: two adults at a low dose, five adults at an intermediate dose, five adults at a high dose, and two pediatric patients at the intermediate dose. Data from the phase I/II BRILLIANCE trial indicated that EDIT-101 was generally safe across all dose levels. Six patients experienced meaningful improvements from baseline in cone-mediated vision, while nine patients (64%) showed meaningful enhancements in BCVA, red light sensitivity, or mobility test scores [123]. However, only three patients exhibited clinically significant visual improvement, with two of these responders having homozygous mutations and just one of the twelve heterozygous patients responding positively. This indicates that EDIT-101 may be more effective in homozygous LCA10 patients compared to heterozygous ones.

Due to the limited number of eligible subjects, Editas Medicine has temporarily halted recruitment for the trial but will continue long-term follow-ups with the treated patients. Despite the challenges of a small patient pool and variable outcomes, EDIT-101 marks a notable advancement in gene editing for ocular diseases. Ongoing research will further refine gene editing methods to enhance their clinical applicability for these conditions.

Innovative Cas-based tools, including base editors (BEs) and prime editors (PEs), have emerged for precise genetic corrections in post-mitotic retinal cells without causing DSBs [125]. BEs utilize a modified catalytically impaired "nicking" Cas nuclease (dead Cas9) linked to DNA-modifying enzymes (cytidine or adenine deaminases) to facilitate the conversion of single bases, such as C to T or A to G [126]. These tools enable the accurate correction of point mutations or single-nucleotide polymorphisms at the target position of genomic DNA [127]. The use of split BE dual AAV vectors has shown therapeutic efficiency in achieving base editing in mouse retinal cells and a rodent model of RP [128–130]. Furthermore, single-AAV adenine base editor (ABE) systems have been created by reducing the size of ABEs and AAV components, thus enhancing in vivo targeting efficiency, editing effectiveness, lowering required AAV doses, and minimizing potential toxicity [131]. Given that many IRDs are caused by point mutations, base editors are expected to be widely applicable for therapeutic use.

Prime editors (PEs) signify a considerable advancement in CRISPR/Cas9-based gene editing. PEs also utilize a modified Cas9, specifically the Cas9-H840A nickase, to cut a single DNA strand and prevent DSB formation. The Cas9 nickase is linked to a reverse transcriptase, which enables the desired edit to be transcribed in reverse at the target site.

PEs use a prime editing guide RNA (pegRNA) that contains both template sequences for reverse transcription and protospacer sequences. The Cas9-H840A nickase cleaves the target DNA strand, allowing the reverse transcriptase to synthesize a template DNA strand, thereby modifying the target site [132]. This method allows PEs to perform targeted genome modifications, including various substitutions and small insertions or deletions (indels), by leveraging cellular DNA repair mechanisms [132].

Several research groups have started using PEs for editing mutations associated with IRDs. In human cells, split PEs delivered via dual AAV1 vectors can mediate insertions and base conversions at multiple endogenous sites [133]. Two independent studies in rd12 mice have evaluated AAV-delivery of PE to edit the RPE65 mutation [134,135]. Subretinal injection of dual-AAVs carrying PE and pegRNA achieved a delivery efficiency of 23% and approximately 6.4% editing efficiency without any detectable indels, unintended substitutions, or off-target effects. Improved dark-adapted ERG responses, reaching up to 67% of the wild-type amplitude, were observed [134]. Another study using an optimized dual-AAV split-PE3 system administered via subretinal injection in rd12 mice showed $11.4 \pm 2.3\%$ editing in RPE cells, the restoration of RPE65 protein levels, and improved photoreceptor function and survival [135]. Although reports on the use of PEs in IRD gene therapy are still limited, they hold substantial promise for ocular gene therapy due to their extensive target site selection, high editing efficiency, and low off-target rates [135].

4. Challenges and Prospectives

As previously mentioned, gene-based therapy has shown great potential for addressing the genetic defects causing IRDs in children with INS. Currently, the majority of clinical trials for IRDs focus on AAV-based gene therapies, which target a variety of retinal conditions such as LCA, X-linked RP, achromatopsia, and X-linked retinoschisis. Innovations in AAV vector engineering, delivery methods, and safety improvements have significantly increased the effectiveness of these therapies, expanding their applicability to a wider range of retinal conditions.

Nevertheless, numerous challenges must be tackled before AAV-based gene therapy can become a truly curative treatment for IRDs [136]. Despite the limited packaging capacity of AAVs, key issues primarily pertain to safety, predictability, and the durability of the gene therapy outcomes. The genetic heterogeneity of IRDs requires the development of mutation-specific therapies, which is both time-consuming and costly. Ensuring long-term systemic efficacy and safety, particularly in terms of immune responses to AAV vectors, is crucial. Additionally, addressing manufacturing and regulatory challenges is essential for making these therapies widely available. Despite these hurdles, rapid progress in gene therapy techniques, coupled with a deeper understanding of the genetic and phenotypic diversity of IRDs, holds great promise for developing effective treatments.

The innovative CRISPR/Cas9 genome editing technology represents a promising approach for treating IRDs. Unlike AAV-mediated gene augmentation, CRISPR/Cas9 has the potential to address a broader spectrum of diseases with more enduring effects. However, efficient in vivo delivery remains a significant challenge for the clinical application of CRISPR/Cas therapeutics. This is because the system often requires two separate AAV vectors—one for the Cas endonuclease and another for the guide RNA (gRNA) expression cassette. Identifying smaller Cas9 orthologs that can be packaged together with the gRNA into a single AAV vector could help overcome this limitation. Additionally, the use of AAV vectors for long-term expression of CRISPR/Cas9 or base editors can result in significant off-target editing, which may reduce the precision of on-target gene edits. To address this, research efforts are increasingly focused on non-viral methods for the transient delivery of the editing machinery, such as nanoparticles [137] or engineered virus-like particles [138]. Furthermore, in vivo genome editing raises various safety and efficacy concerns. Therefore, the risks and benefits of genome editing therapies must be thoroughly evaluated to ensure their clinical viability.

In summary, combining precise genetic diagnostics with advanced AAV-based gene therapy presents a promising strategy for treating INS-associated IRDs. Future research should focus on broadening the range of treatable mutations, optimizing vector design, and guaranteeing the long-term safety and effectiveness of these therapies. By addressing these issues, we can advance toward effectively treating vision loss due to INS-associated IRDs, thus providing hope to affected families and contributing to the overarching goal of eliminating childhood blindness.

5. Conclusions

In conclusion, this review provides a comprehensive overview of the phenotypic and genetic landscape of IRDs, with a particular emphasis on recent advancements in understanding INS-associated IRDs. We have explored the potential of AAV-based gene therapy for retinal diseases, focusing on the clinical trials targeting INS-associated IRDs. A thorough understanding of the fundamental mechanisms of these diseases is essential for developing effective treatment. The diverse genotype–phenotype landscape of these conditions presents both obstacles and opportunities for therapeutic innovation.

High-throughput NGS technologies have significantly enhanced our understanding of the genetic underpinnings of INS-related IRDs, catalyzing the development of gene-based therapies for previously incurable conditions. The approval of Luxturna, the first AAV-based gene therapy, has paved the way for a significant number of gene-based therapies targeting INS-associated IRDs to enter late-stage clinical trials. These trials offer hope for improved vision and quality of life for pediatric patients, with additional approvals of gene-based therapies expected in the near future.

Despite ongoing challenges in AAV-based gene augmentation therapies, the continuous advancement of genetic therapeutic strategies—including gene editing therapies, capsid-engineered vectors, and delivery techniques—will be crucial in overcoming existing obstacles to effective and durable gene therapy for INS-associated IRDs. These advancements are poised to significantly transform the treatment paradigm for INS-associated IRDs.

Supplementary Materials: The following supporting information can be downloaded at: https://www.mdpi.com/article/10.3390/life14111356/s1: Table S1: INS-associated IRDs and their corresponding causative genes. Table S2: INS-associated IRD genes and their corresponding inheritance patterns and phenotypes.

Author Contributions: Conceptualization, X.G. and R.W.H.; Writing—original draft preparation, X.G.; Writing—review and editing, X.G. and R.W.H. All authors have read and agreed to the published version of the manuscript.

Funding: This research received no external funding.

Institutional Review Board Statement: Not applicable.

Informed Consent Statement: Not applicable.

Data Availability Statement: Not applicable.

Acknowledgments: Support for this research was provided by the following divisions of Akron Children's Hospital, Akron, OH: Department of Ophthalmology, and The Rebecca D. Considine Research Institute.

Conflicts of Interest: The authors declare no conflicts of interest.

References

1. Richards, M.D.; Wong, A. Infantile nystagmus syndrome: Clinical characteristics, current theories of pathogenesis, diagnosis, and management. *Can. J. Ophthalmol.* **2015**, *50*, 400–408. [CrossRef] [PubMed]
2. Sarvananthan, N.; Surendran, M.; Roberts, E.O.; Jain, S.; Thomas, S.; Shah, N.; Proudlock, F.A.; Thompson, J.R.; McLean, R.J.; Degg, C.; et al. The prevalence of nystagmus: The Leicestershire nystagmus survey. *Investig. Ophthalmol. Vis. Sci.* **2009**, *50*, 5201–5206. [CrossRef] [PubMed]

3. Nash, D.L.; Diehl, N.N.; Mohney, B.G. Incidence and Types of Pediatric Nystagmus. *Am. J. Ophthalmol.* **2017**, *182*, 31–34. [CrossRef] [PubMed]
4. Papageorgiou, E.; McLean, R.J.; Gottlob, I. Nystagmus in childhood. *Pediatr. Neonatol.* **2014**, *55*, 341–351. [CrossRef] [PubMed]
5. Bertsch, M.; Floyd, M.; Kehoe, T.; Pfeifer, W.; Drack, A.V. The clinical evaluation of infantile nystagmus: What to do first and why. *Ophthalmic Genet.* **2017**, *38*, 22–33. [CrossRef]
6. Weiss, A.H.; Biersdorf, W.R. Visual sensory disorders in congenital nystagmus. *Ophthalmology* **1989**, *96*, 517–523. [CrossRef]
7. Rim, J.H.; Lee, S.T.; Gee, H.Y.; Lee, B.J.; Choi, J.R.; Park, H.W.; Han, S.H.; Han, J. Accuracy of Next-Generation Sequencing for Molecular Diagnosis in Patients With Infantile Nystagmus Syndrome. *JAMA Ophthalmol.* **2017**, *135*, 1376–1385. [CrossRef]
8. Gottlob, I. Nystagmus. *Curr. Opin. Ophthalmol.* **2000**, *11*, 330–335. [CrossRef]
9. Cavuoto, K.M.; Binenbaum, G.; Chang, M.Y.; Heidary, G.; Morrison, D.G.; Trivedi, R.H.; Kim, S.J.; Pineles, S.L. Genetic testing for infantile nystagmus syndrome with or without associated findings. *J. Am. Assoc. Pediatr. Ophthalmol. Strabismus* **2023**, *27*, 259–264. [CrossRef]
10. Brodsky, M.C.; Dell'Osso, L.F. A unifying neurologic mechanism for infantile nystagmus. *JAMA Ophthalmol.* **2014**, *132*, 761–768. [CrossRef]
11. Hertle, R.W. A Story of Discovery and Change: What We Learned from Studying Nystagmus in Infancy and Childhood. *J. Binocul. Vis. Ocul. Motil.* **2022**, *72*, 113–130. [CrossRef] [PubMed]
12. Duncan, J.L.; Pierce, E.A.; Laster, A.M.; Daiger, S.P.; Birch, D.G.; Ash, J.D.; Iannaccone, A.; Flannery, J.G.; Sahel, J.A.; Zack, D.J.; et al. Inherited Retinal Degenerations: Current Landscape and Knowledge Gaps. *Transl. Vis. Sci. Technol.* **2018**, *7*, 6. [CrossRef] [PubMed]
13. Berger, W.; Kloeckener-Gruissem, B.; Neidhardt, J. The molecular basis of human retinal and vitreoretinal diseases. *Prog. Retin. Eye Res.* **2010**, *29*, 335–375. [CrossRef] [PubMed]
14. Self, J.E.; Lee, H. Novel therapeutics in nystagmus: What has the genetics taught us so far? *Ther. Adv. Rare Dis.* **2021**, *2*, 2633004021998714. [CrossRef]
15. Dias, M.F.; Joo, K.; Kemp, J.A.; Fialho, S.L.; da Silva Cunha, A., Jr.; Woo, S.J.; Kwon, Y.J. Molecular genetics and emerging therapies for retinitis pigmentosa: Basic research and clinical perspectives. *Prog. Retin. Eye Res.* **2018**, *63*, 107–131. [CrossRef]
16. Moon, D.; Park, H.W.; Surl, D.; Won, D.; Lee, S.T.; Shin, S.; Choi, J.R.; Han, J. Precision Medicine through Next-Generation Sequencing in Inherited Eye Diseases in a Korean Cohort. *Genes* **2021**, *13*, 27. [CrossRef]
17. He, X.; Fu, Y.; Ma, L.; Yao, Y.; Ge, S.; Yang, Z.; Fan, X. AAV for Gene Therapy in Ocular Diseases: Progress and Prospects. *Research* **2023**, *6*, 0291. [CrossRef]
18. Zhou, R.; Caspi, R.R. Ocular immune privilege. *F1000 Biol. Rep.* **2010**, *2*, 3. [CrossRef]
19. Ghoraba, H.H.; Akhavanrezayat, A.; Karaca, I.; Yavari, N.; Lajevardi, S.; Hwang, J.; Regenold, J.; Matsumiya, W.; Pham, B.; Zaidi, M.; et al. Ocular Gene Therapy: A Literature Review with Special Focus on Immune and Inflammatory Responses. *Clin. Ophthalmol.* **2022**, *16*, 1753–1771. [CrossRef]
20. Ail, D.; Malki, H.; Zin, E.A.; Dalkara, D. Adeno-Associated Virus (AAV)—Based Gene Therapies for Retinal Diseases: Where are We? *Appl. Clin. Genet.* **2023**, *16*, 111–130. [CrossRef]
21. Carss, K.J.; Arno, G.; Erwood, M.; Stephens, J.; Sanchis-Juan, A.; Hull, S.; Megy, K.; Grozeva, D.; Dewhurst, E.; Malka, S.; et al. Comprehensive Rare Variant Analysis via Whole-Genome Sequencing to Determine the Molecular Pathology of Inherited Retinal Disease. *Am. J. Hum. Genet.* **2017**, *100*, 75–90. [CrossRef] [PubMed]
22. Allikmets, R. Leber congenital amaurosis: A genetic paradigm. *Ophthalmic Genet.* **2004**, *25*, 67–79. [CrossRef] [PubMed]
23. Ahmed, E.; Loewenstein, J. Leber congenital amaurosis: Disease, genetics and therapy. *Semin. Ophthalmol.* **2008**, *23*, 39–43. [CrossRef] [PubMed]
24. Koenekoop, R.K. An overview of Leber congenital amaurosis: A model to understand human retinal development. *Surv. Ophthalmol.* **2004**, *49*, 379–398. [CrossRef]
25. Kumaran, N.; Moore, A.T.; Weleber, R.G.; Michaelides, M. Leber congenital amaurosis/early-onset severe retinal dystrophy: Clinical features, molecular genetics and therapeutic interventions. *Br. J. Ophthalmol.* **2017**, *101*, 1147–1154. [CrossRef]
26. Kondkar, A.A.; Abu-Amero, K.K. Leber congenital amaurosis: Current genetic basis, scope for genetic testing and personalized medicine. *Exp. Eye Res.* **2019**, *189*, 107834. [CrossRef]
27. Boye, S.E. Leber congenital amaurosis caused by mutations in GUCY2D. *Cold Spring Harb. Perspect. Med.* **2014**, *5*, a017350. [CrossRef]
28. Sharon, D.; Wimberg, H.; Kinarty, Y.; Koch, K.W. Genotype-functional-phenotype correlations in photoreceptor guanylate cyclase (GC-E) encoded by GUCY2D. *Prog. Retin. Eye Res.* **2018**, *63*, 69–91. [CrossRef]
29. Bouzia, Z.; Georgiou, M.; Hull, S.; Robson, A.G.; Fujinami, K.; Rotsos, T.; Pontikos, N.; Arno, G.; Webster, A.R.; Hardcastle, A.J.; et al. GUCY2D-Associated Leber Congenital Amaurosis: A Retrospective Natural History Study in Preparation for Trials of Novel Therapies. *Am. J. Ophthalmol.* **2020**, *210*, 59–70. [CrossRef]
30. den Hollander, A.I.; Koenekoop, R.K.; Yzer, S.; Lopez, I.; Arends, M.L.; Voesenek, K.E.; Zonneveld, M.N.; Strom, T.M.; Meitinger, T.; Brunner, H.G.; et al. Mutations in the CEP290 (NPHP6) gene are a frequent cause of Leber congenital amaurosis. *Am. J. Hum. Genet.* **2006**, *79*, 556–561. [CrossRef]

31. Leroy, B.P.; Birch, D.G.; Duncan, J.L.; Lam, B.L.; Koenekoop, R.K.; Porto, F.B.O.; Russell, S.R.; Girach, A. Leber congenital amaurosis due to cep290 mutations—Severe vision impairment with a high unmet medical need: A Review. *Retina* **2021**, *41*, 898–907. [CrossRef] [PubMed]
32. Sheck, L.; Davies, W.I.L.; Moradi, P.; Robson, A.G.; Kumaran, N.; Liasis, A.C.; Webster, A.R.; Moore, A.T.; Michaelides, M. Leber Congenital Amaurosis Associated with Mutations in CEP290, Clinical Phenotype, and Natural History in Preparation for Trials of Novel Therapies. *Ophthalmology* **2018**, *125*, 894–903. [CrossRef] [PubMed]
33. Daich Varela, M.; Georgiou, M.; Alswaiti, Y.; Kabbani, J.; Fujinami, K.; Fujinami-Yokokawa, Y.; Khoda, S.; Mahroo, O.A.; Robson, A.G.; Webster, A.R.; et al. CRB1-Associated Retinal Dystrophies: Genetics, Clinical Characteristics, and Natural History. *Am. J. Ophthalmol.* **2023**, *246*, 107–121. [CrossRef] [PubMed]
34. Ehrenberg, M.; Pierce, E.A.; Cox, G.F.; Fulton, A.B. CRB1, one gene, many phenotypes. *Semin. Ophthalmol.* **2013**, *28*, 397–405. [CrossRef]
35. Fahim, A.T.; Bouzia, Z.; Branham, K.H.; Kumaran, N.; Vargas, M.E.; Feathers, K.L.; Perera, N.D.; Young, K.; Khan, N.W.; Heckenlively, J.R.; et al. Detailed clinical characterisation, unique features and natural history of autosomal recessive RDH12-associated retinal degeneration. *Br. J. Ophthalmol.* **2019**, *103*, 1789–1796. [CrossRef]
36. Ba-Abbad, R.; Arno, G.; Robson, A.G.; Bouras, K.; Georgiou, M.; Wright, G.; Webster, A.R.; Michaelides, M. Macula-predominant retinopathy associated with biallelic variants in RDH12. *Ophthalmic Genet.* **2020**, *41*, 612–615. [CrossRef]
37. Falk, M.J.; Zhang, Q.; Nakamaru-Ogiso, E.; Kannabiran, C.; Fonseca-Kelly, Z.; Chakarova, C.; Audo, I.; Mackay, D.S.; Zeitz, C.; Borman, A.D.; et al. NMNAT1 mutations cause Leber congenital amaurosis. *Nat. Genet.* **2012**, *44*, 1040–1045. [CrossRef]
38. Koenekoop, R.K.; Wang, H.; Majewski, J.; Wang, X.; Lopez, I.; Ren, H.; Chen, Y.; Li, Y.; Fishman, G.A.; Genead, M.; et al. Mutations in NMNAT1 cause Leber congenital amaurosis and identify a new disease pathway for retinal degeneration. *Nat. Genet.* **2012**, *44*, 1035–1039. [CrossRef]
39. Dharmaraj, S.R.; Silva, E.R.; Pina, A.L.; Li, Y.Y.; Yang, J.M.; Carter, C.R.; Loyer, M.K.; El-Hilali, H.K.; Traboulsi, E.K.; Sundin, O.K.; et al. Mutational analysis and clinical correlation in Leber congenital amaurosis. *Ophthalmic Genet.* **2000**, *21*, 135–150. [CrossRef]
40. Aboshiha, J.; Dubis, A.M.; Carroll, J.; Hardcastle, A.J.; Michaelides, M. The cone dysfunction syndromes. *Br. J. Ophthalmol.* **2016**, *100*, 115–121. [CrossRef]
41. Hirji, N.; Aboshiha, J.; Georgiou, M.; Bainbridge, J.; Michaelides, M. Achromatopsia: Clinical features, molecular genetics, animal models and therapeutic options. *Ophthalmic Genet.* **2018**, *39*, 149–157. [CrossRef] [PubMed]
42. Thiadens, A.A.; Slingerland, N.W.; Roosing, S.; van Schooneveld, M.J.; van Lith-Verhoeven, J.J.; van Moll-Ramirez, N.; van den Born, L.I.; Hoyng, C.B.; Cremers, F.P.; Klaver, C.C. Genetic etiology and clinical consequences of complete and incomplete achromatopsia. *Ophthalmology* **2009**, *116*, 1984–1989.e1. [CrossRef] [PubMed]
43. Sundaram, V.; Wilde, C.; Aboshiha, J.; Cowing, J.; Han, C.; Langlo, C.S.; Chana, R.; Davidson, A.E.; Sergouniotis, P.I.; Bainbridge, J.W.; et al. Retinal structure and function in achromatopsia: Implications for gene therapy. *Ophthalmology* **2014**, *121*, 234–245. [CrossRef] [PubMed]
44. Aboshiha, J.; Dubis, A.M.; Cowing, J.; Fahy, R.T.; Sundaram, V.; Bainbridge, J.W.; Ali, R.R.; Dubra, A.; Nardini, M.; Webster, A.R.; et al. A prospective longitudinal study of retinal structure and function in achromatopsia. *Investig. Ophthalmol. Vis. Sci.* **2014**, *55*, 5733–5743. [CrossRef] [PubMed]
45. Kohl, S.; Zobor, D.; Chiang, W.C.; Weisschuh, N.; Staller, J.; Gonzalez Menendez, I.; Chang, S.; Beck, S.C.; Garcia Garrido, M.; Sothilingam, V.; et al. Mutations in the unfolded protein response regulator ATF6 cause the cone dysfunction disorder achromatopsia. *Nat. Genet.* **2015**, *47*, 757–765. [CrossRef]
46. Kohl, S.; Varsanyi, B.; Antunes, G.A.; Baumann, B.; Hoyng, C.B.; Jagle, H.; Rosenberg, T.; Kellner, U.; Lorenz, B.; Salati, R.; et al. CNGB3 mutations account for 50% of all cases with autosomal recessive achromatopsia. *Eur. J. Hum. Genet.* **2005**, *13*, 302–308. [CrossRef]
47. Remmer, M.H.; Rastogi, N.; Ranka, M.P.; Ceisler, E.J. Achromatopsia: A review. *Curr. Opin. Ophthalmol.* **2015**, *26*, 333–340. [CrossRef]
48. Burkard, M.; Kohl, S.; Kratzig, T.; Tanimoto, N.; Brennenstuhl, C.; Bausch, A.E.; Junger, K.; Reuter, P.; Sothilingam, V.; Beck, S.C.; et al. Accessory heterozygous mutations in cone photoreceptor CNGA3 exacerbate CNG channel-associated retinopathy. *J. Clin. Investig.* **2018**, *128*, 5663–5675. [CrossRef]
49. Aligianis, I.A.; Forshew, T.; Johnson, S.; Michaelides, M.; Johnson, C.A.; Trembath, R.C.; Hunt, D.M.; Moore, A.T.; Maher, E.R. Mapping of a novel locus for achromatopsia (ACHM4) to 1p and identification of a germline mutation in the alpha subunit of cone transducin (GNAT2). *J. Med. Genet.* **2002**, *39*, 656–660. [CrossRef]
50. Thiadens, A.A.; den Hollander, A.I.; Roosing, S.; Nabuurs, S.B.; Zekveld-Vroon, R.C.; Collin, R.W.; De Baere, E.; Koenekoop, R.K.; van Schooneveld, M.J.; Strom, T.M.; et al. Homozygosity mapping reveals PDE6C mutations in patients with early-onset cone photoreceptor disorders. *Am. J. Hum. Genet.* **2009**, *85*, 240–247. [CrossRef]
51. Pieh, C.; Simonsz-Toth, B.; Gottlob, I. Nystagmus characteristics in congenital stationary night blindness (CSNB). *Br. J. Ophthalmol.* **2008**, *92*, 236–240. [CrossRef] [PubMed]
52. Tsang, S.H.; Sharma, T. Congenital Stationary Night Blindness. *Adv. Exp. Med. Biol.* **2018**, *1085*, 61–64. [PubMed]
53. Zeitz, C.; Robson, A.G.; Audo, I. Congenital stationary night blindness: An analysis and update of genotype-phenotype correlations and pathogenic mechanisms. *Prog. Retin. Eye Res.* **2015**, *45*, 58–110. [CrossRef] [PubMed]

54. Verbakel, S.K.; van Huet, R.A.C.; Boon, C.J.F.; den Hollander, A.I.; Collin, R.W.J.; Klaver, C.C.W.; Hoyng, C.B.; Roepman, R.; Klevering, B.J. Non-syndromic retinitis pigmentosa. *Prog. Retin. Eye Res.* **2018**, *66*, 157–186. [CrossRef]
55. Hamel, C. Retinitis pigmentosa. *Orphanet J. Rare Dis.* **2006**, *1*, 40. [CrossRef]
56. Jordan, S.A.; Farrar, G.J.; Kenna, P.; Humphries, M.M.; Sheils, D.M.; Kumar-Singh, R.; Sharp, E.M.; Soriano, N.; Ayuso, C.; Benitez, J.; et al. Localization of an autosomal dominant retinitis pigmentosa gene to chromosome 7q. *Nat. Genet.* **1993**, *4*, 54–58. [CrossRef]
57. Tee, J.J.; Smith, A.J.; Hardcastle, A.J.; Michaelides, M. RPGR-associated retinopathy: Clinical features, molecular genetics, animal models and therapeutic options. *Br. J. Ophthalmol.* **2016**, *100*, 1022–1027. [CrossRef]
58. Williams, K.M.; Georgiou, M.; Kalitzeos, A.; Chow, I.; Hysi, P.G.; Robson, A.G.; Lingham, G.; Chen, F.K.; Mackey, D.A.; Webster, A.R.; et al. Axial Length Distributions in Patients With Genetically Confirmed Inherited Retinal Diseases. *Investig. Ophthalmol. Vis. Sci.* **2022**, *63*, 15. [CrossRef]
59. De Silva, S.R.; Arno, G.; Robson, A.G.; Fakin, A.; Pontikos, N.; Mohamed, M.D.; Bird, A.C.; Moore, A.T.; Michaelides, M.; Webster, A.R.; et al. The X-linked retinopathies: Physiological insights, pathogenic mechanisms, phenotypic features and novel therapies. *Prog. Retin. Eye Res.* **2021**, *82*, 100898. [CrossRef]
60. Georgiou, M.; Robson, A.G.; Jovanovic, K.; Guimaraes, T.A.C.; Ali, N.; Pontikos, N.; Uwaydat, S.H.; Mahroo, O.A.; Cheetham, M.E.; Webster, A.R.; et al. RP2-Associated X-linked Retinopathy: Clinical Findings, Molecular Genetics, and Natural History. *Ophthalmology* **2023**, *130*, 413–422. [CrossRef]
61. Ghosh, A.K.; Murga-Zamalloa, C.A.; Chan, L.; Hitchcock, P.F.; Swaroop, A.; Khanna, H. Human retinopathy-associated ciliary protein retinitis pigmentosa GTPase regulator mediates cilia-dependent vertebrate development. *Hum. Mol. Genet.* **2010**, *19*, 90–98. [CrossRef] [PubMed]
62. Vervoort, R.; Lennon, A.; Bird, A.C.; Tulloch, B.; Axton, R.; Miano, M.G.; Meindl, A.; Meitinger, T.; Ciccodicola, A.; Wright, A.F. Mutational hot spot within a new RPGR exon in X-linked retinitis pigmentosa. *Nat. Genet.* **2000**, *25*, 462–466. [CrossRef] [PubMed]
63. Shu, X.; McDowall, E.; Brown, A.F.; Wright, A.F. The human retinitis pigmentosa GTPase regulator gene variant database. *Hum. Mutat.* **2008**, *29*, 605–608. [CrossRef] [PubMed]
64. Megaw, R.D.; Soares, D.C.; Wright, A.F. RPGR: Its role in photoreceptor physiology, human disease, and future therapies. *Exp. Eye Res.* **2015**, *138*, 32–41. [CrossRef] [PubMed]
65. Pelletier, V.; Jambou, M.; Delphin, N.; Zinovieva, E.; Stum, M.; Gigarel, N.; Dollfus, H.; Hamel, C.; Toutain, A.; Dufier, J.L.; et al. Comprehensive survey of mutations in RP2 and RPGR in patients affected with distinct retinal dystrophies: Genotype-phenotype correlations and impact on genetic counseling. *Hum. Mutat.* **2007**, *28*, 81–91. [CrossRef]
66. Talib, M.; van Schooneveld, M.J.; Thiadens, A.A.; Fiocco, M.; Wijnholds, J.; Florijn, R.J.; Schalij-Delfos, N.E.; van Genderen, M.M.; Putter, H.; Cremers, F.P.M.; et al. Clinical and genetic characteristics of male patients with RPGR-associated retinal dystrophies: A Long-Term Follow-up Study. *Retina* **2019**, *39*, 1186–1199. [CrossRef]
67. Hadalin, V.; Sustar, M.; Volk, M.; Maver, A.; Sajovic, J.; Jarc-Vidmar, M.; Peterlin, B.; Hawlina, M.; Fakin, A. Cone Dystrophy Associated with a Novel Variant in the Terminal Codon of the RPGR-ORF15. *Genes.* **2021**, *12*, 499. [CrossRef]
68. Yang, L.; Yin, X.; Feng, L.; You, D.; Wu, L.; Chen, N.; Li, A.; Li, G.; Ma, Z. Novel mutations of RPGR in Chinese retinitis pigmentosa patients and the genotype-phenotype correlation. *PLoS ONE* **2014**, *9*, e85752. [CrossRef]
69. Di Iorio, V.; Karali, M.; Melillo, P.; Testa, F.; Brunetti-Pierri, R.; Musacchia, F.; Condroyer, C.; Neidhardt, J.; Audo, I.; Zeitz, C.; et al. Spectrum of Disease Severity in Patients With X-Linked Retinitis Pigmentosa Due to RPGR Mutations. *Investig. Ophthalmol. Vis. Sci.* **2020**, *61*, 36. [CrossRef]
70. Molday, L.L.; Hicks, D.; Sauer, C.G.; Weber, B.H.; Molday, R.S. Expression of X-linked retinoschisis protein RS1 in photoreceptor and bipolar cells. *Investig. Ophthalmol. Vis. Sci.* **2001**, *42*, 816–825.
71. Sikkink, S.K.; Biswas, S.; Parry, N.R.; Stanga, P.E.; Trump, D. X-linked retinoschisis: An update. *J. Med. Genet.* **2007**, *44*, 225–232. [CrossRef] [PubMed]
72. George, N.D.; Yates, J.R.; Bradshaw, K.; Moore, A.T. Infantile presentation of X linked retinoschisis. *Br. J. Ophthalmol.* **1995**, *79*, 653–657. [CrossRef]
73. Hinds, A.M.; Fahim, A.; Moore, A.T.; Wong, S.C.; Michaelides, M. Bullous X linked retinoschisis: Clinical features and prognosis. *Br. J. Ophthalmol.* **2018**, *102*, 622–624. [CrossRef] [PubMed]
74. Peachey, N.S.; Fishman, G.A.; Derlacki, D.J.; Brigell, M.G. Psychophysical and electroretinographic findings in X-linked juvenile retinoschisis. *Arch. Ophthalmol.* **1987**, *105*, 513–516. [CrossRef] [PubMed]
75. Apushkin, M.A.; Fishman, G.A.; Rajagopalan, A.S. Fundus findings and longitudinal study of visual acuity loss in patients with X-linked retinoschisis. *Retina* **2005**, *25*, 612–618. [CrossRef]
76. Wang, D.; Tai, P.W.L.; Gao, G. Adeno-associated virus vector as a platform for gene therapy delivery. *Nat. Rev. Drug Discov.* **2019**, *18*, 358–378. [CrossRef]
77. Bordet, T.; Behar-Cohen, F. Ocular gene therapies in clinical practice: Viral vectors and nonviral alternatives. *Drug Discov. Today* **2019**, *24*, 1685–1693. [CrossRef]
78. Flotte, T.R. Gene therapy progress and prospects: Recombinant adeno-associated virus (rAAV) vectors. *Gene Ther.* **2004**, *11*, 805–810. [CrossRef]

79. Wu, Z.; Asokan, A.; Samulski, R.J. Adeno-associated virus serotypes: Vector toolkit for human gene therapy. *Mol. Ther.* **2006**, *14*, 316–327. [CrossRef]
80. Au, H.K.E.; Isalan, M.; Mielcarek, M. Gene Therapy Advances: A Meta-Analysis of AAV Usage in Clinical Settings. *Front. Med.* **2021**, *8*, 809118. [CrossRef]
81. Kotterman, M.A.; Schaffer, D.V. Engineering adeno-associated viruses for clinical gene therapy. *Nat. Rev. Genet.* **2014**, *15*, 445–451. [CrossRef]
82. Bryant, D.H.; Bashir, A.; Sinai, S.; Jain, N.K.; Ogden, P.J.; Riley, P.F.; Church, G.M.; Colwell, L.J.; Kelsic, E.D. Deep diversification of an AAV capsid protein by machine learning. *Nat. Biotechnol.* **2021**, *39*, 691–696. [CrossRef]
83. Wu, Z.; Yang, H.; Colosi, P. Effect of genome size on AAV vector packaging. *Mol. Ther.* **2010**, *18*, 80–86. [CrossRef]
84. Maddalena, A.; Tornabene, P.; Tiberi, P.; Minopoli, R.; Manfredi, A.; Mutarelli, M.; Rossi, S.; Simonelli, F.; Naggert, J.K.; Cacchiarelli, D.; et al. Triple Vectors Expand AAV Transfer Capacity in the Retina. *Mol. Ther.* **2018**, *26*, 524–541. [CrossRef]
85. Trapani, I.; Tornabene, P.; Auricchio, A. Large gene delivery to the retina with AAV vectors: Are we there yet? *Gene Ther.* **2021**, *28*, 220–222. [CrossRef]
86. Cideciyan, A.V.; Sudharsan, R.; Dufour, V.L.; Massengill, M.T.; Iwabe, S.; Swider, M.; Lisi, B.; Sumaroka, A.; Marinho, L.F.; Appelbaum, T.; et al. Mutation-independent rhodopsin gene therapy by knockdown and replacement with a single AAV vector. *Proc. Natl. Acad. Sci. USA* **2018**, *115*, E8547–E8556. [CrossRef]
87. Lewin, A.S.; Smith, W.C. Gene Therapy for Rhodopsin Mutations. *Cold Spring Harb. Perspect. Med.* **2022**, *12*, a041283. [CrossRef]
88. Leveillard, T.; Fridlich, R.; Clerin, E.; Ait-Ali, N.; Millet-Puel, G.; Jaillard, C.; Yang, Y.; Zack, D.; van-Dorsselaer, A.; Sahel, J.A. Therapeutic strategy for handling inherited retinal degenerations in a gene-independent manner using rod-derived cone viability factors. *Comptes Rendus Biol.* **2014**, *337*, 207–213. [CrossRef]
89. John, M.C.; Quinn, J.; Hu, M.L.; Cehajic-Kapetanovic, J.; Xue, K. Gene-agnostic therapeutic approaches for inherited retinal degenerations. *Front. Mol. Neurosci.* **2022**, *15*, 1068185. [CrossRef]
90. Michalakis, S.; Schon, C.; Becirovic, E.; Biel, M. Gene therapy for achromatopsia. *J. Gene Med.* **2017**, *19*, e2944. [CrossRef]
91. Fischer, M.D.; Michalakis, S.; Wilhelm, B.; Zobor, D.; Muehlfriedel, R.; Kohl, S.; Weisschuh, N.; Ochakovski, G.A.; Klein, R.; Schoen, C.; et al. Safety and Vision Outcomes of Subretinal Gene Therapy Targeting Cone Photoreceptors in Achromatopsia: A Nonrandomized Controlled Trial. *JAMA Ophthalmol.* **2020**, *138*, 643–651. [CrossRef]
92. Cukras, C.; Wiley, H.E.; Jeffrey, B.G.; Sen, H.N.; Turriff, A.; Zeng, Y.; Vijayasarathy, C.; Marangoni, D.; Ziccardi, L.; Kjellstrom, S.; et al. Retinal AAV8-RS1 Gene Therapy for X-Linked Retinoschisis: Initial Findings from a Phase I/IIa Trial by Intravitreal Delivery. *Mol. Ther.* **2018**, *26*, 2282–2294.
93. Petrs-Silva, H.; Dinculescu, A.; Li, Q.; Min, S.H.; Chiodo, V.; Pang, J.J.; Zhong, L.; Zolotukhin, S.; Srivastava, A.; Lewin, A.S.; et al. High-efficiency transduction of the mouse retina by tyrosine-mutant AAV serotype vectors. *Mol. Ther.* **2009**, *17*, 463–471. [CrossRef]
94. Pennesi, M.E.; Yang, P.; Birch, D.G.; Weng, C.Y.; Moore, A.T.; Iannaccone, A.; Comander, J.I.; Jayasundera, T.; Chulay, J.; XLRS-001 Study Group. Intravitreal Delivery of rAAV2tYF-CB-hRS1 Vector for Gene Augmentation Therapy in Patients with X-Linked Retinoschisis: 1-Year Clinical Results. *Ophthalmol. Retin.* **2022**, *6*, 1130–1144. [CrossRef]
95. Hauswirth, W.W.; Aleman, T.S.; Kaushal, S.; Cideciyan, A.V.; Schwartz, S.B.; Wang, L.; Conlon, T.J.; Boye, S.L.; Flotte, T.R.; Byrne, B.J.; et al. Treatment of leber congenital amaurosis due to RPE65 mutations by ocular subretinal injection of adeno-associated virus gene vector: Short-term results of a phase I trial. *Hum. Gene Ther.* **2008**, *19*, 979–990. [CrossRef]
96. Russell, S.; Bennett, J.; Wellman, J.A.; Chung, D.C.; Yu, Z.F.; Tillman, A.; Wittes, J.; Pappas, J.; Elci, O.; McCague, S.; et al. Efficacy and safety of voretigene neparvovec (AAV2-hRPE65v2) in patients with RPE65-mediated inherited retinal dystrophy: A randomised, controlled, open-label, phase 3 trial. *Lancet* **2017**, *390*, 849–860. [CrossRef]
97. Maguire, A.M.; Russell, S.; Chung, D.C.; Yu, Z.F.; Tillman, A.; Drack, A.V.; Simonelli, F.; Leroy, B.P.; Reape, K.Z.; High, K.A.; et al. Durability of Voretigene Neparvovec for Biallelic RPE65-Mediated Inherited Retinal Disease: Phase 3 Results at 3 and 4 Years. *Ophthalmology* **2021**, *128*, 1460–1468. [CrossRef]
98. Wang, X.; Yu, C.; Tzekov, R.T.; Zhu, Y.; Li, W. The effect of human gene therapy for RPE65-associated Leber's congenital amaurosis on visual function: A systematic review and meta-analysis. *Orphanet J. Rare Dis.* **2020**, *15*, 49. [CrossRef]
99. Testa, F.; Bacci, G.; Falsini, B.; Iarossi, G.; Melillo, P.; Mucciolo, D.P.; Murro, V.; Salvetti, A.P.; Sodi, A.; Staurenghi, G.; et al. Voretigene neparvovec for inherited retinal dystrophy due to RPE65 mutations: A scoping review of eligibility and treatment challenges from clinical trials to real practice. *Eye* **2024**, *38*, 2504–2515. [CrossRef]
100. Jacobson, S.G.; Cideciyan, A.V.; Ho, A.C.; Peshenko, I.V.; Garafalo, A.V.; Roman, A.J.; Sumaroka, A.; Wu, V.; Krishnan, A.K.; Sheplock, R.; et al. Safety and improved efficacy signals following gene therapy in childhood blindness caused by GUCY2D mutations. *iScience* **2021**, *24*, 102409. [CrossRef]
101. Jacobson, S.G.; Cideciyan, A.V.; Ho, A.C.; Roman, A.J.; Wu, V.; Garafalo, A.V.; Sumaroka, A.; Krishnan, A.K.; Swider, M.; Mascio, A.A.; et al. Night vision restored in days after decades of congenital blindness. *iScience* **2022**, *25*, 105274. [CrossRef]
102. Cehajic-Kapetanovic, J.; Xue, K.; Martinez-Fernandez de la Camara, C.; Nanda, A.; Davies, A.; Wood, L.J.; Salvetti, A.P.; Fischer, M.D.; Aylward, J.W.; Barnard, A.R.; et al. Initial results from a first-in-human gene therapy trial on X-linked retinitis pigmentosa caused by mutations in RPGR. *Nat. Med.* **2020**, *26*, 354–359. [CrossRef]
103. Martinez-Fernandez De La Camara, C.; Nanda, A.; Salvetti, A.P.; Fischer, M.D.; MacLaren, R.E. Gene therapy for the treatment of X-linked retinitis pigmentosa. *Expert Opin. Orphan Drugs* **2018**, *6*, 167–177. [CrossRef]

104. Gal, A.; Li, Y.; Thompson, D.A.; Weir, J.; Orth, U.; Jacobson, S.G.; Apfelstedt-Sylla, E.; Vollrath, D. Mutations in MERTK, the human orthologue of the RCS rat retinal dystrophy gene, cause retinitis pigmentosa. *Nat. Genet.* **2000**, *26*, 270–271. [CrossRef]
105. Ghazi, N.G.; Abboud, E.B.; Nowilaty, S.R.; Alkuraya, H.; Alhommadi, A.; Cai, H.; Hou, R.; Deng, W.T.; Boye, S.L.; Almaghamsi, A.; et al. Treatment of retinitis pigmentosa due to MERTK mutations by ocular subretinal injection of adeno-associated virus gene vector: Results of a phase I trial. *Hum. Genet.* **2016**, *135*, 327–343. [CrossRef]
106. Saari, J.C.; Crabb, J.W. Focus on molecules: Cellular retinaldehyde-binding protein (CRALBP). *Exp. Eye Res.* **2005**, *81*, 245–246. [CrossRef]
107. McLaughlin, M.E.; Ehrhart, T.L.; Berson, E.L.; Dryja, T.P. Mutation spectrum of the gene encoding the beta subunit of rod phosphodiesterase among patients with autosomal recessive retinitis pigmentosa. *Proc. Natl. Acad. Sci. USA* **1995**, *92*, 3249–3253. [CrossRef]
108. Pichard, V.; Provost, N.; Mendes-Madeira, A.; Libeau, L.; Hulin, P.; Tshilenge, K.T.; Biget, M.; Ameline, B.; Deschamps, J.Y.; Weber, M.; et al. AAV-mediated Gene Therapy Halts Retinal Degeneration in PDE6beta-deficient Dogs. *Mol. Ther.* **2016**, *24*, 867–876. [CrossRef]
109. Cideciyan, A.V.; Jacobson, S.G.; Drack, A.V.; Ho, A.C.; Charng, J.; Garafalo, A.V.; Roman, A.J.; Sumaroka, A.; Han, I.C.; Hochstedler, M.D.; et al. Effect of an intravitreal antisense oligonucleotide on vision in Leber congenital amaurosis due to a photoreceptor cilium defect. *Nat. Med.* **2019**, *25*, 225–228. [CrossRef]
110. Russell, S.R.; Drack, A.V.; Cideciyan, A.V.; Jacobson, S.G.; Leroy, B.P.; Van Cauwenbergh, C.; Ho, A.C.; Dumitrescu, A.V.; Han, I.C.; Martin, M.; et al. Intravitreal antisense oligonucleotide sepofarsen in Leber congenital amaurosis type 10, a phase 1b/2 trial. *Nat. Med.* **2022**, *28*, 1014–1021. [CrossRef]
111. Jacobson, S.G.; Cideciyan, A.V.; Sumaroka, A.; Roman, A.J.; Charng, J.; Lu, M.; Choi, W.; Sheplock, R.; Swider, M.; Kosyk, M.S.; et al. Outcome Measures for Clinical Trials of Leber Congenital Amaurosis Caused by the Intronic Mutation in the CEP290 Gene. *Investig. Ophthalmol. Vis. Sci.* **2017**, *58*, 2609–2622. [CrossRef] [PubMed]
112. Slijkerman, R.W.; Vache, C.; Dona, M.; Garcia-Garcia, G.; Claustres, M.; Hetterschijt, L.; Peters, T.A.; Hartel, B.P.; Pennings, R.J.; Millan, J.M.; et al. Antisense Oligonucleotide-based Splice Correction for USH2A-associated Retinal Degeneration Caused by a Frequent Deep-intronic Mutation. *Mol. Ther. Nucleic Acids* **2016**, *5*, e381. [CrossRef] [PubMed]
113. Tsai, Y.T.; Wu, W.H.; Lee, T.T.; Wu, W.P.; Xu, C.L.; Park, K.S.; Cui, X.; Justus, S.; Lin, C.S.; Jauregui, R.; et al. Clustered Regularly Interspaced Short Palindromic Repeats-Based Genome Surgery for the Treatment of Autosomal Dominant Retinitis Pigmentosa. *Ophthalmology* **2018**, *125*, 1421–1430. [CrossRef] [PubMed]
114. Li, P.; Kleinstiver, B.P.; Leon, M.Y.; Prew, M.S.; Navarro-Gomez, D.; Greenwald, S.H.; Pierce, E.A.; Joung, J.K.; Liu, Q. Allele-Specific CRISPR-Cas9 Genome Editing of the Single-Base P23H Mutation for Rhodopsin-Associated Dominant Retinitis Pigmentosa. *CRISPR J.* **2018**, *1*, 55–64. [CrossRef]
115. Ruan, G.X.; Barry, E.; Yu, D.; Lukason, M.; Cheng, S.H.; Scaria, A. CRISPR/Cas9-Mediated Genome Editing as a Therapeutic Approach for Leber Congenital Amaurosis 10. *Mol. Ther.* **2017**, *25*, 331–341. [CrossRef]
116. Kleinstiver, B.P.; Prew, M.S.; Tsai, S.Q.; Topkar, V.V.; Nguyen, N.T.; Zheng, Z.; Gonzales, A.P.; Li, Z.; Peterson, R.T.; Yeh, J.R.; et al. Engineered CRISPR-Cas9 nucleases with altered PAM specificities. *Nature* **2015**, *523*, 481–485. [CrossRef]
117. Hu, J.H.; Miller, S.M.; Geurts, M.H.; Tang, W.; Chen, L.; Sun, N.; Zeina, C.M.; Gao, X.; Rees, H.A.; Lin, Z.; et al. Evolved Cas9 variants with broad PAM compatibility and high DNA specificity. *Nature* **2018**, *556*, 57–63. [CrossRef]
118. Vazquez-Dominguez, I.; Garanto, A.; Collin, R.W.J. Molecular Therapies for Inherited Retinal Diseases-Current Standing, Opportunities and Challenges. *Genes* **2019**, *10*, 654. [CrossRef]
119. Cai, Y.; Cheng, T.; Yao, Y.; Li, X.; Ma, Y.; Li, L.; Zhao, H.; Bao, J.; Zhang, M.; Qiu, Z.; et al. In vivo genome editing rescues photoreceptor degeneration via a Cas9/RecA-mediated homology-directed repair pathway. *Sci. Adv.* **2019**, *5*, eaav3335. [CrossRef]
120. Bohrer, L.R.; Wiley, L.A.; Burnight, E.R.; Cooke, J.A.; Giacalone, J.C.; Anfinson, K.R.; Andorf, J.L.; Mullins, R.F.; Stone, E.M.; Tucker, B.A. Correction of NR2E3 Associated Enhanced S-cone Syndrome Patient-specific iPSCs using CRISPR-Cas9. *Genes* **2019**, *10*, 278. [CrossRef]
121. Jo, D.H.; Song, D.W.; Cho, C.S.; Kim, U.G.; Lee, K.J.; Lee, K.; Park, S.W.; Kim, D.; Kim, J.H.; Kim, J.S.; et al. CRISPR-Cas9-mediated therapeutic editing of Rpe65 ameliorates the disease phenotypes in a mouse model of Leber congenital amaurosis. *Sci. Adv.* **2019**, *5*, eaax1210. [CrossRef] [PubMed]
122. Maeder, M.L.; Stefanidakis, M.; Wilson, C.J.; Baral, R.; Barrera, L.A.; Bounoutas, G.S.; Bumcrot, D.; Chao, H.; Ciulla, D.M.; DaSilva, J.A.; et al. Development of a gene-editing approach to restore vision loss in Leber congenital amaurosis type 10. *Nat. Med.* **2019**, *25*, 229–233. [CrossRef] [PubMed]
123. Pierce, E.A.; Aleman, T.S.; Jayasundera, K.T.; Ashimatey, B.S.; Kim, K.; Rashid, A.; Jaskolka, M.C.; Myers, R.L.; Lam, B.L.; Bailey, S.T.; et al. Gene Editing for CEP290-Associated Retinal Degeneration. *N. Engl. J. Med.* **2024**, *390*, 1972–1984. [CrossRef] [PubMed]
124. Quinn, J.; Musa, A.; Kantor, A.; McClements, M.E.; Cehajic-Kapetanovic, J.; MacLaren, R.E.; Xue, K. Genome-Editing Strategies for Treating Human Retinal Degenerations. *Hum. Gene Ther.* **2021**, *32*, 247–259. [CrossRef]
125. Kantor, A.; McClements, M.E.; MacLaren, R.E. CRISPR-Cas9 DNA Base-Editing and Prime-Editing. *Int. J. Mol. Sci.* **2020**, *21*, 6240. [CrossRef]
126. Gaudelli, N.M.; Komor, A.C.; Rees, H.A.; Packer, M.S.; Badran, A.H.; Bryson, D.I.; Liu, D.R. Programmable base editing of A*T to G*C in genomic DNA without DNA cleavage. *Nature* **2017**, *551*, 464–471. [CrossRef]

127. Huang, T.P.; Newby, G.A.; Liu, D.R. Precision genome editing using cytosine and adenine base editors in mammalian cells. *Nat. Protoc.* **2021**, *16*, 1089–1128. [CrossRef]
128. Levy, J.M.; Yeh, W.H.; Pendse, N.; Davis, J.R.; Hennessey, E.; Butcher, R.; Koblan, L.W.; Comander, J.; Liu, Q.; Liu, D.R. Cytosine and adenine base editing of the brain, liver, retina, heart and skeletal muscle of mice via adeno-associated viruses. *Nat. Biomed. Eng.* **2020**, *4*, 97–110. [CrossRef]
129. Wu, Y.; Wan, X.; Zhao, D.; Chen, X.; Wang, Y.; Tang, X.; Li, J.; Li, S.; Sun, X.; Bi, C.; et al. AAV-mediated base-editing therapy ameliorates the disease phenotypes in a mouse model of retinitis pigmentosa. *Nat. Commun.* **2023**, *14*, 4923. [CrossRef]
130. Su, J.; She, K.; Song, L.; Jin, X.; Li, R.; Zhao, Q.; Xiao, J.; Chen, D.; Cheng, H.; Lu, F.; et al. In vivo base editing rescues photoreceptors in a mouse model of retinitis pigmentosa. *Mol. Ther. Nucleic Acids* **2023**, *31*, 596–609. [CrossRef]
131. Davis, J.R.; Wang, X.; Witte, I.P.; Huang, T.P.; Levy, J.M.; Raguram, A.; Banskota, S.; Seidah, N.G.; Musunuru, K.; Liu, D.R. Efficient in vivo base editing via single adeno-associated viruses with size-optimized genomes encoding compact adenine base editors. *Nat. Biomed. Eng.* **2022**, *6*, 1272–1283. [PubMed]
132. Anzalone, A.V.; Randolph, P.B.; Davis, J.R.; Sousa, A.A.; Koblan, L.W.; Levy, J.M.; Chen, P.J.; Wilson, C.; Newby, G.A.; Raguram, A.; et al. Search-and-replace genome editing without double-strand breaks or donor DNA. *Nature* **2019**, *576*, 149–157. [PubMed]
133. Zhi, S.; Chen, Y.; Wu, G.; Wen, J.; Wu, J.; Liu, Q.; Li, Y.; Kang, R.; Hu, S.; Wang, J.; et al. Dual-AAV delivering split prime editor system for in vivo genome editing. *Mol. Ther.* **2022**, *30*, 283–294. [CrossRef] [PubMed]
134. Jang, H.; Jo, D.H.; Cho, C.S.; Shin, J.H.; Seo, J.H.; Yu, G.; Gopalappa, R.; Kim, D.; Cho, S.R.; Kim, J.H.; et al. Application of prime editing to the correction of mutations and phenotypes in adult mice with liver and eye diseases. *Nat. Biomed. Eng.* **2022**, *6*, 181–194. [CrossRef] [PubMed]
135. She, K.; Liu, Y.; Zhao, Q.; Jin, X.; Yang, Y.; Su, J.; Li, R.; Song, L.; Xiao, J.; Yao, S.; et al. Dual-AAV split prime editor corrects the mutation and phenotype in mice with inherited retinal degeneration. *Signal Transduct. Target. Ther.* **2023**, *8*, 57. [CrossRef]
136. Colella, P.; Ronzitti, G.; Mingozzi, F. Emerging Issues in AAV-Mediated In Vivo Gene Therapy. *Mol. Ther. Methods Clin. Dev.* **2018**, *8*, 87–104. [CrossRef]
137. Kabra, M.; Shahi, P.K.; Wang, Y.; Sinha, D.; Spillane, A.; Newby, G.A.; Saxena, S.; Tong, Y.; Chang, Y.; Abdeen, A.A.; et al. Nonviral base editing of KCNJ13 mutation preserves vision in a model of inherited retinal channelopathy. *J. Clin. Investig.* **2023**, *133*, e171356. [CrossRef]
138. An, M.; Raguram, A.; Du, S.W.; Banskota, S.; Davis, J.R.; Newby, G.A.; Chen, P.Z.; Palczewski, K.; Liu, D.R. Engineered virus-like particles for transient delivery of prime editor ribonucleoprotein complexes in vivo. *Nat. Biotechnol.* **2024**, *42*, 1526–1537.

life

MDPI

Review

Low Vision Rehabilitation and Eye Exercises: A Comprehensive Guide to Tertiary Prevention of Diabetic Retinopathy

Tibor Rák [1,2], Andrea Kovács-Valasek [3,4], Etelka Pöstyéni [3], Róbert Gábriel [3,4,*] and Adrienne Csutak [1]

1 Department of Ophthalmology, Medical School, University of Pécs, Szigeti út 12, 7624 Pécs, Hungary; rak.tibor@pte.hu (T.R.); csutak.adrienne@pte.hu (A.C.)
2 Department of Pharmacognosy, Faculty of Pharmacy, University of Pécs, Rókus utca 2, 7624 Pécs, Hungary
3 Department of Neurobiology, Faculty of Sciences, University of Pécs, Ifjúság útja 6, 7624 Pécs, Hungary; valasek@gamma.ttk.pte.hu (A.K.-V.); etelka14@gmail.com (E.P.)
4 János Szentágothai Research Centre, University of Pécs, Ifjúság útja 20, 7624 Pécs, Hungary
* Correspondence: gabriel@gamma.ttk.pte.hu

Abstract: Diabetic retinopathy (DR) is a leading cause of vision loss in patients with diabetes. While medical treatments like retinal laser photocoagulation, anti-VEGF therapy, and vitrectomy are primary, complementary therapies are gaining increasing attention. Based on the existing literature, a healthy lifestyle, including a balanced diet, stress management techniques, and regular physical activity targeting DR, can help regulate blood sugar levels and improve overall physical and mental health to reduce complications. This article explores physical activities and visual training methods related to DR, emphasizing complementary therapies, even though some of these practices are currently not fully integrated into evidence-based ophthalmology. Low vision exercises and aids help patients make the most of their remaining vision, improving their ability to perform everyday tasks, reducing the impact of vision loss, and promoting independence. There is some evidence that eye-related physiotherapy can improve the quality of life for patients with DR, although selection bias cannot be excluded in the presented studies. Consistent physical activity promotes holistic health, and therapies should be regularly monitored by ophthalmologists. This review further helps integrative healthcare professionals in offering appropriate therapies for rehabilitation purposes in the treatment of ophthalmic diseases, particularly DR.

Keywords: diabetic retinopathy; low vision rehabilitation; visual training; Bates–Schneider method; animal-assisted therapy; ocular yoga; orthoptic exercises; physical exercise

1. Introduction

Diabetic retinopathy (DR) is a leading cause of vision loss in people with diabetes, affecting millions worldwide [1,2]. According to precise estimates by the Vision Loss Expert Group of the Global Burden of Disease Study and the Global Burden of Disease 2019 Blindness and Vision Impairment Collaborators, approximately 1.07 million people were affected by blindness and nearly 3.28 million had moderate to severe visual impairment globally due to DR [3]. Proper management of diabetes, including blood sugar control, stress management, and physical activity, is crucial in preventing the progression of this condition [4,5]. This ocular condition is primarily managed through medical interventions such as retinal laser photocoagulation, anti-VEGF therapy, and vitrectomy [6]. A significant number of patients with diabetic retinopathy experience varying degrees of vision impairment, the management of which presents a considerable challenge to ophthalmology [7]. However, treatments such as anti-VEGF therapy and certain surgeries can improve vision in many cases. Recent studies have demonstrated the efficacy of anti-VEGF therapies and certain surgical interventions in improving vision in patients with severe visual impairment [7,8]. The Diabetic Retinopathy Clinical Research (DRCR) Network found that intravitreal anti-VEGF injections significantly improve vision in diabetic macular edema patients [8]. Furthermore, systematic reviews of cataract surgeries have shown substantial visual improvement, particularly in high-income countries [9,10]. Combined cataract surgery and visual rehabilitation have also been reported to enhance visual acuity and quality of life in visually impaired individuals [9].

However, complementary therapies like meditation, yoga, and mindfulness are gaining attention for their potential benefits in managing diabetes and its complications, including DR [6]. Evidence suggests that the adoption of a healthy lifestyle, including a balanced diet rich in antioxidants and regular physical activity, can significantly impact the progression of DR [5,6,11]. Furthermore, stress management techniques, such as yoga and meditation, can help control blood sugar levels and reduce the risk of complications [6]. According to the results of Amore et al. using data from the Italian Ministry of Health, 19% of low vision specialists were orthoptists [12]. Italian statistics also demonstrate a slightly higher involvement of orthoptists (2.6%) than ophthalmologists (1.2%) in low vision rehabilitation care [12]. This ratio may be comparable to that of other developed countries, indicating that visually impaired DR patients are likely to utilize visual training services more frequently, although precise statistics are unfortunately lacking in several areas. Fitzmaurice noted that in Australia, orthoptists are skilled in providing low vision rehabilitation training, and patients experience high levels of both subjective and objective satisfaction [13]. Raphanel et al. mentioned examples of actual low vision practice in other countries: despite the current recommendations outlined by the French High Health Authority, only 10–15% of French visually impaired patients are referred to a low vision professional, while it is estimated that 90% of patients could benefit from this service [14]. In India, only 30% of eligible patients are referred for low vision rehabilitation [14]. In the UK, there is often a mismatch between what treatments eye care professionals believe are available and what visually impaired patients actually receive [14]. In Canada, there is neither a standardized low vision rehabilitation model nor a consistent referral system across different regions [14]. Based on the above data, there is a significant global demand for accessible rehabilitation options for DR patients. Additionally, given the high workload of ophthalmologists, orthoptists and visual therapists can play a crucial role in the rehabilitation and mental well-being of patients. Collaboration between other interdisciplinary healthcare professionals is advisable. As a result of teamwork, lifestyle suggestions can be planned and maintained together with the involvement of family members and friends [15]. Integrating a multifaceted lifestyle medicine approach, which includes medical treatments, lifestyle changes, and the use of low vision aids and visual exercises, can significantly enhance the quality of life for patients by maximizing their remaining vision and promoting independence.

In the following sections, the best known physical activities and visual training practices and their relation to DR are explained. As the practice of eye-related exercises has become very popular, it is important that ophthalmologists, visual trainers (or orthoptists), opticians, and optometrists are familiar with complementary therapies. In the following sections, the authors elaborate on different methods and their relevance in ophthalmic diseases.

2. Methods

2.1. Search Strategy

A systematic literature review and meta-analysis were conducted to assess the efficacy and scope of complementary and rehabilitative interventions in the tertiary prevention of DR. The search was performed in four major databases for studies published between 1969 and 2025: PubMed, Scopus, Web of Science, and Google Scholar. The search strategy employed both MeSH terms and free-text keywords derived from the concepts covered in this review article. The final search string included combinations of the following keywords: ("diabetic retinopathy" OR "DR") AND ("low vision rehabilitation" OR "visual training" OR "orthoptic exercises" OR "ocular physiotherapy" OR "ocular yoga" OR "animal-assisted therapy" OR "guide dog" OR "diabetic alert dog" OR "hippotherapy" OR "Bates method" OR "Schneider method" OR "capillary exercise" OR "Chinese eye exercise" OR "office exercises" OR "Trataka" OR "Kolpakov gymnastics" OR "Katsuzo Nishi" OR "Zalmanov principle" OR "eurhythmy"/"Eurhythmie" OR "mind-body therapy" OR "meditation" OR "Shinrin-Yoku"/"forest bathing" OR "psychosomatic"). In addition, a hand search of references in key review articles was performed to identify any studies not captured by the database searches.

2.2. Study Selection and Data Extraction

Two independent reviewers screened the titles and abstracts. After duplicate removal, the full texts of potentially relevant articles were assessed. Disagreements were resolved through discussion or consultation with a third expert.

Inclusion criteria were as follows:

- Clinical studies (RCTs, controlled before–after studies, and cohort studies) evaluating low vision rehabilitation or complementary mind–body therapies in patients with DR or diabetic visual impairment;
- Studies reporting visual, psychological, or metabolic outcomes;
- Interventions including eye-focused exercises, physical activity, animal-assisted rehabilitation, yoga, or integrated lifestyle medicine protocols.

Exclusion criteria were as follows:

- Studies focused exclusively on pharmacologic or surgical management (e.g., anti-VEGF and laser therapy);
- Non-clinical studies, editorials, conference abstracts, or letters;
- Populations with non-diabetic causes of visual impairment.

Data extraction included the following:

- Study design and country;
- Sample size and patient demographics;
- Type, duration, and frequency of intervention;
- Outcome measures, e.g., visual acuity (VA), quality of life (QoL), blood glucose metrics (HbA1c), contrast sensitivity, and psychosocial scales;
- Results and reported effect sizes.

When necessary, corresponding authors were contacted for missing data.

The methodological quality of included studies was assessed descriptively. Particular attention was paid to selection bias, especially in studies involving lifestyle interventions or non-blinded designs (e.g., yoga, dog-assisted therapy, and office exercises). Studies with a high risk of bias were excluded.

3. Evidence-Based Low Vision Rehabilitation in Diabetic Vision Loss

Visual impairment or low vision (caused by DR) is a type of vision loss that cannot be further corrected with standard glasses, contact lenses, medical treatment, or surgery, and it significantly and permanently affects an individual's daily life in most cases [16]. Low vision rehabilitation methods (Table 1) are designed to help individuals with visual impairments make the most of their remaining vision as tertiary prevention. These exercises can improve visual function, enhance contrast sensitivity, and reduce strain on the eyes. For DR patients, low vision exercises can be particularly beneficial in maintaining visual acuity and adapting to changes in vision [17]. Contrast sensitivity training that involves distinguishing between different levels of contrast can help improve an individual's ability to see in low-light conditions and recognize objects against varying backgrounds [17]. This is crucial for patients with DR, who often experience reduced contrast sensitivity. For patients with central vision loss, training the eccentric and peripheral vision can help improve overall visual awareness [17]. Exercises that encourage the use of peripheral vision can aid in navigating environments and performing daily tasks more effectively. Low vision aids are specialized devices designed to assist individuals with visual impairments in maximizing their remaining vision [16,17]. These aids can significantly improve the ability to perform daily activities, such as reading, writing, and recognizing faces, which are challenging for DR patients [16]. Handheld or stand magnifiers and telescopic devices can enlarge text and images, making it easier for patients to read and perform detailed tasks. Electronic magnifiers, which offer adjustable magnification and contrast settings, provide additional flexibility and convenience [16,17]. For example, a clip-on magnifier lens allows both hands of the patient to be free while self-dosing the prescribed insulin [17]. These are also particularly useful for activities such as watching television, attending events, or navigating outdoor environments. Electronic aids such as screen readers, text-to-speech software, and electronic magnifiers can assist patients in accessing digital content and performing computer-based tasks [17]. These aids can enhance independence and productivity in both personal and professional settings. Proper adaptive lighting is essential for maximizing visual function: adjustable lamps, task lighting, and devices that enhance contrast can help reduce eye strain and improve visibility in various environments [17]. The integration of low vision exercises and aids into management plans for DR patients can have a profound impact on their quality of life. These interventions not only help patients adapt to their visual impairments but also empower them to maintain independence and engage in daily activities.

Table 1. Classification of low vision rehabilitation modalities.

Low Vision Rehabilitation Modalities [1]	
Optical aids	For near and intermediate activities For distant activities
Non-optical aids	Environmental modifications Digital devices Non-electric aids Animal-assisted therapy
Psychotherapy	Psychological consultation Mind–body practices
Neurofunctional rehabilitation	Biofeedback training Vision therapy

Table 1. *Cont.*

Low Vision Rehabilitation Modalities [1]	
Prostheses	Implantable intraocular devices (telescopic IOL [2], bionic retina, etc.) Extraocular devices (orbital and ocular prostheses, prosthetic contact lenses, etc.)
Cellular and gene therapy	Stem cell therapy

[1] Self-edited table adapted from multiple sources: Cooke et al. (2001), Muhsin et al. (2024), Németh et al. (2024), O'Loughlin et al. (2024), and Vingolo et al. (2015) [16–20]. [2] IOL: intraocular lens.

3.1. *Animal-Assisted Therapy in Blindness from Diabetic Retinopathy*

According to Bassan et al., there are approximately 500,000 service dogs in the United States; however, only about 2% of blind or visually impaired individuals keep them, highlighting the need for increased access to and availability of these trained animals [21]. The border between animal-assisted therapy interventions and assistance animals is not consistently clear due to subjective and psychosocial factors. The former has therapeutic value, while the latter is a low vision rehabilitation method [22]. A scoping review highlighted that animal-assisted interventions can enhance well-being and function in hospital rehabilitation settings. These interventions, which often involve officially certified dogs, have shown improvements in social and emotional well-being, ambulation, motor skills, and verbal communication [20,22,23]. Dogs generally provide companionship and emotional support, which can significantly reduce feelings of isolation and depression often experienced by individuals with visual impairments. Lundquist et al. stated that dog-assisted interventions can have positive effects in healthcare settings, including for patients with cognitive and psychiatric conditions [24]. Guide dogs are trained to help navigate obstacles, enhancing the mobility and independence of their handlers [23]. The findings of Glenk et al. show that blind people, compared to non-dog owners, with a guide dog in Austria are more likely to believe that service animals provide more benefits regarding their health and psychosocial status [23]. A total of 93% of respondents reported that their guide dog's importance was the same as that of their relatives, and they were generally satisfied with the dog's work [23]. It is also known in the literature that visually impaired individuals, regardless of age, prefer to keep guide dogs despite the availability of the most modern digital navigation devices or mobility aids [23]. Although the literature on guide dogs does not specifically address connections with patients who are blind due to diabetes, diabetic alert dogs offer a more interesting solution for them [25]. Keeping these dogs can be beneficial for minors and adolescents with type 1 diabetes and older patients with type 2 diabetes as they not only read their owner's physical body signals but also olfactorily detect hypoglycemia (0.29–0.80 g/dL) and hyperglycemia (0.49–0.96 g/dL) [25]. Lippi and Lebani concluded that the current published evidence may contribute to undeniable health and psychological benefits for diabetic people. However, their efficiency in detecting harmful, even life-threatening, blood glucose variations remain questionable [25]. Additionally, there is no data in the literature on diabetic retinopathy and diabetic alert dogs. Equine-assisted therapy (or hippotherapy) has shown promising benefits for individuals with blindness or visual impairments. This intervention can provide a multisensory experience that stimulates various senses, compensating for the lack of visual input. The rhythmic and repetitive movements of the horse can help improve balance, coordination, and core strength, which are often areas of concern for those with visual impairments [26]. Additionally, the bond formed between the patient and the horse can foster a sense of trust, confidence, and emotional stability. This connection is particularly beneficial for individuals who may experience anxiety or depression due to their visual impairment. Moreover, it can enhance spatial awareness and orientation skills as patients learn to navigate and interact with their environment in new ways. The therapeutic setting also offers a safe and supportive space for patients to explore their capabilities and build resilience [26]. The authors did not find any studies in the literature describing the real connection between hippotherapy and DR. However, Klimova et al. conducted a unique study with children aged 7–13 years with type 1 diabetes [27]. According to their results, during hippotherapy rehabilitation performed twice a week, the children's blood glucose levels significantly normalized after an initial increase [27]. The psychological assessment also showed that children were emotionally bonded with the animals, which could make intervention beneficial. Based on their results, Klimova et al. found hippotherapy to be more advantageous for young diabetic patients compared to other rehabilitation methods [27]. Although canines and horses are commonly used for therapeutic purposes, several other animal species can also be utilized (e.g., feline-assisted therapy or ornithotherapy). These benefits can extend to ophthalmic patients, especially DR patients, by providing emotional support and improving their overall well-being. While there is growing interest in animal assisted therapy, some studies highlight the need for more rigorous research to fully understand its benefits and mechanisms on DR patients and integrating these therapies into standard care practices. Another limitation of this intervention is its high financial burden. The training, certification, and living costs for a guide dog range between EUR 34,000 and 40,000 [23], while the costs of a diabetic alert dog depend on their training organization, ranging between EUR 1500 and 18,500 [25].

3.2. Orthoptic Exercises in Diabetic Eyes

Orthoptic exercises were originally evaluated for the management of heterophoria, intermittent strabismus, convergence insufficiency, and presbyopic accommodative disorders [28], but the following causes give a rationale in diabetic patients too. Studies indicate that approximately 1–14% of diabetics experience ocular motor nerve palsies during the course of their disease [29]. This condition is significantly more common in diabetic individuals compared to non-diabetic individuals, with the incidence being 5–10 times higher [29]. Individuals with DR or diabetic nephropathy face a heightened risk of developing ophthalmoplegia [30]. While DR primarily affects the retina, the same vascular complications can also impact the nerves controlling the extraocular muscles, leading to conditions such as cranial mononeuropathy [30]. Isolated cranial nerve palsy is approximately seven times more prevalent among diabetic patients, likely due to microvascular infarction [30,31]. The third cranial nerve is most commonly affected [29,31,32], but the fourth and sixth cranial nerves can also be involved [29,31], with functional recovery typically occurring within six months [33]. Although multiple cranial nerve palsies are rare in diabetes, they necessitate neuroimaging to exclude compressive lesions [30]. Oculomotor nerve palsies in diabetic patients can lead to symptoms such as diplopia and ptosis [17,29]. Idiopathic eyelid ptosis (a disorder of the third cranial nerve) is linked to an increased risk of insulin resistance [34]. Sensory neuropathy of the fifth cranial nerve can cause corneal hypesthesia, heightening the risk of dry eye and neurotrophic keratitis [35]. Regular blinking exercises can help patients become more aware of their blinking, supplemented by artificial eye drops [36]. Raskind reports the main characteristics of diabetes-induced convergence insufficiency causing diplopia and other visual disturbances [37]. The results of Rundström and Eperjesi indicate the necessity of evaluating the binocular vision of low vision patients (e.g., DR patients) to detect orthoptic disorders [38]. Treatment of convergence insufficiency aims to improve convergence exercises by pencil push-ups (accommodative orthoptic exercise by using a pencil), near-distance fixation changes, stereograms, etc. [32]. According to Raskind, systemic convergence insufficiency should primarily be managed using prismatic and additive lenses [37]. Orthoptic exercises and strabismus surgery is indicated if the misalignment is stable and long-standing and if the patients do not tolerate prismatic glasses [32].

Macfarlene et al. reports that there are limited data on the convergence training of stroke patients in the literature. It is important to mention this neurovascular condition because they also emphasize that diabetic patients, especially with visual impairments such as DR or cataract, are generally more vulnerable to stroke [39]. According to their study, the outcomes support the benefits of orthoptic interventions; therefore, a visual trainer/orthoptist is an essential part in teamwork [39]. Maagard et al. reported that vergence exercises (e.g., Brock's string fusion, picture fusion using an aperture ruler, etc.) induce faster recovery of convergence insufficiency than accommodation exercises (e.g., distance-to-near accommodative visual acuity cards, etc.) in school children [40]. Horwood and Toor found that separating convergence and accommodation exercises seemed more effective than training both functions concurrently [28]; therefore, it is wiser to choose one type of training method to improve the patient's orthoptic skills. Yadav et al. compared the efficacy of pencil push-up vs. office-based orthoptic therapy on patients with asthenopic symptoms due to convergence insufficiency [41]. Unfortunately, they excluded diabetes-related convergence insufficiency from their study, although useful conclusions could have been drawn.

Orthoptic exercises could have limitations and complications, such as pain, subconjunctival suffusion, and suture rupture in postoperative cases [42]. Zeng et al. used postoperative rehabilitation exercises in a study on postoperative patients with blow-out orbital fracture with retrobulbar anesthesia [42], but this can be interpreted on pars plana vitrectomy on DR. Therefore, we discourage vitrectomy patients from performing orthoptic exercises postoperatively to avoid severe complications. Another limitation of these exercises is the clinically rare phenomenon, termed horror fusionis, derived from Bielschowsky, which refers to a specific clinical syndrome where patients' extraocular muscles actively avoid bifoveal fixation despite all attempts to achieve stereoscopic vision. It is an acquired disorder of central fusion, which causes diplopia in all positions of gaze, leading the patient to experience anxiety because even with correction, both eyes could neither fuse nor suppress images. Orthoptic visual therapists should warn their patients not to perform any orthoptic exercises on their own without medical professional guidance and diagnosis [43].

4. Physical Activity—Evidence in Ophthalmology

DR is a common complication of diabetes, characterized by damage to the retinal blood vessels due to prolonged hyperglycemia [6]. Among these, whole-body exercises and physical activity play a crucial role in both the prevention and management of DR [6,44]. These interventions directly impact ocular health through several physiological mechanisms: Regular physical activity enhances cardiovascular health, leading to better blood flow and oxygen delivery to the retina. Improved circulation helps maintain the integrity of retinal blood vessels and reduces the risk of ischemic damage [44]. Regular exercise improves endothelial function and increases ocular perfusion, which can have direct benefits for eye health by enhancing blood flow to ocular tissues [45]. Physical activity increases insulin sensitivity and helps regulate blood glucose levels. Better glycemic control is essential for preventing the progression of DR as high blood sugar levels are a primary cause of retinal damage [5,6,44]. Exercise induces the production of antioxidant enzymes and reduces systemic inflammation. Lower levels of oxidative stress and inflammation protect retinal cells from damage and slow the progression of DR [6,46]. Regular physical activity improves mitochondrial function and energy production in retinal cells. Enhanced mitochondrial function supports cellular health and resilience against metabolic stress [44]. Over time, diabetes damages peripheral circulation, thus causing changes in the eyes, kidneys, extremities, brain, etc. Based on Zalmanov's "peripheral disease" principle, it is important for diabetic patients to perform capillary exercise every day, as well as exercises that stimulate the circulation of the limbs and help preserve long-term visual functions. Whole-body vibration increases the expression of fibronectin type III domain-containing protein 5 and brain-derived neurotrophic factor (BDNF) transcriptional factors, which are neuroprotective and enhance muscle contraction and relaxation [6,47]. These factors play a role in maintaining retinal health and preventing neurodegeneration [5,6,44]. Maintaining a healthy weight and cardiovascular health through exercise reduces the risk of comorbid conditions, such as hypertension and hyperlipidemia, which can exacerbate DR [6]. The Centers for Disease Control and Prevention, the World Health Organization (WHO), and the American Heart Association (AHA) all recommend 150 min of moderate aerobic exercise per week, which is equivalent to about 30 min per day for five days per week. This can include walking, cycling, swimming, dancing, and even active gardening [48].

Whole-body exercises and physical activity are integral components of a comprehensive approach to managing DR. Incorporating regular exercise into management plans for individuals with diabetes can significantly reduce the risk and progression of DR, ultimately preserving vision and enhancing quality of life.

4.1. *Office-Based Exercises and Diabetic Retinopathy*

The modern sedentary lifestyle and workplace environment are posing new difficulties to people's health, notably their eyes and visual function. Prolonged near work and screen time in the office put significant strain on the eyes, resulting in a variety of health complications [49,50]. Continuous near concentrating can lead to eye fatigue, dry eye syndrome, and visual impairment [49,50]. In the worst-case scenario, both extended hours of near labor and the resulting chronic dry eye have been linked to an increased risk of depression and, in severe cases, suicide [36,51]. These rates are similarly higher among individuals with DR compared to the general population [52,53]. A study from the Kangbuk Samsung Cohort found that adults with prediabetes who worked more than 52 h per week had a significantly higher risk (hazard ratio of 2.00) of developing type 2 diabetes compared to those working 35–40 h per week [54]. Research published in Diabetes Care indicates that individuals with diabetes are less likely to be employed and, if employed, tend to have more work-loss days and health-related work limitations. This study did not find a significant change in weekly hours worked due to diabetes [55]. Another study highlighted that certain work-related factors, such as long working hours and shift work, could increase the risk of cardiovascular diseases in workers with diabetes, which can be related to diabetic complications like retinopathy [56]. Recently, increasing regional and international recommendations for various office gymnastics have been developed with the primary aim of reducing eye and mental fatigue. Additionally, these exercises have the secondary goal of alleviating stress caused by prolonged screen work, contributing to overall physical and mental well-being [36,57], which is also essential to diabetic and DR patients. The exercises of the Bates–Schneider Method include palming (covering the eyes with the palms) and regular blinking, which are basic relaxation techniques [57,58]. Occupational health workers and the American Academy of Ophthalmology (AAO) recommend the 20/20/20 rule for computer workers [36,59,60], which involves taking a 20 s break to look at something 20 feet (6 m) away every 20 min, and the regular use of preservative-free artificial tears [36], although there are controversial results in its efficacy [60]. In Hungary and Austria, "12 screen Tibetans", developed by Martin Donner, an orthopedic surgeon, was promoted in online recommendations and publications for occupational health to interrupt computer office work by performing simple gymnastics to counteract the lack of movement and prevent neck and back pain. These practices are also promoted by the Chamber of Labor for Workers and Employees in Vienna, Austria (Kammer für Arbeiter und Angestellte für Wien), and Hungarian occupational health experts. In Thailand, a study was conducted on a series of relaxing exercises prescribed for office workers, incorporating complex eye, head, and neck movements [61]. Kolpakov and colleagues patented several eye health-focused movement and massage forms 36 years ago following their research [6,36,62,63]. These exercises (Figures S1–S17), which can be performed both at home and at work, aim to enhance overall body and eye blood circulation, thereby preserving eye health. Similarly, the lifestyle program developed by Japanese Katsuzō Nishi [6,64], which includes capillary exercises (Figures S18 and S19) targeting both body and eye health based on the mZalmanov principle [6,65], follows the same principles. A recent Japanese study also incorporated cognitive behavioral therapy-based exercises, including classic sports warm-ups, sit-ups, crunches, squats, arm and leg lifts, etc. [50]. Chinese eye exercises, based on massaging periocular anatomical points with the fingertips according to traditional Chinese medicine, aim to strengthen the eye muscles, improve blood circulation, and reduce eye fatigue [66]. Nevertheless, it proved effective for dry eye and asthenopic complaints observed during office work [57,67–69]. Overall, the authors believe that eye-relaxing office exercises targeting the entire body and head–neck region with active movement and muscle work could be beneficial for diabetic patients with a sedentary lifestyle to reduce the risk of DR [49,50,54,61]. It is important to highlight that no recommendations or studies have been made regarding some of these exercises for DR, but among these, Kolpakov et al. recommended and patented their gymnastics for preventing DR and enhancing retinal health [6].

4.1.1. Bates–Schneider Method

The Bates–Schneider method, including ocular and visual training, as well as ocular yoga procedures, is based on the Bates technique, which is widely promoted by behavioral optometrists and visual therapists (Figures S20 and S21) [57,58]. These oculomotor muscle exercises should be performed together with appropriate lifestyle management and diet counseling based on international recommendations in order to improve the patients' quality of life as holistically as possible [70]. For the proper functioning of the extraocular muscles, the visual trainer compiles the exercises in such a way that all eye muscles can be dynamically exercised, thereby increasing the muscles' need for oxygen [70]. The exercises and professional experiences were published in 1891 and 1911 by William Horatio Bates, who attributed the origin of visual problems to mental stress, wrong gazing habits, and extraocular and facial muscle strain [57,71]. His technique emphasizes the original visual organ, the brain, instead of the eye, referring to the vision as mental process. Additionally, repetitive relaxation of the internal and external ocular muscles can enhance subjective visual function by relieving unnecessary mental and ocular stress. Meir Schneider is the founder of the self-healing method based on Bates' visual exercises promoting active "body–mind work". This method offers combinations of massage, physical exercises, visualization, breathing exercises, and a complete lifestyle program. Schneider believes that all patients can learn techniques to empower themselves to improve their own health, mostly focusing on rehabilitation [72]. The Bates–Schneider method is generally misinterpreted in the laymen and even scientific opinions because the original observations of Bates are different regarding visual performance than what the popular culture currently refers to. The application of this method in the refractive error's viewpoint can be harmful without an ophthalmologist's professional control.

The Bates–Schneider method can be useful for patients with visual impairment (such as DR) in increasing self-confidence and reducing their fears and insecurities associated with their eye disease. Integrating these techniques into a low vision rehabilitation approach can support patients with DR by improving their overall well-being and increasing focus on ocular health. However, it is crucial to note that the effectiveness of the Bates–Schneider method in treating refractive errors has been largely debunked by critics and scientific reviews, with no credible evidence supporting its use for DR management. Anecdotal evidence suggests that visual improvements in these cases are based on short-term blurry adaptation and perceptual learning rather than actual objective vision improvement. Therefore, further well-designed clinical studies are needed to provide objective scientific validation towards DR.

Sunning

Along with the skin of the human body, the eyes are the organs most exposed to ultraviolet radiation [73]. Sunning, a Bates visual exercise, means enjoying the light of the sun with closed eyes (to avoid retinal photodamage, e.g., solar maculopathy) and performing rhythmic movements of the head and hands (Figure S20B–D) [57,74]. From an ophthalmological point of view, sunning can be a method for slowing down the progression of myopia, but the scientific literature does not directly indicate that sunshine has a positive effect on DR. If the authors had to mention a possible positive outcome of the sunning method as a rehabilitation technique in relation to DR, it would be the following: Payne et al. assessed the significant relationship between insufficient serum vitamin D in the case of DR, especially in patients with proliferative DR, compared to a non-diabetic population [75]. There is some evidence suggesting that vitamin D, which is synthesized in the skin through exposure to sunlight, may have beneficial effects on diabetes and its complications. Vitamin D is known to reduce inflammation and oxidative stress, both of which are key factors in the development and progression of DR [76]. Adequate levels of vitamin D (and its oral supplementation) may help in maintaining better glycemic control and reducing the risk of complications associated with diabetes [76]. According to professional recommendation, approximately 5–30 min of direct sunlight on the face, arms, legs, or back twice weekly can provide the amount of vitamin D necessary to protect bones from the development of rachitis and/or osteomalacia. The consumption of 1800 IU (international unit) of vitamin D per day improves retinal blood flow [6,77]. Other sunlight-dependent biochemical pathways lead to the synthesis of alpha melanocyte-stimulating hormone, calcitonin gene-related peptide, neuropeptide substance P, and endorphins; the latter explains the psychologically beneficial effect of sunbathing [73]. Based on the above, the Mediterranean lifestyle takes on a new meaning; in addition to high-quality red wine with a high resveratrol content and a Mediterranean diet rich in antioxidants, a sufficient amount of sunlight is also considered a protective factor [46,78,79]. Emphasis must be placed on a moderate healthy amount of sunlight exposure: according to Simó and Hernández, this method maintains vitamin D biosynthesis, but sunlight exposure for $\geq$5 h a day is significantly associated with an increased risk of DR [80,81].

Swaying and Swinging

Swaying and swinging are underestimated and neglected components of the Bates–Schneider method [57]. During the rhythmic swinging of the head or the whole body [57,74], the eyes are forced to focus on a specific target object that is aligned stationarily on the fovea (Figure S2). Visual information from the environment is gathered through these quick eye movements, which consist of a series of saccades and fixations [82]. Saccadic movements are also important in the abovementioned "sunning" technique. Fixation saccade training as a gait rehabilitation strategy can be an important therapeutic option to improve movement coordination of patients with progressive visual loss, including those with DR. The applied elements of this rehabilitation strategy are balance training, foot stepping, and eye movement exercises [82]. Based on this hypothesis, the Bates method of swaying–swinging may be incorporated into low vision rehabilitation to lower the risk of falls and difficulties of a low quality of life. These exercises can help to enhance the coordination and balance of patients, which is crucial for those with visual impairments. Incorporating such activities into the route of individuals with DR may aid in maintaining better ocular health and overall well-being, supporting the principles of lifestyle medicine.

4.2. General and Ocular Yoga

Yoga (Sanskrit term meaning "to connect") has become increasingly popular, with its regular practice doubled within a 10-year period, increasing from 0.46% in 1999 to 1.11% in 2008 [83]. According to a 2022 report from the National Health Interview Survey, 16.9% of adults aged 18 and older in the United States practiced yoga in the past 12 months [84]. The principal elements of yoga are asanas and mudras (leg and hand positions) with breathing exercises and meditation [83]. Considering the fact that vision loss and psychosocial emotional responses go together, relaxation, psychotherapy, or other stress reduction programs should be helpful in reducing the impact of low vision [85]. The regular practice of yoga results in increased alpha and theta brain activity, thus alleviating autonomic functions [83]. Yoga has been mentioned to help in managing type 2 diabetes mellitus, which can indirectly benefit DR by improving glycemic control, reducing stress, and enhancing overall health. Improved glycemic control can help prevent or slow the progression of DR [86]. Nathani and Nanduri reported in a case study that a patient with DR experienced significant improvement in vision and overall diabetes management after following Yoga Prana Vidya healing protocols. The patient showed reduced blurriness in vision and better control of diabetes [87]; however, since this is evidence from a single case, its scientific power is very low. Yogic ocular exercises, including extraocular muscle exercise (palming, blinking, gazing in nine directions, focusing on near and distant points, accommodative exercises, etc.), Trāṭaka Kriya, and 45 min of daily meditation, also significantly reduce intraocular pressure and anxiety.

Ocular yoga, an advanced version of the Bates method stemming from Indian Ayurvedic medicine, includes exercises such as circular eye movements, focusing on objects at different distances, and resting the eyes [88,89]. A specific subtype is the meditative practice Trāṭaka kriya (yogic gazing) [88], which has been studied as a complementary treatment for glaucoma patients, but its role in DR is unknown. Ocular yoga exercises also affect the average retinal thickness of the macula in a favorable way and increase oxygen saturation in the blood through deep breathing [89]. Kumar et al. reported a 3-month Ayurvedic treatment protocol containing Rasayana yoga aimed at ameliorating DR. According to their analysis, they achieved the following primary outcome: the trial group showed statistically significant improvement in superficial and dot-blot retinal hemorrhages, hard exudates, and visual acuity. However, the secondary outcome, hypoglycemia, was not statistically significant compared to the control group [90]. According to the AAO's recommendations, patients with known associated primary or secondary glaucoma or ocular hypertension should be advised against practicing yoga exercises with 10 head-down positions (Śīrṣāsana, Adho Mukha Śvānāsana, Uttānāsana, Halāsana, Viparīta Karaṇī, Kākāsana, Vṛścikāsana, Adho-mukha-vṛksāsana, Piñcha-mayūrāsana, and Sarvāṅgāsana), which are associated with a transient 2-fold increase in intraocular pressure in glaucomatous and healthy eyes. Regular practicing of the previously described positions caused visual field progression [91,92]. Further limitations in yogic practice are that diabetic patients (similarly to myopic patients) have been shown to have an altered composition of their vitreous body, which may predispose them to vitreous degeneration or syneresis. During yoga exercises, postural changes, especially when transitioning from an upright position to a head-down position and vice versa, may cause abrupt shifts in the vitreous gel. Vitreous hemorrhage may be seen in acute posterior vitreous detachment due to the spontaneous rupture of small retinal capillaries [93]. Not only can vitreous shear forces cause retinopathy, but the Valsalva mechanism induced by a closed glottis during breathing exercises can also contribute to this condition. Recently, Kaushal et al. documented the first case of Valsalva retinopathy resulting from 30 min of daily Anuloma pranayama (a type of Hatha yoga) over a one-month period [94]. The retinas of patients with DR are particularly susceptible to pressure changes associated with the Valsalva maneuver. Altogether, yoga is considered a safe and effective complementary practice affecting DR. However, based on the AAO's recommendations, DR patients should disclose their activity to their treating ophthalmologist and other specialists evaluating the risks and benefits regarding their ocular condition.

5. Psychosomatic Correlations of Eye Diseases

Patients with severe vision loss from DR face significant challenges when performing daily tasks, driving, and moving [85,95], which affects their psychosocial well-being. While there is a wide range of studies in the literature on the quality-of-life assessment of, e.g., DR patients, there is less information about exact methods to improve it. Educating the patient about the disease and encouraging two-way communication between the patient and physician are known to reduce anxiety and stress. Inadequate information and uncertainty due to fear-inducing diagnosis and prognosis also make healthcare personnel responsible [85,95]. Important parts of the treatment process are empathy and taking the time to explain the illness to the patients, who can often become isolated without much understanding from medical professionals and/or their relatives. Later, patients feel that their disease is untreatable, which leads to a decrease in compliance and discontinuation of treatment [96]. It is important as therapists to explain to patients that their ocular disease is chronic and that the severity of symptoms may fluctuate depending on the body's internal conditions and reactions to environmental conditions [96]. Even though the eye is a relatively small organ of the human body, the term "sick eye in a sick body" syndrome is increasingly appropriate, as Dada et al. states, as it can be associated with disorders of the cardiovascular, central nervous, and endocrine systems. Excessive stress exposure of the brain to glucocorticosteroids can become toxic to neurons and even retinal tissue [46,85]. Psychological stress can also activate inflammatory responses through the neuronal activation of signaling pathways, which ensures increased nuclear factor kappa B (NF-κB) production [46,85]. In glaucomatous, diabetic, and healthy eyes, mental stress is associated with an increase in IOP and vasoregulatory disturbances caused by glucocorticoids, proinflammatory cytokines, and endothelin-1 [46,85]. Tumor necrosis factor alpha (TNF-α) and interleukin 6 (IL-6) levels also increase in the aqueous humor. These factors can all contribute to the loss of metabolic control. Thus, if metabolic shift slows down, degenerative processes gain ground, causing diseases of the retina [46,85]. These processes can be further aggravated by multiple stressors, and therefore, it is necessary for ophthalmologists and other health professionals to understand the usefulness of stress-reducing therapies in ophthalmology, e.g., for diabetic patients [97]. Complementary therapies such as yoga, meditation, autogenic training, music therapy, and psychotherapy support have been shown to reduce stress—related to vision loss and fear of blindness—and improve quality of life [85,95]. Depression and suicide rates are notably higher among individuals with DR compared to the general population. According to a study by Bao et al., the prevalence of depression in patients with moderate to severe non-proliferative DR or proliferative DR was significantly higher (14.3%) compared to those with mild retinopathy or no retinopathy (6.9% and 7.0%, respectively) [52]. Additionally, a study highlighted by Ha et al. found that individuals with sight-threatening eye diseases, including DR, had a higher incidence rate ratio of suicide compared to those without such diagnoses. Specifically, DR was associated with the highest suicide deaths among sight-threatening eye diseases, accounting for 57% of cases [53].

The holistic model of healthcare emphasizes the integration of physical, mental, emotional, and spiritual needs of patients. This model suggests that addressing these aspects in a balanced manner can lead to better health outcomes [98]. Research by Sridhar showed that spirituality and religion can play a significant role in coping with diabetes. Spiritual practices and beliefs can help manage the emotional and psychological stresses associated with diabetes, contributing to overall well-being [99]. Another study highlighted the significant emotional and social strain experienced by individuals with DR. This strain can affect their overall health and well-being, indicating the importance of addressing emotional and social factors in managing the condition [100]. Furthermore, spiritual intelligence and mindfulness can positively influence the mental well-being of individuals with diabetes. These factors can help manage emotional dysregulation and depression, which are common in diabetes [101,102]. These findings underscore the importance of comprehensive care that includes mental health support for individuals with DR to address the increased risk of depression and suicide.

5.1. Mind–Body Therapies (Meditation and Visualization)

Mindfulness-Based Stress Reduction (MBSR) is a structured program that combines mindfulness meditation and gentle yoga. Developed by Jon Kabat-Zinn, MBSR aims to cultivate nonjudgmental awareness of the present moment, which can help individuals manage stress and improve overall well-being [102]. For people with diabetes, MBSR has been shown to reduce stress, improve glycemic control, and enhance quality of life [102]. By reducing stress and promoting relaxation, MBSR may indirectly benefit eye health and help manage DR. A bibliometric study by Jiang et al. highlighted the increasing interest in mindfulness interventions for diabetes. While the primary focus is on mental health and metabolic control, these interventions could potentially benefit overall eye health and reduce the risk of complications like DR [103]. Other mind–body therapies, including meditation, yoga, and other relaxation techniques, are promising options for managing diabetes [104]. These therapies aim to counteract the stress response and promote a state of relaxation, which can help regulate cortisol and other stress hormones [104]. Chronic stress is known to negatively impact blood glucose control and contribute to the development of diabetic complications [6,104]. By reducing stress and improving metabolic control, mind–body therapies may help prevent or slow the progression of DR. Meditation and yoga are traditional practices that focus on mental and physical well-being. Meditation involves focused attention and diaphragmatic breathing, which can help reduce stress and improve emotional health [104].

In conclusion, the direct impacts of meditation, yoga, and mindfulness on DR require further research. Incorporating these complementary therapies into a comprehensive diabetes management plan may help improve overall health and potentially benefit eye health in individuals with DR.

5.2. Effects of Eye Eurhythmy on Diabetic Retinopathy

Eurhythmy therapy [Ancient Greek: εὖ (good) + ῥυθμός (regular, symmetric motion)] is an active exercise of anthroposophic medicine, introduced by the Austro-Hungarian Rudolf Steiner in 1911 [105–107]. According to the principles of anthroposophy, a person's physical, emotional, and spiritual–individual levels interact harmoniously, and their imbalance is a characteristic of illness. Therefore, the central idea of eurhythmy is to restore the balance of health with meditative movement exercises [108]. Speech movements are transposed into exercises, targeting the patient's self-expression and self-healing abilities. In general, eurhythmy improves the breathing pattern and posture, strengthens muscle tone, and increases physical vitality, so it is particularly beneficial in the complementary treatment of neurological locomotor diseases [105–107]. The question of whether eurhythmy cannot also be used for eye diseases due to the neural connection may arise. The first steps toward the development of eurhythmy exercises for ophthalmic diseases (in German, Augeneurhythmie) was first invented and standardized by Ilse Knauer between 1938 and 1971 [107]. Margret Thiersch reports on the symbolism of eurythmic body movements, which can basically help the patient to perceive and visualize their eyes and vision more consciously [106]. Paul Blok und Ralf Brukart advise patients with DR to perform therapeutic eurhythmy exercises with the "D", "T", "I", "M", and "S" sound series daily in addition to relaxation techniques and breathing exercises. According to the principles of eurhythmy and with the addition of the authors' opinion, "D" symbolizes the response to external stimuli, and its downward movement imitates the reduction in blood sugar levels. The "T" sequence, with its analogy of moving from top to bottom and from outside to inside, models cellular processes, such as glucose uptake by the cell through the GLUT4 (Glucose Transporter Type 4) transporter. The stretching exercises of the "I" sequence focus on the patient's individuality and symbolize self-reflection, projecting the direction of vision. The "M" sequence represents the bidirectional metabolic processes occurring in the human eye and body. In terms of vision, it serves to raise awareness of DR and the acceptance of sometimes incurable conditions, alongside the "groping" in darkness associated with blindness and low vision. The complexity of the "S" sequence not only illustrates the dynamism of the retina and the body's blood circulation but also serves to relax the patient, as order emerges from chaos [106,109]. Anger suggests that eurythmy can serve as a secondary prevention intervention for diabetic patients [110]. However, the psycho-philosophical concepts underlying the two types of diabetes differ: In type 2 diabetes, the condition is attributed to a "weakened Ego". In type 1 diabetes, the "Ego" has withdrawn, leading to an autoimmune response [110]. Hilgard suggested eurhythmy as complementary therapy among children and adolescents with type 1 diabetes. According to her recommendations, the cornerstones of childhood diabetes management are glycemic control with insulin substitution and antidiabetics, which should be completed with eurhythmy, art therapy, and individual psychological care [111]. During eurhythmy, participants engage in skill-based activities and rhythmic exercises. The primary objective is to cultivate lifelong somatic awareness in adolescents with type 1 diabetes, ensuring meticulous regulation of glucose homeostasis [111]. In the context of DR, direct supporting evidence is limited and warrants further research; although eye eurhythmy provides a tool for a patient in accepting their disease and reducing anxiety, there is no available placebo-controlled clinical study. After eurhythmy treatment, increased autonomic nervous system-associated stress adaptation can be expected, as well as general improvement in cardiovascular parameters, ultimately leading to a better quality of life. Adult and pediatric patients generally report good cooperation and satisfaction without any known documented side effects [112]. Based on several clinical results, it significantly improves the sense of self-competence in patients with chronic diseases; improves posture and breathing and heart rate patterns; and strengthens muscle tone [108,113].

5.3. Shinrin-Yoku

The Japanese term Shinrin-Yoku (kanji: 森林浴; "forest bathing") was first introduced by the Ministry of Agriculture, Forestry, and Fisheries of Japan in 1982 [114]. Studies have been conducted to investigate the effects of forest environments on human physical and mental health since 2004 [115]. Recently, Li established the term "Forest Medicine" in 2022, referring to a potential healing and rehabilitation technique with a "prevention is better than cure" lifestyle approach [114,115]. Dynamic forest bathing refers to patients participating in forest environment interventions (e.g., alpinism, hiking, yoga, etc.) and effectively enhances the activity of the parasympathetic nervous system while reducing the activity of the sympathetic nervous system. Other physiologic benefits are the increase in human natural killer cells and the intracellular levels of anti-cancer proteins, as well as reductions in blood pressure and heart rate [114]. From an ophthalmologic point of view, forest landscapes with naturally occurring colors (green, yellow, and red) potentially stimulate the visual pathway and photoreceptors. Everyday technostress (electric devices, artificial lighting, etc.) affecting the visual system ultimately leads to mental stress, anxiety, depression, computer vision syndrome (headaches, mental fatigue, eye, and neck strain), and insomnia [115]. Ohtsuka et al. found that Shinrin-Yoku significantly decreased blood glucose levels in non-insulin-dependent diabetic patients. The study involved 87 diabetic patients who participated in forest bathing sessions for over six years. The results showed a significant reduction in mean blood glucose levels and glycated hemoglobin (HbA1c) after the forest bathing sessions [116]. These findings suggest that Shinrin-Yoku can be an effective complementary therapy for managing blood glucose levels in diabetic patients, likely due to the combined effects of physical activity, stress reduction, and the natural environment. According to the studies of Marigold and Hollands, it can be useful to guide patients on a therapeutic tour (e.g., Shinrin-Yoku and nature therapy) for rehabilitation purposes, as walking on uneven ground or terrain requires visual information to guide foot placement, and anxiety is reduced in a green environment [82]. Due to possible indirect neuroprotective effects, it may have complementary benefits for patients with DR by reducing stress, improving mental health, and potentially enhancing overall ocular health. Although more scientific evidence needs to be collected and evaluated, incorporating Shinrin-Yoku into a lifestyle medicine approach could support traditional treatments for DR.

6. Clinical Considerations and Recommendations

Patients with DR or glaucoma suffer from greater visual field loss, altered color vision perception, and decreased stereoacuity due to the apoptosis and autophagy of retinal ganglion cells, which all affect activities of daily life [46,117,118].

Eye movement disorders caused by strabismus and eye muscle paralysis of various origins (e.g., diabetic neuropathy, etc.) [42,119] cause diplopia and other visual disturbances in addition to esthetic problems. Vision deterioration caused by advanced DR and post-operative complications, e.g., progressive proliferative diabetic vitreoretinopathy even after successful pars plana vitrectomy, makes patients visually impaired with an ever-decreasing quality of life.

While in the case of the former diseases, the functions of visual perception and facial and eye muscles would be partially or optimally fully restored, in the latter case, the aim would be to develop functional vision that can be developed even with "impaired vision". Integrating lifestyle medicine and low vision rehabilitation principles, such as a balanced diet, regular physical activity, stress management, and proper eye hygiene, into patients' care plan can support overall eye health and improve quality of life (Figure 1). Current recommendations for DR management include strict control of blood sugar levels, blood pressure, and cholesterol, along with regular eye examinations and timely treatment interventions or medications, laser therapy, and surgical interventions. These are supported by a wealth of evidence for their efficacy [46]. It is also essential for healthcare providers to clearly evaluate the ophthalmic conditions of their patients to select complementary therapy or even discourage certain practices. Patients with DR may have other ocular comorbidities like glaucoma or high-grade myopia, increasing the risk of vitreous liquefaction, vitreous hemorrhage, tractional retinal tear, or retinal detachment [6,93]. The tractional movement of the vitreous body causes the spontaneous rupture of the small capillaries of the retina, so, due to the aforementioned risks, according to the authors' professional opinion, Kolpakov's gymnastics, Nishi's capillary exercises, and Bates' swaying–swinging exercises are not recommended in order to avoid vitreoretinal injuries [6,93,120]. Prevention, information, and advice are also provided by other specialists, e.g., opticians, optometrists, and rehabilitation visual trainers, to facilitate the work of ophthalmologists [121]. Therapies must be regularly checked by an ophthalmologist and modified if necessary. Although longer treatment times and patience are required, these therapies significantly improve the quality of life [96].

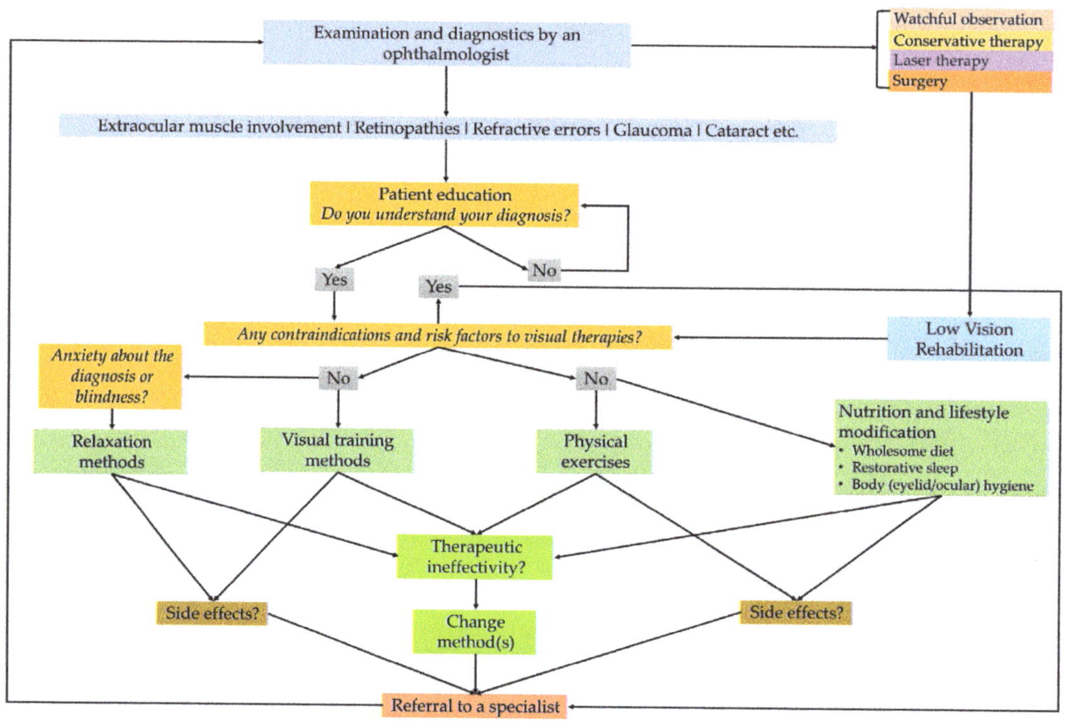

Figure 1. A flowchart of clinical steps in integrative medical care helping ophthalmologists, visual therapists, and behavioral optometrists in eye-related lifestyle interventions. The chart emphasizes the importance of correct diagnosis and strict medical treatments based on the recommendations of a specialist ophthalmologist (self-edited).

A limitation of this study is the limited scope of direct evidence in the scientific literature. There are few or no high-impact, peer-reviewed studies directly correlating specific complementary therapies with measurable improvements in DR outcomes. Most of the presented practices could indirectly or directly improve the systemic conditions caused by DR and diabetes, reducing their progression, and this study can help initiate scientific dialog and promote evidence and research in this field in the future. Future research may provide more insights into their potential roles. For those interested in complementary therapies, it is essential to consult with healthcare professionals to ensure any adjunctive treatments are safe and do not interfere with conventional treatments. Prevention, information, and advice are also provided by other specialists, e.g., opticians, optometrists, and rehabilitation visual trainers, to facilitate the work of ophthalmologists [121]. Therapies must be regularly checked by an ophthalmologist and modified if necessary (Figure 2).

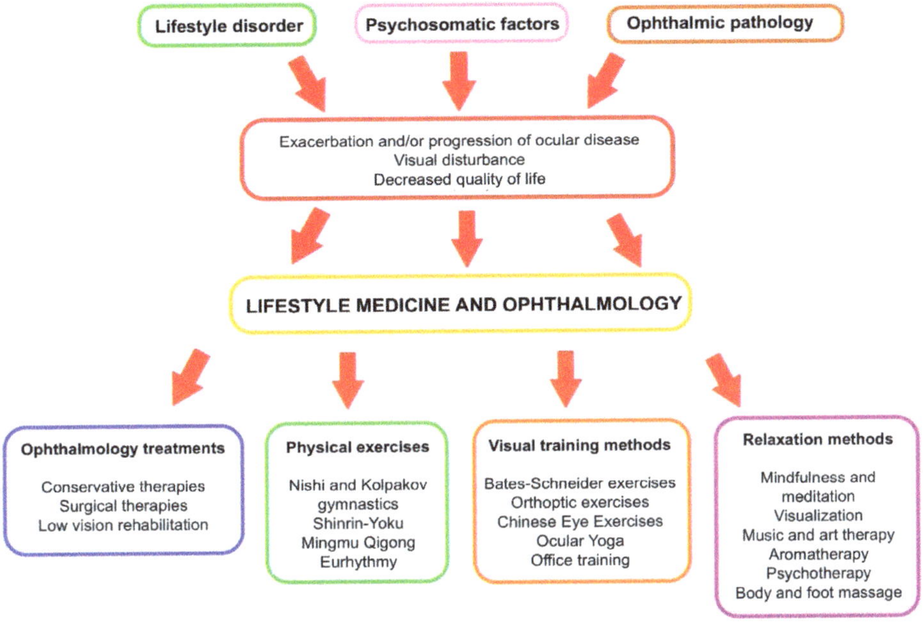

Figure 2. The role of lifestyle medicine in the system of ophthalmology treatments. If a specialist has identified multifactorial causes of the patient's eye disease impairing quality of life, a therapeutic decision can be made in addition to the professional treatment of the pathophysiological cause, as well as personalized complementary therapy involving lifestyle change (self-edited). The red arrow in the flowchart represents a transition point where the negative outcomes—exacerbation or progression of ocular disease, visual disturbance, and decreased quality of life—lead to the implementation of lifestyle medicine and ophthalmology interventions.

7. Conclusions

Low vision exercises and aids can help patients make the most of their remaining vision, improving their ability to perform everyday tasks and reducing the impact of vision loss on their lives. By providing tools and techniques to enhance visual function, low vision aids and exercises enable patients to maintain a higher level of independence. This can lead to improved self-esteem and a greater sense of control over their condition. Vision loss can be associated with an increased risk of depression and anxiety. By improving visual function and reducing the challenges associated with vision impairment, low vision aids and exercises can contribute to better mental health and overall well-being. In conclusion, low vision exercises and aids play a crucial role in the comprehensive management of DR. These interventions help patients maximize their remaining vision, maintain independence, and improve their quality of life. Incorporating these strategies into the care plan for DR patients can lead to better outcomes and a more positive outlook on managing their condition.

Supplementary Materials: The following supporting information can be downloaded at: https://www.mdpi.com/article/10.3390/life15060857/s1. Figures S1–S7: Kolpakov's industrial (wrokplace) gymnastics; Figures S8–S17: Kolpakov's hygienic (domestic) gymnastics; Figures S18–S19: Katsuzo Nishi's workout; Figures S20–S21: Bates-Schneider exercises.

Author Contributions: Conceptualization, T.R., R.G. and A.C.; writing—original draft preparation, T.R.; writing—review and editing, A.K.-V., E.P., R.G. and A.C.; visualization, T.R. and E.P.; supervision, R.G. and A.C. All authors have read and agreed to the published version of the manuscript.

Funding: This research was financed by the Thematic Excellence Program 2021 Health Sub-program of the Ministry for Innovation and Technology in Hungary within the framework of the EGA-16 project of the Pécs of University. The publication of this article was also supported by the authors' scientific and research recognition grant provided by the University of Pécs (University of Pécs Research Fellow Grant (EKÖP), grant no. EKÖP-24-4-I-PTE-170).

Institutional Review Board Statement: Not applicable.

Informed Consent Statement: Informed consent was received from the participants involved in demonstrating the Supplementary Material Figures.

Data Availability Statement: No new data were created or analyzed in this study. Data sharing is not applicable to this article.

Acknowledgments: The authors would like to acknowledge the photographers and subjects who took part in formatting the supplementary material figures: Dalma Komjáti (University of Pécs, Pécs, Hungary) and István Szabó (University of Pécs, Pécs, Hungary). The authors would like to express their gratitude to Lilla Ungváry (Főnixmed, Budapest, Hungary) and Anna Nagyné Halász (Konnektív Kft., Miskolc, Hungary), visual trainers and co-workers of the "Ki-Látás" Foundation (Pécs, Hungary), for sharing their expertise in the field of visual training exercises and low vision rehabilitation.

Conflicts of Interest: The authors declare no conflicts of interest.

Abbreviations

AAO	American Academy of Ophthalmology
AHA	American Heart Association
BDNF	Brain-Derived Neurotrophic Factor
DR	Diabetic Retinopathy
GLUT4	Glucose Transporter Type 4
IL	Interleukin
IU	International Unit
MBSR	Mindfulness-Based Stress Reduction
NF-κB	Nuclear factor kappa B
TNF-α	Tumor Necrosis Factor Alpha (α)
WHO	World Health Organization

References

1. Tao, Z.; Shi, A.; Zhao, J. Epidemiological Perspectives of Diabetes. *Cell Biochem. Biophys.* **2015**, *73*, 181–185. [CrossRef] [PubMed]
2. Wang, W.; Lo, A.C.Y. Diabetic Retinopathy: Pathophysiology and Treatments. *Int. J. Mol. Sci.* **2018**, *19*, 1816. [CrossRef] [PubMed]
3. Curran, K.; Peto, T.; Jonas, J.B.; Friedman, D.; Kim, J.E.; Leasher, J.; Tapply, I.; Fernandes, A.G.; Cicinelli, M.V.; Arrigo, A.; et al. Global Estimates on the Number of People Blind or Visually Impaired by Diabetic Retinopathy: A Meta-Analysis from 2000 to 2020. *Eye* **2024**, *38*, 2047–2057. [CrossRef]
4. Chong, D.D.; Das, N.; Singh, R.P. Diabetic Retinopathy: Screening, Prevention, and Treatment. *Cleve. Clin. J. Med.* **2024**, *91*, 503–510. [CrossRef] [PubMed]
5. AlQabandi, Y.; Nandula, S.A.; Boddepalli, C.S.; Gutlapalli, S.D.; Lavu, V.K.; Abdelwahab, R.A.M.; Huang, R.; Potla, S.; Bhalla, S.; Hamid, P. Physical Activity Status and Diabetic Retinopathy: A Review. *Cureus* **2022**, *14*, e28238. [CrossRef]
6. Rák, T.; Kovács-Valasek, A.; Pöstyéni, E.; Csutak, A.; Gábriel, R. Complementary Approaches to Retinal Health Focusing on Diabetic Retinopathy. *Cells* **2023**, *12*, 2699. [CrossRef]
7. Sun, J.K.; Liu, D. Challenges in the Clinical Management of Proliferative Diabetic Retinopathy: Treatment Choice and Follow-Up. *JAMA Ophthalmol.* **2023**, *141*, 46–47. [CrossRef]
8. Glassman, A.R. Anti-Vascular Endothelial Growth Factor Options and Questions for Diabetic Eye Disease Treatment. *JAMA Ophthalmol.* **2025**, *143*, 336–337. [CrossRef]

9. Limoli, P.G.; Limoli, C.; Nebbioso, M. Potential Guidelines for Cataract Surgery and Rehabilitation in Visually Impaired Patients: Literature Analysis. *Aging Med.* **2024**, *7*, 802–812. [CrossRef]
10. Han, X.; Zhang, J.; Liu, Z.; Tan, X.; Jin, G.; He, M.; Luo, L.; Liu, Y. Real-World Visual Outcomes of Cataract Surgery Based on Population-Based Studies: A Systematic Review. *Br. J. Ophthalmol.* **2023**, *107*, 1056–1065. [CrossRef]
11. Bryl, A.; Mrugacz, M.; Falkowski, M.; Zorena, K. The Effect of Diet and Lifestyle on the Course of Diabetic Retinopathy—A Review of the Literature. *Nutrients* **2022**, *14*, 1252. [CrossRef] [PubMed]
12. Amore, F.; Silvestri, V.; Turco, S.; Fortini, S.; Giudiceandrea, A.; Cruciani, F.; Mariotti, S.P.; Antonini, D.; Rizzo, S. Vision Rehabilitation Workforce in Italy: A Country-Level Analysis. *BMC Health Serv. Res.* **2024**, *24*, 1323. [CrossRef] [PubMed]
13. Fitzmaurice, K. The Role of the Orthoptist in Visual Rehabilitation: An Australian Perspective. *Am. Orthopt. J.* **1996**, *46*, 159–166. [CrossRef]
14. Raphanel, M.; Shaughness, G.; Seiple, W.H.; Arleo, A. Current Practice in Low Vision Rehabilitation of Age-Related Macular Degeneration and Usefulness of Virtual Reality as a Rehabilitation Tool. *J. Aging Sci.* **2018**, *6*, 194. [CrossRef]
15. Lianov, L.; Johnson, M. Physician Competencies for Prescribing Lifestyle Medicine. *JAMA* **2010**, *304*, 202–203. [CrossRef]
16. Németh, J.; Barcsay, G.; Barcsay-Veres, A.; Nagy, Z.Z. The Ophthalmologist's Duties in Vision Rehabilitation. [A Szemészorvos Feladatai a Látásrehabilitációban]. *Szemeszet* **2024**, *161*, 11–17. [CrossRef]
17. Cooke, J.B.; Cochrane, A.L. A Practical Guide to Low Vision Management of Patients with Diabetes. *Clin. Exp. Optom.* **2001**, *84*, 155–161. [CrossRef]
18. Muhsin, Z.J.; Qahwaji, R.; Ghanchi, F.; Al-Taee, M. Review of Substitutive Assistive Tools and Technologies for People with Visual Impairments: Recent Advancements and Prospects. *J. Multimodal User Interfaces* **2024**, *18*, 135–156. [CrossRef]
19. Vingolo, E.M.; De Rosa, V.; Domanico, D.; Anselmucci, F. Low Vision Rehabilitation: Current Perspectives. *Clin. Optom.* **2015**, *7*, 53–58. [CrossRef]
20. O'Louglin, M.; Edwards, R.; Bould, E.; Devine, S.; Downing, S. Animal-Assisted Interventions in Adult Hospital Rehabilitation Settings: A Scoping Review. *Nurs. Health Sci.* **2024**, *26*, e13138. [CrossRef]
21. Bassan, E.; Mair, A.; De Santis, M.; Bugianelli, M.; Loretti, E.; Capecci, A.; Mutinelli, F.; Contalbrigo, L. An Overview of the Literature on Assistance Dogs Using Text Mining and Topic Analysis. *Front. Vet. Sci.* **2024**, *11*, 1463332. [CrossRef] [PubMed]
22. Mittly, V.; Fáy, V.; Dankovics, N.; Pál, V.; Purebl, G. The Role of Dog Therapy in Clinical Recovery and Improving Quality of Life: A Randomized, Controlled Trial. *BMC Complement. Med. Ther.* **2024**, *24*, 229. [CrossRef] [PubMed]
23. Glenk, L.M.; Weissenbacher, K.; Přibylová, L.; Stetina, B.U.; Demirel, S. Perceptions on Health Benefits of Guide Dog Ownership in an Austrian Population of Blind People with and without a Guide Dog. *Animals* **2019**, *9*, 428. [CrossRef]
24. Lundqvist, M.; Carlsson, P.; Sjödahl, R.; Theodorsson, E.; Levin, L.Å. Patient Benefit of Dog-Assisted Interventions in Health Care: A Systematic Review. *BMC Complement. Altern. Med.* **2017**, *17*, 358. [CrossRef]
25. Lippi, G.; Plebani, M. Diabetes Alert Dogs: A Narrative Critical Overview. *Clin. Chem. Lab. Med.* **2019**, *57*, 452–458. [CrossRef]
26. Ramos, M.M.; Nabeiro, M. The Influence of Equine-Assisted Services on the Balance of a Participant with Visual Impairment and Autism Characteristics. *J. Bodyw. Mov. Ther.* **2022**, *31*, 57–61. [CrossRef]
27. Klimova, V.K.; Strelkova, Y.A.; Klimova, M.V.; Kholodova, O.A. Physical Rehabilitation of Patients with Diabetes Using Hippotherapy. [Иппотерапия Как Оздоровительная Технология При Заболевании Сахарным Диабетом 1 Типа]. *Teor. I Prakt. Fiz. Kult.* **2011**, *10*, 55–57.
28. Horwood, A.; Toor, S. Clinical Test Responses to Different Orthoptic Exercise Regimes in Typical Young Adults. *Ophthalmic Physiol. Opt.* **2014**, *34*, 250–262. [CrossRef] [PubMed]
29. Chebel, S.; Bouatay, A.B.; Ammar, M.; Ben-Yahia, S.; Khairallah, M.; Ayed, M.F. Diabetes Mellitus-Associated Ocular Motor Nerve Palsies. *Neurosciences* **2009**, *4*, 386–388.
30. Lassie, N.; Ashan, H.; Triola, S.; Widiastuti, W. Risk Factors of Opthalmoplegia in Diabetes Mellitus. *J. Penelit. Pendidik. IPA* **2023**, *9*, 868–875. [CrossRef]
31. Watanabe, K.; Hagura, R.; Akanuma, Y.; Takasu, T.; Kajinuma, H.; Kuzuya, N.; Irie, M. Characteristics of Cranial Nerve Palsies in Diabetic Patients. *Diabetes Res. Clin. Pract.* **1990**, *10*, 19–27. [CrossRef]
32. Iliescu, D.A.; Timaru, C.M.; Alexe, N.; Gosav, E.; De Simone, A.; Batras, M.; Stefan, C. Management of Diplopia. *Rom. J. Ophthalmol.* **2017**, *61*, 166. [CrossRef]
33. Burde, R.M. Neuro-Ophthalmic Associations and Complications of Diabetes Mellitus. *Am. J. Ophthalmol.* **1992**, *114*, 498–501. [CrossRef] [PubMed]
34. Bosco, D.; Costa, R.; Plastino, M.; Branca, D.; Cotronei, P.; Sperlì, T.; Santacroce, N.; Siniscalchi, A.; Consoli, D.; Ceccotti, C.; et al. Glucose Metabolism in the Idiopathic Blepharoptosis: Utility of the Oral Glucose Tolerance Test (OGTT) and of the Insulin Resistance Index. *J. Neurol. Sci.* **2009**, *284*, 24–28. [CrossRef]
35. Rogell, G.D. Corneal Hypesthesia and Retinopathy in Diabetes Mellitus. *Ophthalmology* **1980**, *87*, 229–233. [CrossRef] [PubMed]
36. Rák, T.; Csutak, A. Exploring Novel Pharmacological Trends: Natural Compounds in Dry Eye Disease Management. *Acta Pharm.* **2024**, *74*, 383–404. [CrossRef] [PubMed]

37. Raskind, R.H. Problems at the Reading Distance. *Am. Orthopt. J.* **1976**, *26*, 53–59. [CrossRef]
38. Rundström, M.M.; Eperjesi, F. Is There a Need for Binocular Vision Evaluation in Low Vision? *Ophthalmic Physiol. Opt.* **1995**, *15*, 525–528. [CrossRef]
39. Macfarlane, A.; Jolly, N.; Ma, D.; Thompson, K. Orthoptic Interventions in Stroke Patients. *Aust. Orthopt. J.* **2009**, *43*, 17–23.
40. Maagaard, M.L.; Nisted, I.; Bek, T. Vergence Exercises for Six Weeks Induce Faster Recovery of Convergence Insufficiency Than Accommodation Exercises in School Children. *Investig. Ophthalmol. Vis. Sci.* **2021**, *62*, 23. [CrossRef]
41. Yadav, S.; Singh, A.; Agrawal, A.; Mittal, S.; Panyala, R.; Kumar, B. Pencil Push-up Therapy vs. Office-Based Orthoptic Therapy in Emmetropes with Asthenopic Symptoms Due to Convergence Insufficiency: A Randomized Controlled Trial. *Himal. J. Ophthalmol.* **2022**, *16*, 4. [CrossRef]
42. Zeng, C.; Fan, C.; Liu, J.; Xiao, Q.; Zhu, Y.; Song, X.; Chen, H. Gradual Oculomotor Training in Blow-out Orbital Fracture Reconstruction Recovery. *J. Int. Med. Res.* **2019**, *48*, 300060519893846. [CrossRef] [PubMed]
43. Bixenman, W.W. Central Fusion Disruption Is Not Horror Fusionis. *Arch. Ophthalmol.* **2010**, *128*, 648–649. [CrossRef]
44. Zhang, Q.; Jiang, Y.; Deng, C.; Wang, J. Effects and Potential Mechanisms of Exercise and Physical Activity on Eye Health and Ocular Diseases. *Front. Med.* **2024**, *11*, 1353624. [CrossRef] [PubMed]
45. Gale, J.; Wells, A.P.; Wilson, G. Effects of Exercise on Ocular Physiology and Disease. *Surv. Ophthalmol.* **2009**, *54*, 349–355. [CrossRef]
46. Kovács-Valasek, A.; Rák, T.; Pöstyéni, E.; Csutak, A.; Gábriel, R. Three Major Causes of Metabolic Retinal Degenerations and Three Ways to Avoid Them. *Int. J. Mol. Sci.* **2023**, *24*, 8728. [CrossRef]
47. Ren, J.; Xiao, H. Exercise for Mental Well-Being: Exploring Neurobiological Advances and Intervention Effects in Depression. *Life* **2023**, *13*, 1505. [CrossRef]
48. Yang, Y.J. An Overview of Current Physical Activity Recommendations in Primary Care. *Korean J. Fam. Med.* **2019**, *40*, 135. [CrossRef]
49. Wolkoff, P. "Healthy" Eye in Office-like Environments. *Environ. Int.* **2008**, *34*, 1204–1214. [CrossRef]
50. Sano, K.; Kawashima, M.; Takechi, S.; Mimura, M.; Tsubota, K. Exercise Program Improved Subjective Dry Eye Symptoms for Office Workers. *Clin. Ophthalmol.* **2018**, *12*, 307. [CrossRef]
51. Jeong, N.R.; Lee, S.H.; Kim, Y.J.; Lee, J.G.; Yi, Y.H.; Tak, Y.J.; Hwang, H.R.; Kim, G.L.; Lee, S.Y.; Cho, Y.H.; et al. Association between Near Work Time and Depression among Workers in South Korea. *Korean J. Fam. Med.* **2021**, *42*, 390. [CrossRef] [PubMed]
52. Bao, Y.; Cope, S.; Gaddis, M.; Drees, B. Prevalence and Predictors of Depression in Patients with Diabetic Retinopathy in a Nationally Representative Sample. *Investig. Ophthalmol. Vis. Sci.* **2020**, *61*, 1902.
53. Ha, A.; Kim, S.H.; Kang, G.; Yoon, H.J.; Kim, Y.K. Association between Sight-Threatening Eye Diseases and Death by Suicide in South Korea: A Nationwide Population-Based Cohort Study. *Ophthalmology* **2023**, *130*, 804–811. [CrossRef]
54. Seo, E.; Lee, Y.; Mun, E.; Kim, D.H.; Jeong, Y.; Lee, J.; Jeong, J.; Lee, W. The Effect of Long Working Hours on Developing Type 2 Diabetes in Adults with Prediabetes: The Kangbuk Samsung Cohort Study. *Ann. Occup. Environ. Med.* **2022**, *34*, e4. [CrossRef]
55. Tunceli, K.; Bradley, C.J.; Nerenz, D.; Williams, L.K.; Pladevall, M.; Lafata, J.E. The Impact of Diabetes on Employment and Work Productivity. *Diabetes Care* **2005**, *28*, 2662–2667. [CrossRef]
56. Saif-Ur-Rahman, K.M.; Mamun, R.; Li, Y.; Matsunaga, M.; Ota, A.; Yatsuya, H. Work-Related Factors among People with Diabetes and the Risk of Cardiovascular Diseases: A Systematic Review. *J. Occup. Health* **2021**, *63*, 12278. [CrossRef] [PubMed]
57. Szczygieł, E.; Fudacz, N.; Berus, T.; Rojek, J.; Golec, E. Physiotherapy in Ophthalmology—A Literature Review. *Rehabil. Med.* **2023**, *27*, 41–48. [CrossRef]
58. Di Noto, P.; Uta, S.; DeSouza, J.F.X. Eye Exercises Enhance Accuracy and Letter Recognition, but Not Reaction Time, in a Modified Rapid Serial Visual Presentation Task. *PLoS ONE* **2013**, *8*, e59244. [CrossRef]
59. Alghamdi, W.M.; Alrasheed, S.H. Impact of an Educational Intervention Using the 20/20/20 Rule on Computer Vision Syndrome. *African Vis. Eye Health* **2020**, *79*, a554. [CrossRef]
60. Johnson, S.; Rosenfield, M. 20-20-20 Rule: Are These Numbers Justified? *Optom. Vis. Sci.* **2023**, *100*, 52–56. [CrossRef]
61. Lertwisuttipaiboon, S.; Pumpaibool, T.; Neeser, K.J.; Kasetsuwan, N. Effectiveness of a Participatory Eye Care Program in Reducing Eye Strain among Staff Computer Users in Thailand. *Risk Manag. Healthc. Policy* **2017**, *10*, 71. [CrossRef]
62. Kolpakov, S.P.; Rumiantseva, A.G. [Experience with the Use of a Complex Method of Correcting the Psychophysiologic State of Humans Working with Constant Visual Strain]. *Fiziol. Cheloveka* **1987**, *13*, 42–49. [PubMed]
63. Gumeniuk, V.A.; Klassina, S.I.; Orbachevskaia, G.N.; Kolpakov, S.P. [Massage as a Means for Correcting Visual Perception and Improving the Physiological Functions of the Working Man]. *Gig. Tr. Prof. Zabol.* **1990**, *10*, 50–52.
64. Venslauskas, M.; Ostasevičius, V.; Marozas, V. Limb's Vibrations Exercise Monitoring with MEMS Accelerometer to Identify Influence of Cardiovascular System. In Proceedings of the Vibroengineering procedia 15th Int. Conf. "Vibroengineering—2013", Druskininkai, Lithuania, 17–19 September 2013; Volume 1, pp. 48–52.

65. Eremin, M.S.; Shevchenko, L.I.; Korgun, Z.F.; Eremin, S.M.; Pegova, L.A.; Bazilevskaia, T.N. [Hydrotherapy According to the Method of I. Gillershtein and A.S. Zalmanov]. *Vopr. Kurortol. Fizioter. Lech. Fiz. Kult.* **1969**, *34*, 467–468. [PubMed]
66. Wang, H.; Qian, Y.; Congdon, N.; Boswell, M.; Rozelle, S.; Ma, X. Effect of Chinese Eye Exercises on Change in Visual Acuity and Eyeglasses Wear among School-Aged Children in Rural China: A Propensity-Score-Matched Cohort Study. *BMC Complement. Med. Ther.* **2020**, *20*, 82. [CrossRef]
67. Li, Y.; Zhu, L.; Wang, R.; Yang, X.; Jiang, X.; Lu, T. Guided Meditation for Vision Acuity Training on Adolescent Myopia: Study Protocol for an Open-Label, Prospective, Multicenter, Randomized Controlled Trial. *Trials* **2022**, *23*, 16. [CrossRef]
68. Li, S.M.; Kang, M.T.; Peng, X.X.; Li, S.Y.; Wang, Y.; Li, L.; Yu, J.; Qiu, L.X.; Sun, Y.Y.; Liu, L.R.; et al. Efficacy of Chinese Eye Exercises on Reducing Accommodative Lag in School-Aged Children: A Randomized Controlled Trial. *PLoS ONE* **2015**, *10*, e0117552. [CrossRef]
69. Lin, Z.; Vasudevan, B.; Jhanji, V.; Gao, T.Y.; Wang, N.L.; Wang, Q.; Wang, J.; Ciuffreda, K.J.; Liang, Y.B. Eye Exercises of Acupoints: Their Impact on Refractive Error and Visual Symptoms in Chinese Urban Children. *BMC Complement. Altern. Med.* **2013**, *13*, 306. [CrossRef]
70. Gupta, S.K.; Aparna, S. Effect of Yoga Ocular Exercises on Eye Fatigue. *Int. J. Yoga* **2020**, *13*, 76. [CrossRef]
71. Poulere, E.; Moschandreas, J.; Kontadakis, G.A.; Pallikaris, I.G.; Plainis, S. Effect of Blur and Subsequent Adaptation on Visual Acuity Using Letter and Landolt C Charts: Differences between Emmetropes and Myopes. *Ophthalmic Physiol. Opt.* **2013**, *33*, 130–137. [CrossRef]
72. Robb, W.J.W. Self-Healing: A Concept Analysis. *Nurs. Forum* **2006**, *41*, 60–77. [CrossRef] [PubMed]
73. Mead, M.N. Benefits of Sunlight: A Bright Spot for Human Health. *Environ. Health Perspect.* **2008**, *116*, A160–A167. [CrossRef]
74. Gopinathan, G.; Dhiman, K.S.; Manjusha, R. A Clinical Study to Evaluate the Efficacy of Trataka Yoga Kriya and Eye Exercises (Non-Pharmocological Methods) in the Management of Timira (Ammetropia and Presbyopia). *Ayu* **2012**, *33*, 543. [CrossRef]
75. Payne, J.F.; Ray, R.; Watson, D.G.; Delille, C.; Rimler, E.; Cleveland, J.; Lynn, M.J.; Tangpricha, V.; Srivastava, S.K. Vitamin D Insufficiency in Diabetic Retinopathy. *Endocr. Pract.* **2012**, *18*, 185–193. [CrossRef] [PubMed]
76. Vasdeki, D.; Tsamos, G.; Dimakakos, E.; Patriarcheas, V.; Koufakis, T.; Kotsa, K.; Cholewka, A.; Stanek, A. Vitamin D Supplementation: Shedding Light on the Role of the Sunshine Vitamin in the Prevention and Management of Type 2 Diabetes and Its Complications. *Nutrients* **2024**, *16*, 3651. [CrossRef] [PubMed]
77. Razzaque, M.S. Sunlight Exposure: Do Health Benefits Outweigh Harm? *J. Steroid Biochem. Mol. Biol.* **2018**, *175*, 44–48. [CrossRef]
78. Moïse, M.M.; Benjamin, L.-M.; Doris, T.M.; Dalida, K.N.; Augustin, N.O.; Moïse, M.M.; Benjamin, L.-M.; Doris, T.M.; Dalida, K.N.; Augustin, N.O. Role of Mediterranean Diet, Tropical Vegetables Rich in Antioxidants, and Sunlight Exposure in Blindness, Cataract and Glaucoma among African Type 2 Diabetics. *Int. J. Ophthalmol.* **2012**, *5*, 231–237. [CrossRef]
79. Lindqvist, P.G.; Epstein, E.; Landin-Olsson, M. Sun Exposure—Hazards and Benefits. *Anticancer Res.* **2022**, *42*, 1671–1677. [CrossRef]
80. Simó, R.; Hernández, C. What Else Can We Do to Prevent Diabetic Retinopathy? *Diabetologia* **2023**, *66*, 1614–1621. [CrossRef]
81. Lee, H.J.; Kim, C.O.; Lee, D.C. Association between Daily Sunlight Exposure Duration and Diabetic Retinopathy in Korean Adults with Diabetes: A Nationwide Population-Based Cross-Sectional Study. *PLoS ONE* **2020**, *15*, e0237149. [CrossRef]
82. Srivastava, A.; Ahmad, O.F.; Pacia, C.P.; Hallett, M.; Lungu, C. The Relationship between Saccades and Locomotion. *J. Mov. Disord.* **2018**, *11*, 93–106. [CrossRef] [PubMed]
83. Kumar, S.N.; Venu, A.; Jaya, M.H. Effect of Yoga Mudras in Improving the Health of Users: A Precautionary Measure Practice in Daily Life for Resisting the Deadly COVID-19 Disease. In *Lessons from COVID-19 Impact on Healthcare Systems and Technology*; Academic Press: Cambridge, MA, USA, 2022; pp. 41–59. [CrossRef]
84. Elgaddal, N.; Weeks, J.D. Yoga Among Adults Age 18 and Older: United States, 2022. *NCHS Data Brief* **2024**, *501*, 1–11. [CrossRef]
85. Sabel, B.A.; Wang, J.; Cárdenas-Morales, L.; Faiq, M.; Heim, C. Mental Stress as Consequence and Cause of Vision Loss: The Dawn of Psychosomatic Ophthalmology for Preventive and Personalized Medicine. *EPMA J.* **2018**, *9*, 133. [CrossRef]
86. Metri, K.; Nagaratna, R.; Singh, A. Role of Yoga in Prevention and Management of Type 2 Diabetes Mellitus (T2DM) and Its Complications. *Princ. Pract. Yoga Cardiovasc. Med.* **2022**, 197–203. [CrossRef]
87. Nathani, P.; Nanduri, V.S. A Case of Diabetic Retinopathy with Blurred Vision Healed Successfully Using Yoga Prana Vidya (Ypv) Healing Protocols. *Indian J. Clin. Exp. Ophthalmol.* **2023**, *9*, 269–273. [CrossRef]
88. Sankalp; Dada, T.; Yadav, R.K.; Sharma, H.B.; Netam, R.K.; Kochhar, K.P. Effect of Tratak (Yogic Ocular Exercises) on Intraocular Pressure in Glaucoma: An RCT. *Int. J. Yoga* **2022**, *15*, 59. [CrossRef] [PubMed]
89. Galina, D.; Etsuo, C.; Takuhei, S.; Kanno, J.; Antonela, L.; Olivera, L.; Ana, G.; Dushan, K. Immediate Effect of Yoga Exercises for Eyes on the Macular Thickness. *Int. J. Yoga* **2020**, *13*, 223. [CrossRef]
90. Kumar, V.K.; Singh, B.V.D.; Manjusha, R. Add-on Effect of Ayurvedic Treatment Protocol for Diabetic Retinopathy. *AYU (An Int. Q. J. Res. Ayurveda)* **2021**, *42*, 118–129. [CrossRef] [PubMed]

91. Jasien, J.V.; Jonas, J.B.; Gustavo De Moraes, C.; Ritch, R. Intraocular Pressure Rise in Subjects with and without Glaucoma during Four Common Yoga Positions. *PLoS ONE* **2015**, *10*, e0144505. [CrossRef]
92. Bertschinger, D.R.; Mendrinos, E.; Dosso, A. Yoga Can Be Dangerous—Glaucomatous Visual Field Defect Worsening Due to Postural Yoga. *Br. J. Ophthalmol.* **2007**, *91*, 1413. [CrossRef]
93. Chong, S.Y.; Fhun, L.C.; Tai, E.; Chong, M.F.; Teo, K.S.S. Posterior Vitreous Detachment Precipitated by Yoga. *Cureus* **2018**, *10*, e2109. [CrossRef] [PubMed]
94. Kaushal, D.; Kapila, S.; Gupta, A.S.; Rattan, A. Yoga-Induced Valsalva Retinopathy. *Delhi J. Ophthalmol.* **2024**, *34*, 49–51. [CrossRef]
95. Maeng, K.J.; Lee, K.; Kim, S.; Park, C.K.; Kim, E.W.; Lee, S.Y.; Bae, H.W.; Seong, G.J.; Kim, C.Y. Effects of Glaucoma Medication on Dry Eye Syndrome and Quality of Life in Patients with Glaucoma. *Korean J. Ophthalmol.* **2021**, *35*, 467. [CrossRef]
96. Aragona, P.; Giannaccare, G.; Mencucci, R.; Rubino, P.; Cantera, E.; Rolando, M. Modern Approach to the Treatment of Dry Eye, a Complex Multifactorial Disease: A P.I.C.A.S.S.O. Board Review. *Br. J. Ophthalmol.* **2021**, *105*, 446–453. [CrossRef] [PubMed]
97. Dada, T.; Bhai, N.; Midha, N.; Shakrawal, J.; Kumar, M.; Chaurasia, P.; Gupta, S.; Angmo, D.; Yadav, R.; Dada, R.; et al. Effect of Mindfulness Meditation on Intraocular Pressure and Trabecular Meshwork Gene Expression: A Randomized Controlled Trial. *Am. J. Ophthalmol.* **2021**, *223*, 308–321. [CrossRef]
98. Steele, L. Holistic Well-Being: Mental, Physical, and Spiritual. In *Good Health and Well-Being*; Springer: Berlin/Heidelberg, Germany, 2020; pp. 373–382. [CrossRef]
99. Sridhar, G.R. Diabetes, Religion and Spirituality. *Int. J. Diabetes Dev. Ctries.* **2013**, *33*, 5–7. [CrossRef]
100. Fenwick, E.; Rees, G.; Pesudovs, K.; Dirani, M.; Kawasaki, R.; Wong, T.Y.; Lamoureux, E. Social and Emotional Impact of Diabetic Retinopathy: A Review. *Clin. Experiment. Ophthalmol.* **2012**, *40*, 27–38. [CrossRef]
101. Ajele, W.K.; Oladejo, T.A.; Akanni, A.A.; Babalola, O.B. Spiritual Intelligence, Mindfulness, Emotional Dysregulation, Depression Relationship with Mental Well-Being among Persons with Diabetes during COVID-19 Pandemic. *J. Diabetes Metab. Disord.* **2021**, *20*, 1705–1714. [CrossRef]
102. Whitebird, R.R.; Kreitzer, M.J.; O'Connor, P.J. Mindfulness-Based Stress Reduction and Diabetes. *Diabetes Spectr.* **2009**, *22*, 226–230. [CrossRef]
103. Jiang, S.; Pan, X.; Li, H.; Su, Y. Global Trends and Developments in Mindfulness Interventions for Diabetes: A Bibliometric Study. *Diabetol. Metab. Syndr.* **2024**, *16*, 43. [CrossRef]
104. DiNardo, M.M. Mind-Body Therapies in Diabetes Management. *Diabetes Spectr.* **2009**, *22*, 30–34. [CrossRef]
105. Büssing, A.; Ostermann, T.; Majorek, M.; Matthiessen, P.F. Eurythmy Therapy in Clinical Studies: A Systematic Literature Review. *BMC Complement. Altern. Med.* **2008**, *8*, 8. [CrossRef]
106. Thiersch, M. How Ocular Eurhythmy Works. [Wirkensweise Der Augenheileurythmie]. *Der Merkurstab* **2007**, *2*, 116–130. [CrossRef]
107. Lötzke, D.; Heusser, P.; Büssing, A. A Systematic Literature Review on the Effectiveness of Eurythmy Therapy. *J. Integr. Med.* **2015**, *13*, 217–230. [CrossRef]
108. Kanitz, J.L.; Pretzer, K.; Reif, M.; Witt, K.; Reulecke, S.; Voss, A.; Längler, A.; Henze, G.; Seifert, G. The Impact of Eurythmy Therapy on Fatigue in Healthy Adults—A Controlled Trial. *Eur. J. Integr. Med.* **2012**, *4*, e289–e297. [CrossRef]
109. Roemer, F. *Therapy Concepts of Anthroposophic Medicine, Step-by-Step Plans with Differential Diagnostics. [Therapiekonzepte Der Anthroposophischen Medizin Stufenplane Mit Differenzialdiagnostik]*, 2nd ed.; Thieme Medical and Scientific Publishers Pvt. Ltd.: Leipzig, Germany, 2018; ISBN 9783132420434.
110. Anger, C. Heileurythmische Behandlung Bei Diabetes Mellitus Typ 1. *Der Merkurstab* **2002**, *55*, 56–59. [CrossRef]
111. Hilgard, D. Kooperative Behandlungsansätze Bei Diabeteskranken Kindern. *Der Merkurstab* **2002**, *55*, 65–73. [CrossRef]
112. Meier-Girard, D.; Ribi, K.; Gerstenberg, G.; Ruhstaller, T.; Wolf, U. Eurythmy Therapy versus Slow Movement Fitness in the Treatment of Fatigue in Metastatic Breast Cancer Patients: Study Protocol for a Randomized Controlled Trial. *Trials* **2020**, *21*, 612. [CrossRef]
113. Logtenberg, R. The Effect of Eurhythmy Therapy on Self-Determination, Health Complaints and Psychological Symptoms: A Non-Randomised Trial. *Complement. Ther. Med.* **2020**, *49*, 102347. [CrossRef]
114. Wen, Y.; Gu, X.; Deng, W.; Zou, Q.; Hu, Y.; Yan, Q.; Pan, Y.; Wen, Z.; Wan, R.; Sheng, G.; et al. The Effects of Dynamic and Static Forest Bathing (Shinrin-Yoku) on Physiological and Psychological Health in Males and Females. *Forests* **2023**, *14*, 1592. [CrossRef]
115. Li, Q. Effects of Forest Environment (Shinrin-Yoku/Forest Bathing) on Health Promotion and Disease Prevention—The Establishment of "Forest Medicine". *Environ. Health Prev. Med.* **2022**, *27*, 43. [CrossRef] [PubMed]
116. Ohtsuka, Y.; Yabunaka, N.; Takayama, S. Shinrin-Yoku (Forest-Air Bathing and Walking) Effectively Decreases Blood Glucose Levels in Diabetic Patients. *Int. J. Biometeorol.* **1998**, *41*, 125–127. [CrossRef] [PubMed]
117. Zwierko, T.; Jedziniak, W.; Lesiakowski, P.; Śliwiak, M.; Kirkiewicz, M.; Lubiński, W. Eye–Hand Coordination Impairment in Glaucoma Patients. *Int. J. Environ. Res. Public Health* **2019**, *16*, 4332. [CrossRef] [PubMed]

118. Lakshmanan, Y.; George, R.J. Stereoacuity in Mild, Moderate and Severe Glaucoma. *Ophthalmic Physiol. Opt.* **2013**, *33*, 172–178. [CrossRef]
119. Yoon, K.H.; Lee, S.; Lim, J.S.; Cho, Y.E.; Lee, H.J.; Kim, J.H.; Kang, J.W.; Lee, S.H. Experience of Bell's Palsy Patients on Facial Exercise and Efficient Educational Program: A Qualitative Study. *J. Acupunct. Res.* **2015**, *32*, 67–78. [CrossRef]
120. Maggiano, J.; Yu, M.C.M.; Chen, S.; You, T.; Rathod, R. Retinal Tear Formation after Whole-Body Vibration Training Exercise. *BMC Ophthalmol.* **2020**, *20*, 37. [CrossRef]
121. Welp, A.; Woodbury, R.B.; McCoy, M.A.; Teutsch, S.M. (Eds.) *The Role of Public Health and Partnerships to Promote Eye and Vision Health in Communities*; National Academies Press (US): Washington, DC, USA, 2016; ISBN 978-0-309-44001-1.

MDPI AG
Grosspeteranlage 5
4052 Basel
Switzerland
Tel.: +41 61 683 77 34

Life Editorial Office
E-mail: life@mdpi.com
www.mdpi.com/journal/life

www.ingramcontent.com/pod-product-compliance
Lightning Source LLC
Chambersburg PA
CBHW061617210926
51850CB00009B/2347
9783725857852